Theory of Suboptimal Decisions

Decomposition and Aggregation

Mathematics and Its Applications (*Soviet Series*)

Theory of Suboptimal Decisions

Decomposition and Aggregation

A. A. Pervozvanskii

Faculty of Physics and Mechanics,
Leningrad Polytechnical Institute, U.S.S.R.

and

V. G. Gaitsgori

Faculty of Mathematics and Mechanics,
Jadjic University, Dushambe, U.S.S.R.

KLUWER ACADEMIC PUBLISHERS
DORDRECHT / BOSTON / LONDON

THE INTERNATIONAL INSTITUTE FOR APPLIED SYSTEMS ANALYSIS

Library of Congress Cataloging in Publication Data

Pervozvanskii, A. A. (Anatolii Arkad'evich)
[Dekompozitsiia, agregirovanie i priblizhennaia optimizatsiia. English]
Theory of suboptimal decisions decomposition and aggregation / A.A. Pervozvanskii, V.G. Gaitsgori.
p. cm. -- (Mathematics and its applications (Soviet series))
Translation of: Dekompozitsiia, agregirovanie i priblizhennaia optimizatsiia.
Includes index.
ISBN 9027724016
1. Mathematical optimization. 2. System analysis. I. Gaitsgori, Vladimir Grigor'evich. II. Title. III. Series: Mathematics and its applications (D. Reidel Publishing Company). Soviet series.
QA402.5.P4313 1988
519--dc19 88-4643
CIP

ISBN 90-277-2401-6

Published by Kluwer Academic Publishers,
P.O. Box 17, 3300 AA Dordrecht, The Netherlands.

Kluwer Academic Publishers incorporates
the publishing programmes of
D. Reidel, Martinus Nijhoff, Dr W. Junk and MTP Press.

Sold and distributed in the U.S.A. and Canada
by Kluwer Academic Publishers,
101 Philip Drive, Norwell, MA 02061, U.S.A.

In all other countries, sold and distributed
by Kluwer Academic Publishers Group,
P.O. Box 322, 3300 AH Dordrecht, The Netherlands.

Printed in The Netherlands

THE INTERNATIONAL INSTITUTE FOR APPLIED SYSTEMS ANALYSIS

is a nongovernmental research institution, bringing together scientists from around the world to work on problems of common concern. Situated in Laxenburg, Austria. IIASA was founded in October 1972 by the academies of science and equivalent organizations of twelve countries. Its founders gave IIASA a unique position outside national, disciplinary, and institutional boundaries so that it might take the broadest possible view in pursuing its objectives:

To promote international cooperation in solving problems arising from social, economic, technological, and environmental change

To create a network of institutions in the national member organization countries and elsewhere for joint scientific research

To develop and formalize systems analysis and the sciences contributing to it, and promote the use of analytical techniques needed to evaluate and address complex problems

To inform policy advisors and decision makers about the potential application of the Institute's work to such problems

The Institute now has national member organizations in the following countries:

Austria
The Austrian Academy of Sciences

Bulgaria
The National Committee for Applied Systems Analysis and Management

Canada
The Canadian Committee for IIASA

Czechoslovakia
The Committee for IIASA of the Czechoslovak Socialist Republic

Finland
The Finnish Committee for IIASA

France
The French Association for the Development of Systems Analysis

German Democratic Republic
The Academy of Sciences of the German Democratic Republic

Federal Republic of Germany
Association for the Advancement of IIASA

Hungary
The Hungarian Committee for Applied Systems Analysis

Italy
The National Research Council

Japan
The Japan Committee for IIASA

Netherlands
The Foundation IIASA – Netherlands

Poland
The Polish Academy of Sciences

Sweden
The Swedish Council for Planning and Coordination of Research

Union of Soviet Socialist Republics
The Academy of Sciences of the Union of Soviet Socialist Republics

United States of America
The American Academy of Arts and Sciences

SERIES EDITOR'S PREFACE

Approach your problems from the right end and begin with the answers. Then one day, perhaps you will find the final question.

'The Hermit Clad in Crane Feathers' in R. van Gulik's *The Chinese Maze Murders.*

It isn't that they can't see the solution. It is that they can't see the problem.

G.K. Chesterton. *The Scandal of Father Brown* 'The point of a Pin'.

Growing specialization and diversification have brought a host of monographs and textbooks on increasingly specialized topics. However, the "tree" of knowledge of mathematics and related fields does not grow only by putting forth new branches. It also happens, quite often in fact, that branches which were thought to be completely disparate are suddenly seen to be related.

Further, the kind and level of sophistication of mathematics applied in various sciences has changed drastically in recent years: measure theory is used (non-trivially) in regional and theoretical economics; algebraic geometry interacts with physics; the Minkowsky lemma, coding theory and the structure of water meet one another in packing and covering theory; quantum fields, crystal defects and mathematical programming profit from homotopy theory; Lie algebras are relevant to filtering; and prediction and electrical engineering can use Stein spaces. And in addition to this there are such new emerging subdisciplines as "experimental mathematics", "CFD", "completely integrable systems", "chaos, synergetics and large-scale order", which are almost impossible to fit into the existing classification schemes. They draw upon widely different sections of mathematics. This programme, Mathematics and Its Applications, is devoted to new emerging (sub)disciplines and to such (new) interrelations as exempla gratia:

- a central concept which plays an important role in several different mathematical and/or scientific specialized areas;
- new applications of the results and ideas from one area of scientific endeavour into another;
- influences which the results, problems and concepts of one field of enquiry have and have had on the development of another.

The Mathematics and Its Applications programme tries to make available a careful selection of books which fit the philosophy outlined above. With such books, which are stimulating rather than definitive, intriguing rather than encyclopaedic, we hope to contribute something towards better communication among the practitioners in diversified fields.

Because of the wealth of scholarly research being undertaken in the Soviet Union, Eastern Europe, and Japan, it was decided to devote special attention to the work emanating from these particular regions. Thus it was decided to start three regional series under the umbrella of the main MIA programme.

The world is full of very large and complex systems. Indeed, at the individual particle, molecule, economic agent, ... level practically all (dynamic) systems, optimization problems are large, usually far too large to handle either analytically or numerically. The problem is, of course, recognized and finding ways to deal with such large complex problems is a relatively new and fast-growing branch of mathematics. Two ways of approach are decomposition and aggregation, the topic of this book. Of course science abounds with aggregate variables such as pressure, national income, amounts of chemical reactants, surface tension, ... These work after a fashion as far as we know, but still an enormous amount of work will have to be done to justify them and to derive the macroscopic equations they satisfy (if any). Major work in this area is e.g. going on in chemistry and in infinity particle system theory.

Of course, exact aggregation and decomposition is too much to hope for. More than approximate aggregation with perhaps a good limit behaviour is not to be expected. Thus approximation and mathematical modelling itself get inextricably mixed up with aggregation and decomposition ideas. And that then is the subject of this book which presents all these matters in an integrated whole.

The unreasonable effectiveness of mathematics in science ...

Eugene Wigner

Well, if you know of a better 'ole, go to it.

Bruce Bairnsfather

What is now proved was once only imagined.

William Blake

As long as algebra and geometry proceeded along separate paths, their advance was slow and their applications limited.

But when these sciences joined company they drew from each other fresh vitality and thenceforward marched on at a rapid pace towards perfection.

Joseph Louis Lagrange.

Bussum, October 1987 **Michiel Hazewinkel**

Table of Contents

Preface

Decomposition and aggregation are the main approaches to the analysis of complex systems. Decomposition consists of partitioning the original problem into a set of independent problems, while aggregation means that a particular set of system state variables is substituted by a single state variable — a so-called aggregate. There has been much research concerned with developing methods for introducing such transformations in optimal decision making problems. However, this research is based on the requirement of full equivalence between the original and transformed problems. Such a condition greatly limits the class of cases for which such transformations are possible at all; moreover, because of this condition the degree of complexity of the transformation procedure becomes compatible with that of direct solution of the original problem. At the same time, decomposition and aggregation are commonly used on the heuristic level, especially in economic studies. Moreover, development of the original model itself is impossible without rational exclusion of certain connections between the process under consideration and its environment and without rational disregard for some small distinctions between elements of the process.

This book is devoted mainly to the development of mathematically rigorous procedures for decomposition and aggregation based on the rejection of the full equivalence requirement. Allowing for approximate correspondence leads naturally to use of the small parameter method (perturbation theory) as a formal basis of the procedures proposed. Thus, this book is concerned with two aspects. First, the consistent derivation of the small parameter method for different types of optimization problems is presented. Second, concrete schemes for approximate decomposition and aggregation of different problems are developed.

In this Introduction the general philosophy of the approach is outlined. The first chapter presents the main results of perturbation theory for finite-dimensional optimization problems. The second chapter is totally concerned with application of these results to different types of economic and engineering problems. In the third chapter, a certain singular case of the perturbation method is considered where transition to a cruder description in which we neglect small factors leads to drastic changes in the results of optimization. The fourth chapter is concerned with certain problems of stochastic systems optimization, once again with some applications. The fifth and the sixth chapters contain results on approximate optimization of dynamic systems with application to suboptimal hierarchical control systems design.

As is indicated by this short survey, the contents of the book will be of interest to specialists in mathematical programming (Chapters 1 and 3), mathematical economics and industrial control (Chapters 1, 2, and 4), and automatic control theory (Chapters 5 and 6).

The basic material is presented in a mathematically rigorous way, though a knowledge of classical analysis and matrix algebra would be enough for readers (except for Chapter 6).

At the same time, readers interested only in applications may without detriment omit those parts of the text (appearing in small print) that are concerned with more complex proofs.

Moreover we can recommend an abridged reading of this book consisting of the Introduction, Sections 1.1–1.3 of Chapter 1; Chapter 2, Sections 4.1 and 4.2; and Chapter 5. A reader who wishes to master the practical side as quickly as possible may wish to concentrate on these sections.

The original plan of this book was approved by S. Emelianov. The constant support of E. Ershov had an important rôle in its creation. Material, obligingly provided by N. Viltshevski, V. Kolmanovski, O. Kultshitski, V. Plotnikov, and V. Savtshkov was very useful. Professor V. Findeisen and his colleagues have to be mentioned especially. With their assistance a brief course and useful discussions were organized in Polytechnika Warszawska. Critical remarks by E. Piatnetski and L. Rosonoer, who were reviewers of the original text, are taken into account with gratitude.

Special thanks to our wives and colleagues T. Pervozvanskaja and B. Gaitsgori, who not only suffered with us but also were largely responsible for our task of assembling all of the material.

The original text of the book was published in Russian in 1979. However, the USSR Committee for System Analysis, as early as in 1978, had initiated talks with IIASA on simultaneous publication of this book in English in the international series of books on systems analysis. These talks continued for rather a long time, so that the final decision on inclusion of the book in the series was reached after the Russian edition had already been published. This delay in fact turned out to be very useful. It permitted us to completely overhaul the main text and supplement it by certain new results.

Chapter 1 has been written anew with mathematical results presented in a more effective form, and the proofs made easier. It also became possible to find an important systems analytic interpretation of the main decomposition and aggregation scheme, as a tool for developing suboptimal decisions in hierarchical systems. In Chapter 3 an interesting scheme for the regulatization of singular problems was included. The material concerned with synthesis of suboptimal linear regulators was extended and presented separately in Chapter 5. Of course, the necessity of preserving a reasonable size of the book dictated the removal of many cumbersome proofs; perhaps this has made the book better for those readers who are not mathematicians.

The comments by reviewers, particularly by Dr. Geoffrion, were very useful in preparing the new text, which was delivered to IIASA in 1981. IIASA's Editorial

Committee provided invaluable support in preparing the text, which was directed it to the D. Reidel Publishing Company in 1984. Some new changes were introduced thereafter. Editing notes by Dr. J.F. Bonnans and Dr. F. Delebecque were essential, although full responsibility for any English language problems is taken by the authors, for whom it was the first serious trial.

A.A. Pervozvanskii
V.G. Gaitsgori

Leningrad
January 1987

Introduction

The perturbation method is one of the main tools of applied mathematics. Its fundamental idea suggests that the description of a complicated system is divided into the basic ("framework") structure and the detailed ("fine") structure. The detailed structure is then treated as a perturbation of the basic reduced scheme, and the behavior of a system as a whole is approximately estimated by certain corrections to the framework scheme analysis; calculation of these corrections is much simpler than direct investigation of a complicated system.

In theoretical physics, immense experience of applications of the perturbation method has been accumulated. It is this method that has permitted us to overcome in many cases the barrier between highly idealized models and those of realistic complexity. It should also be noted that, notwithstanding the standard nature of adopted procedures, the applications of the perturbation method are by no means standard because deep penetrations into the essence of a problem is required to distinguish between the main and secondary factors. The basic structure is usually outlined intuitively or from experience (which itself is, to some extent, heuristic); but the schemes of the perturbation method allow us to gauge the power of these heuristic notions and, whenever necessary, to obtain improved results.

The problems of theoretical physics investigated with the aid of perturbation method are usually of a descriptive character. Mathematically, they reduce to the requirement of solving some equation. In view of this, the perturbation method has been the subject of as intensive development as that of construction of approximate solutions of equations. At the same time, the "cybernetics" approach to engineering, economic, and other problems necessitated investigation of a new type of mathematical problem — that of optimization on closed sets; such problems do not reduce to the analysis of equations.

The relevant topics are considered in mathematical programming and in the mathematical theory of optimal control. Historically, these disciplines arose independently, having been connected with different practical problems. The former was stimulated by problems of planning economic and military operations; and the latter, by problems of control of engineering systems. Nowadays, the theoretical foundations of both disciplines have almost completely coalesced, and it is reasonable to speak about a united mathematical trend — the theory of optimization. some differences have, however, remained both in the sphere of applications and in the very style of investigation.

Mathematical programming developed first as a numerical, not analytical, discipline. Its elements are usually described algebraically, mainly in finite-dimensional representations. On the other hand, the mathematical theory of optimal control arose as an analytical discipline, closely connected with the theory of differential equations. These circumstances have to some extent predestined the role of the perturbation method in the development of both disciplines.

The numerical preoccupations of mathematical programming have resulted in a certain underestimation of the value of analytical approaches in general and the perturbation method in particular. The growing power of computational techniques promotes the idea of the possibility and expediency of direct numerical solution of arbitrarily complicated problems. It is of practical importance, however, to suggest solutions that allow analytical representation, even if it is incomplete and approximate. In particular, these solutions considerably simplify implementation of algorithms for automatic systems where optimal solutions need to be worked out in real time in accordance with rapidly changing conditions. Moreover, within the framework of a purely numerical scheme, application of the classical procedures often becomes impossible without taking into account the intrinsic features of a problem, and one has to resort to various simplifications and use some heuristic recipes. These remarks relate particularly to nonlinear problems.

Even taking account of the great variety of procedures, one may say that almost all of them rest on a few central ideas which may be formulated briefly as follows:

(a) to perform decomposition, neglecting weak connections between particular subsystems;
(b) to aggregate these subsystems, using the smallness of differences among them; and
(c) taking strong differences into account, to outline the "bottlenecks", having withdrawn unimportant constraints on the basis of *a priori* estimates.

One must bear in mind that these ideas come into play even before the application of any numerical procedures at the stage of initial modeling. For any engineering or economic problem, the starting point in modeling is classification of the products and resources (see, e.g., Chapter 1.4 herein).

The introduction of a finite set of nomenclatures is itself a result of neglecting small differences among the different types of products and resources (for example, all the processing units may be considered as one and the same resource, irrespective of the differences in their periods of exploitation). Taking account of only some of the external and internal factors restricting the capabilities of a system is undoubtedly connected with the intuitive prognosis of the bottlenecks.

Last, the elements of optimization themselves may be distinguished from a set of genuinely interacting parts of the global system only when we neglect, for heuristic reasons, certain mutual connections, or when these connections depend weakly on the decision taken.

On the other hand, one may suppose that the initial description of a system is given with maximal completeness and detail, and that a reasonable simplification is brought about by following specific procedures motivated by intuition, containing the outlined heuristic ideas and at the same time allowing distinct logical justification within the framework of the method of perturbations.

These considerations justify wide application of the method of perturbations even to finite-dimensional problems of mathematical programming; this point, however, has only recently come to be appreciated.

A different situation has arisen in optimal control theory. The application of the method of perturbations (the small parameter method) is traditional for the theory of differential equations, and it is natural that this method should rapidly penetrate optimization theory. However, serious interest in it arose only very recently in connection with problems concerning really complicated systems, particularly technological systems.

It is necessary to note that the division of dynamics into "fast" and "slow" is the basic heuristic principle in the theory of dynamic systems. This method is intrinsically connected with the procedures of aggregation and decomposition as described. In fact, the supposition that the motion is "slow" corresponds to the hypothesis of small differences between the values of variables describing the consecutive states of a system. To a first approximation, therefore, these variables may be replaced by a single aggregated variable, which is the same as substituting the constant state for a varying one.

On the other hand, the hypothesis that the motion is "fast" obviously coincides with that of weakness of connections between consecutive states, so that we are free to treat them as independent to a first approximation. The resemblance in the underlying heuristic ideas results in the similarity of procedures employed in the different branches of perturbation theory, though this similarity is not so obvious and has remained unnoticed for a long time.

The use of perturbation theory in solving optimal control problems allows considerable expansion of the scope for finding optimal trajectories as well as for solving the problem of control synthesis. The very structure of the feedback is well-matched to the special structural and dynamic features of the system, and this simplifies practical implementation.

In conclusion, it is necessary to point out that the method of perturbations in the theory of optimization is not only based upon wide use of intuitive ideas and concepts at the formulation stage of a problem, but also stimulates intuitive understanding of the special features of a solution.

CHAPTER 1

The Perturbation Method in Mathematical Programming

First of all, a general description of the perturbation method is given here. The main idea of the method consists of dividing an original program into a simple reduced structure and into perturbations. The perturbations are supposed to be small, and their level is characterized by a parameter ϵ. A goal is to construct a suboptimal solution to the original program using information about the reduced solution and, to some degree, properties of the perturbations. In this connection, the meaning of the terms "simplicity" and "smallness" ("proximity") is discussed in a semi-formal way in Section 1.1.

Important classes of reduced programs, called directly decomposable and aggregatable classes, are introduced. After that a procedure of suboptimal decision making in hierarchical organizations is described as a formal consequences of the perturbation method.

The general conception outlined in Section 1.1 is basic to the application of the perturbation method to any optimization problem, although the presentation given here is for finite-dimensional mathematical programs (MP) only.

Later, the principal problem of the proximity of the reduced and perturbed solutions is investigated more formally and systematically, it being supposed that the descriptions of the reduced and perturbed programs coincide within terms that are proportional to the small parameter ϵ.

The formal presentation opens in Section 1.2 with linear programs as the simplest case. To begin with, the dependence of the perturbed solution with respect to ϵ is described in the simple situation when the solutions to both the reduced program and its dual are unique. Thereafter, more general situations are considered, in which either the reduced solution or the dual solution is nonunique. An auxiliary program is constructed to isolate an asymptotically optimal solution to the perturbed program among the reduced solution set. As a consequence, a modified simplex procedure is obtained for the numerical solution of linear programs. Section 1.2 concludes with a general theorem whose conditions permit nonuniqueness both for the primal reduced solution and for the dual solution. However, an essential hypothesis is used: the dual solution set is bounded. An example shows that the violation of this condition may lead to perturbed dual solutions that tend to ∞ for $\epsilon \to 0$; which, in turn, implies a gap in the perturbed optimal value at the point $\epsilon = 0$, and hence the correspondence between the perturbed and reduced solutions vanishes.

Section 1.3 is devoted to the simplest nonlinear programs in which only the objective function is perturbed. At first, some simple estimates on suboptimality of the reduced solution are constructed, which are applicable for any ϵ. After that, the convergence of the perturbed solutions (as $\epsilon \to 0$) to the reduced solution is established if the latter is unique. It is also shown that the perturbed solution converges toward the auxiliary program solution if the reduced solution is nonunique. To justify the convergence one needs to use a boundedness condition on the perturbed solution set. The importance of this condition is illustrated by an example and, after that, restrictions that are necessary and sufficient for boundedness are formulated. In conclusion, an optimal numerical search procedure based on the perturbation method is described. This procedure may be effective in overcoming difficulties connected with "ridges". Note that familiarity with Sections 1.1–1.3 is sufficient for an understanding of the applied problems described in Chapter 2. Additional sections contain more complicated material, which will be mainly of interest to specialists in mathematical programming.

Section 1.4 is devoted to perturbed programs with associated convex reduced programs. In the first part, some classical results concerning necessary and sufficient conditions for an extremum are formulated. Various types of regularity conditions, which ensure the existence of Lagrange multipliers at the extreme, are introduced. One of them is equivalent to nonemptiness and unboundedness of the set of Lagrange multipliers. Supposing that this regularity condition is met and that the reduced solution is unique, the convergence of the perturbed solution to the reduced one is established. After that, some attention is given to constructing correction terms in the form of ϵ-expansions. The last part of Section 1.4 contains a general theorem giving conditions for convergence, in situations where both the primal and dual reduced solutions are nonunique.

Perturbed programs having nonconvex reduced programs are considered in Section 1.5. The convergence of the perturbed solution to the auxiliary program solution is proved here, too. In conclusion some precise estimates are constructed.

1.1. Formulation and peculiarities of problems

Consider a mathematical program

$$\max \{f_0(x) \,/\, g_0(x) \leq 0\} \triangleq F^*(0) \tag{1.1}$$

where x is an n-vector and $g_0(x)$ is an m-vector function. Along with this consider the family of programs

$$\max\{f_0(x) + \epsilon f_1(x) \,/\, g_0(x) + \epsilon g_1(x) \le 0\} \triangleq F^*(\epsilon) \tag{1.2}$$

characterized by a scalar non-negative parameter ϵ. The program (1.1) is called the reduced program with respect to the family (1.2), which, in turn, is called the perturbed program. The functions $f_1(x)$, $g_1(x)$ determine the structure of the perturbations, and the parameter ϵ determines their intensity.

The problem of how to relate solutions to (1.1) and to (1.2) arises in the following situations.

(a) Let the program (1.1) be solved. One needs to have a description of the solutions to the family (1.2) without solving each program for particular values of the parameter. It is assumed implicitly that a complexity of a solution to (1.2) is the same for any fixed value of ϵ, including $\epsilon = 0$. Such problems are covered by the theory of *parameter programming.* Note that, as a rule, parametric programming investigates (1.2) globally, i.e., for a large interval of ϵ value.

(b) Let the functions $f_0(x)$, $g_0(x)$ be given, but let the nature of $f_1(x)$, $g_1(x)$ be unknown and, to a considerable extent, arbitrary. Then the correctness problem arises for (1.1). It is unclear whether the solution to the perturbed program (1.2) is close to the reduced program solution if the intensity of perturbation is small.

Such problems are the subject of the theory of stability of solutions to mathematical programs. An important aspect of this theory is the regularization problem: when (1.1) has a nonunique solution, perturbations are introduced artificially to separate out a solution having given properties.

(c) Let one need to solve the perturbed program for some particular value ϵ_0 of the parameter ϵ. Suppose that a direct solution to that program is difficult, but that a solution to the reduced one is comparatively simple. Then it is reasonable to solve the original program by means of a successive approximations procedure, where the first phase is the solution to the reduced program and the second phase is the computation of a correction to it.

For that purpose, one also needs to investigate the solution behavior over some interval of the parameter ϵ, but this investigation is carried out with the object of solving the problem for $\epsilon = \epsilon_0$, i.e., the original one.

Using the terminology of R. Bellman, one can say that the original program is imbedded in a family of programs distinguished by their ϵ values.

This third situation refers to perturbation theory proper and is the main subject of this book.

An explicit description in the form (1.2) may be not given, and this is typical in applications of the perturbation theory. If the original problem is presented in the general form

$$\max\{f(x)/g(x) \le 0\} \tag{1.3}$$

one can write it in the form (1.2) by setting:

$$\epsilon f_1(x) = f(x) - f_0(x) \ ; \ \epsilon g_1(x) = g(x) - g_0(x) \tag{1.4}$$

The way one introduces functions $f_0(x)$, $g_0(x)$ and states the value of ϵ is arbitrary, but, of course, the real success of the perturbation method procedure is essentially determined by the following two factors: the first one is simplicity of the reduced program, and the second one is proximity of the functions describing the original program and the reduced program.

The terms "simplicity" and "proximity" are used informally here, and the selection of a reduced program from the original one is just taken to be a heuristic procedure.

It is necessary to explain in detail what "simplicity" means. We shall not try to provide any rigorous formalization of simplicity for mathematical programs. However, we note that linear programs constitute the simplest class. Complexity increases as we consider convex programs, concave programs, and arbitrary multi-extremal problems.

Within the limits of each class, complexity is determined by the program's size. But, for any fixed size, a solution can be simplified if one uses the specific structure of the programs considered.

Let us examine some ideal types of structure admitting direct simplification of programs.

1. Let a program have the form

$$\max\{F(z_1(x^1),\dots,z_k(x^k)) \,/\, x^s \in X_s \;,\; s=1,2,\dots,k\} \tag{1.5}$$

where $F(z_1,\dots,z_k)$ is a monotonically increasing function of its arguments, x^s are n_s-vectors, and the X_s's are subsets of n_s-dimensional euclidean spaces. To solve this program it is sufficient to solve k independent programs

$$\max\{z_s(x^s) \,/\, x^s \in X_s\} \;,\; s = 1,2,\dots,k \tag{1.6}$$

If a program has the structure (1.6) one can call it a *directly decomposable program.*

2. Let a program have the form

$$\max\{F[z(x)] \,/\, g[z(x)] \le 0 \;,\; x \in X\} \tag{1.7}$$

where $z(x)$ is a scalar function of the n-vector x, $X \subset E^n$, and $F(z)$ is an arbitrary function. Suppose one can solve the aggregated program

$$\max\{F(x) \,/\, g(z) \le 0 \;,\; \check{z} \le z \le \hat{z}\} \tag{1.8}$$

where

$$\check{z} \triangleq \min\{z(x)/x \in X\} \;,\; \hat{z} \triangleq \max\{z(x)/x \in X\}$$

If z^* is a solution to (1.8) and there exists a solution x^* to the equation

$$z(x) = z^* \tag{1.9}$$

then x^* is a solution to (1.7), too. Programs in the form of (1.7) may be named *directly aggregatable.*

3. Let us consider a program

$$\max F[z_1(x^1),\ldots,z_k(x^k)] \tag{1.10}$$

such that

$$g[z_1(x^1),\ldots,z_k(x^k)] \leq 0 \ , \ x^s \in X_s \ , \ s = 1,\ldots,k$$

where notation is in accordance with (1.5) but $F(z_1,\ldots,z_k)$ is not necessarily monotonic. One can obtain solutions z_s^* to the aggregated program

$$\max \{F(z_1,\ldots,z_k) \,/\, g(z_1,\ldots,z_k) \leq 0 \ , \ \check{z}_s \leq z_s \leq \hat{z}_s \ , \ s = 1,\ldots,k\} \tag{1.11}$$

where

$$\check{z}_s \triangleq \min \{z_s(x^s) \,/\, x^s \in X_s\} \ , \ \hat{z}_s \triangleq \max \{z_s(x^s) \,/\, x^s \in X_s\}$$

if there exist† solutions x^{s*} of the equations

$$z_s(x^s) = z_s^*$$

then $x^* = (x^{s*})$ is a solution to (1.10), too. Evidently, (1.10) is a generalization of (1.5) and (1.7). One can call (1.10) a *directly aggregatable and decomposable program.*

In what follows we suppose that a reduced program differs from the original program in its "simplicity": either a reduced program belongs to a less general class (e.g., the original program is nonlinear, but the reduced one is linear), or the reduced program belongs to the same class as the original program, but admits direct aggregation and or decomposition.

† This is evidently true if the X_s's are any connected sets and $z_s(x^s)$ are any continuous functions.

We shall now explain some formal and informal aspects of the notion of "proximity".

As a formal measure of the proximity of the original program to the reduced program, one can use a cost

$$\Delta \triangleq \sup \{\sup_{x \in \bar{X}} |f(x) - f_o(x)| \ , \sup_{x \in \bar{X}} |g_l(x) - g_{0l}(x)| \ , l = 1,\ldots,\omega.\} \tag{1.12}$$

where $\bar{X}$ includes the feasible set Θ of the original program. If Δ is finite, then it can be viewed as a measure of perturbation intensity in this program setting $\Delta = \epsilon_0$. Then by introducing

$$f_1(x) = \frac{1}{\epsilon_0}[f(x) - f_0(x)] \ ; \ g_1(x) = \frac{1}{\epsilon_0}[g(x) - g_0(x)] \tag{1.13}$$

one can write the perturbed program (1.12), which coincides with the original one for $\epsilon = \epsilon_0$.

Generally speaking, the less Δ is, the closer the solution to the original program will be to that of the reduced one. The simplest accuracy estimates (see below, Section 1.2, and also [115–118]) permit one to express errors of the solution approximation as functions of the objectives' proximity. Therefore, it seems reasonable to separate the reduced program by means of a previous procedure of objective approximation, i.e., to choose functions $f_0(x)$, $g_0(x)$ such that they are closer to $f(x)$, $g(x)$, the "simplicity" conditions being satisfied at the same time.

Some examples of effective approximation schemes for specific problems are shown below (see Sections 1.2 and 2.2). In general, however, the approximation problem is difficult (especially if it is stated as an optimal one) and may be as complex as the original one. Moreover, it should not be supposed that significant deviations in the objectives always imply correspondingly significant errors in the resulting approximate solutions. This is obvious in *Figure 1*, where we have two one-dimensional programs with widely different objectives $f(x)$, $g(x)$, and in *Figure 2*, where one can see two programs with identical objective $f(x)$ and strongly differing feasible sets Θ, Θ_0. In both cases, the optimal solutions to the different programs coincide.

Since *a priori* estimates are, as a rule, very rough, they are seldom used. This is also true of formal schemes of (1.12) and (1.13) type.

In many practical cases the perturbed structure is constructed heuristically without any estimates. The parameter ϵ_0 is chosen as a specific numerical parameter of a problem, e.g., if one has the program

$$\max \, [x_1 + x_2 + 0.1x_1^2] \tag{1.14}$$

such that

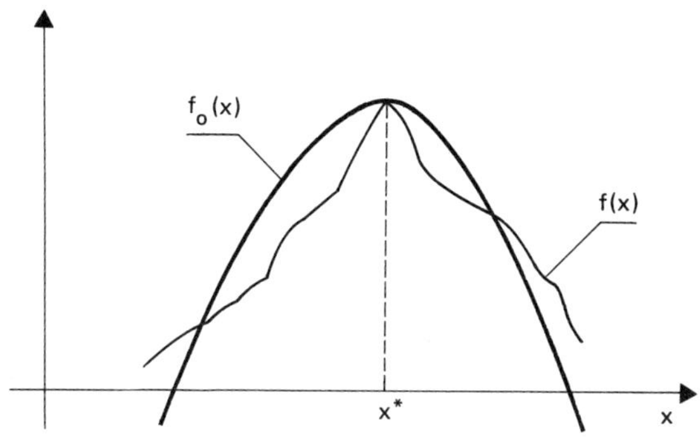

Figure 1.

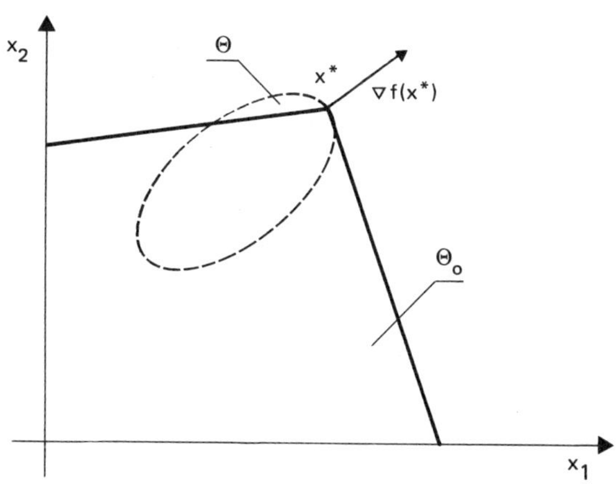

Figure 2.

$$x_1 + x_2 + 0.2x_2^2 \leq 1 \ ; \ x_1 \geq 0 \ ; \ x_2 \geq 0$$

one can put $\epsilon_0 = 0.1$ and investigate the perturbed program

$$\max \, [x_1 + x_2 + \epsilon x_1^2]$$

such that

$$x_1 + x_2 + 2\epsilon x_2^2 \leq 1 \ ; \ x_1 \geq 0 \ , \ x_2 > 0$$

One can check whether a solution that is found heuristically is optimal or suboptimal *a posteriori*. Moreover, if the obtained accuracy is not satisfactory, the perturbation method allows scope for improvement. [In particular, the perturbation method gives the precise solution of (1.14) by some standard correction scheme.]

It is now clear that, although the perturbation method procedure is theoretically applicable to almost any mathematical program, its successful application depends on the specific features of the original program.

The perturbation theory cannot give universal algorithms for mathematical programs. However, it is clear that many practical large-scale programs may be adapted to the perturbation theory approach. This statement is based on the knowledge of "weak connections" and "small differences" effects, mentioned in the Introduction, which are specific features of a variety of models of real large-scale systems.

Mathematical models of large-scale systems optimization are close to the structure of (1.5) if the state of each subsystem is determined by its "own" vector x^s and if the activities of each subsystem depend on those of other subsystems only weakly. If a mathematical model is close to the structure exhibited by (1.6), then the total interaction of subsystems does not depend on the details of large-system behavior but only on the aggregated components. A more complete description of the various types of problem that are amenable to the perturbation method will be given in Chapter 2, following a presentation of some basic mathematical results.

However, before this presentation, let us establish an important feature of programs that are close to being directly aggregatable and decomposable.

A formal description of such programs has the form

$$\max \, \{F_0[z_1(x^1),\ldots,z_k(x^k)] + \epsilon F_1(x)\} \tag{1.15}$$

such that

$$g_0[z_1(x^1),\ldots,z_k(x^k)] + \epsilon g_1(x) \leq 0 \tag{1.16}$$

$$x = (x^s \ , \ s = 1,\ldots,k) \qquad x^s \in X_s \tag{1.17}$$

This description can be interpreted in terms of the theory of hierarchical organization. Let the performance index of an organization in the large depend on the aggregated indices of the individual activities of subsystems. Let the general constraints be expressed chiefly in the aggregated indices allowing for some weak direct dependence on the activities of subsystems. These activities are restricted, however, by the industrial capacity of each subsystem, which is expressed by conditions (1.17).

The reduced program corresponding to (1.15)–(1.17) has the form (1.10) and can be aggregated into that of program (1.11). Its solution can be naturally interpreted as the optimal schedule (plan) of a "coordinating center", and it is expressed in terms of the aggregated indices. Note that the "coordinator" must obtain information about limiting feasible values of these indices, $\check{z}_s$, $\hat{z}_s$, which are numerical measures of the subsystems' capabilities.

The central schedule does not determine the subsystems' activities completely since the same value of the index z_s^* can be obtained at different values of the vector x^s satisfying the equation

$$z_s(x^s) = z_s^*$$

It is not clear, however, whether all these values are equivalent or there exists a value that is closer to the total program solution. The perturbation method allows one to obtain the necessary correction. As will be shown later, one needs to solve an auxiliary program

$$\max \{F_1(x) - \Lambda^T g_1(x) \,/\, z(x) = z^* \ , \ x^s \in X_s\} \tag{1.18}$$

where Λ is the vector of Lagrange multipliers corresponding to the aggregated program's constraints. (This result is valid under some additional requirements on the perturbed program's properties, which have been omitted in the previous discussion.) The solution x^* to the auxiliary program determines, as a rule, an element of the aggregated problem solution set that is close to the perturbed program solution for small ϵ. (The theory also gives the decrease in the performance index if x^* is used in place of the optimal solution to the total program.)

The meaning of the auxiliary program is clear. To choose the best from the aggregated problem set, one needs to improve the resolution of the method by taking into account small quantities omitted for the construction of the aggregated program but which now play the main role. Small perturbations of the performance index are directly introduced in the auxiliary program objective function, and the significance of perturbations arising in the constraints is taken into account by means of Lagrange multipliers. This second term in the objective function can be interpreted as the cost of resource expenditures owing to additional activities. This interpretation gives rise to a traditional presentation of Lagrange multipliers as some objectively stipulated estimates of resources, which was introduced by L. Kantrovich for an economic interpretation of linear programs.

If the functions $F_1(x)$, $g_1(x)$ are separable, i.e., if they may be written in the form

$$F_1(x) = \sum_s F_{1s}(x^s) \quad ; \quad g_1(x) = \sum_s g_{1s}(x^s)$$

then the correction for each subsystem can be made independently by the solutions to the following programs

$$\max \left[F_{1s}(x^s) - \Lambda^T g_{1s}(x^s)\right]$$

such that

$$z_s(x^s) = z_s^* \quad ; \quad x^s \in X_s$$

It seems, in practice, that the procedure described above is a reasonable mathematical model of some scheduling procedures used in large organizations. In the first phase, aggregated knowledge of subsystems' industrial capacities is prepared on the "micro-level" and this information is then transmitted to the coordinating center. The second phase is the "macro-level" scheduling, where the analysis is made using the aggregated indices only. Their optimal values and estimates of the resources are communicated to the subsystems. In the third phase, the subsystems create their detailed schedules on the basis of those data and their own information.

Recall also that a practical coordination procedure does not pretend any rigorous optimality. In practice, one needs a "good solution" that is in some sense close to the optimal one.

In conclusion, we make some remarks about connections between the decomposition approach and the aggregation approach, which are given by the perturbation method, with well-known iterative schemes initially suggested by the pioneering work by G. Dantzig and F. Wolfe [23]. To consider this problem here in detail would be unreasonable, since there are many summary presentations (see, e.g., [23, 48, 64, 141, 126]).

Some comparisons will be given, in connection with more specific problems, in Chapters 2 and 3; but it is already clear that the iterative procedures have more powerful computational scope and are necessary to obtain precise solutions. At the same time, the perturbation method approach is simpler, and is applicable to many practical problems. Its computational efficiency can be excellent under some specific conditions (see below, Section 2.2). On the other hand, the perturbation method procedure can be more attractive to researchers in system analysis since the three-phase scheme that was obtained above as a direct consequence of this method might be used as a model of real decision processes, and it seems that this model is closer to reality than the infinite-step iterative schemes.

1.2. Perturbations in linear programs

Linear programming problems are the most deeply investigated and possibly the most important problems in optimization theory. It is natural to begin a formal description of the perturbation theory approach with these problems, especially as one can then obtain the most complete and simple results. In correspondence with the general definition, we introduce the perturbed linear program

$$\max \{c^T(\epsilon)x \,/\, A(\epsilon)x = b(\epsilon) \ ; \ x \geq 0\} \triangleq F^*(\epsilon) \tag{2.1}$$

and the reduced one

$$\max \{c_0^T x \,/\, A_0 x = b_0 \ ; \ x \geq 0\} \triangleq F^*(0) \tag{2.2}$$

where $A(\epsilon)$ is an $(m \times n)$-matrix, $b(\epsilon)$ is an m-vector, $c(\epsilon)$ is an n-vector and

$$A_0 \triangleq \{A_{0j} \,,\, j = 1,\ldots,n\} = A(0)$$
$$b_0 \triangleq (b_{0i} \,,\, i = 1,\ldots,m) = b(0)$$
$$c_0 \triangleq (c_{oj} \,,\, j = 1,\ldots,n) = c(0)$$

We shall investigate the behavior of solutions $x^*(\epsilon) \triangleq (x_j^*(\epsilon) \,,\, j = 1,\ldots,n)$ to (2.1) for small $\epsilon > 0$ and their connection with solutions $x_0^* \triangleq (x_{0j}^* \,,\, j = 1,\ldots,n)$ to the program (2.2). To begin with, consider the well-known simplest case studied as far back as in [144]. Let the solution x_0^* to the reduced program be unique and, moreover, let the solution λ^* to the dual program

$$\min \{b_0^T\lambda / A_0^T\lambda \geq c_0\} \tag{2.3}$$

also be unique.

In this case, there exists (see, e.g., [21, 23]) a nonsingular $(m \times n)$-submatrix A_b of the matrix A_0 such that the following conditions are fulfilled:

$$A_b x_b^* = b_0 \tag{2.4}$$
$$x_b^* > 0 \tag{2.5}$$
$$A_b^T\lambda^* = c_b \tag{2.6}$$
$$A_{\bar{b}}^T\lambda^* > c_{\bar{b}} \tag{2.7}$$

Here, as usual, x_b^* is an m-vector of *basic variables* such that

$$x_b^* = (x_j^* \ , j \in J_0) \ ; \ x_j^* = 0 \ ; \ j \bar{\in} J_0$$

J_0 is the optimal *basic index set*, so that

$$A_b \triangleq \{A_{0j} \ , j \in J_0\}$$
$$A_{\bar{b}} \triangleq \{A_{0j} \ , j \bar{\in} J_0\}$$
$$c_b \triangleq \{c_{0j} \ , j \in J_0\} \ ; \ c_{\bar{b}} \triangleq \{c_{0j} \ , j \bar{\in} J_0\}$$

The condition (2.7) guarantees the uniqueness of the solution and is equivalent to the requirement that all *column estimates*

$$\Delta_j \triangleq A_{0j}^T \lambda^* - c_{0j} \tag{2.8}$$

corresponding to the nonbasic columns, $j \bar{\in} J_0$, are strictly positive.

The condition (2.5) is called the *nondegeneracy* condition and guarantees the uniqueness of the solution λ^* to the dual program since the *complementarity slackness* conditions

$$x_j^* \Delta_j = 0 \tag{2.9}$$

must be fulfilled for all j. [The equation (2.6) determining λ^* uniquely follows from (2.5) and (2.9)].

Let us write the optimality conditions for the perturbed program: $x(\epsilon)$ is an optimal solution to (2.1) if and only if there exists such a vector $\lambda(\epsilon)$ that

$$A(\epsilon)x(\epsilon) = b(\epsilon) \ ; \ x(\epsilon) \geq 0 \tag{2.10}$$

$$A^T(\epsilon)\lambda(\epsilon) \geq c(\epsilon) \tag{2.11}$$

$$x^T(\epsilon)[A^T(\epsilon)\lambda(\epsilon) - c(\epsilon)] = 0 \tag{2.12}$$

[The conditions (2.10) and (2.11), respectively, express the primal and dual feasibility; (2.12) is the complementarity slackness condition.]

Let us seek $x(\epsilon)$, $\lambda(\epsilon)$ in the form

$$\begin{aligned} x(\epsilon) &= [x_b^*(\epsilon), x_{\bar{b}}^*(\epsilon)] \ ; \\ x_b^*(\epsilon) &\triangleq [x_j^*(\epsilon), j \in J_0] = P(\epsilon)b(\epsilon) \ ; \ x_{\bar{b}}^*(\epsilon) = 0 \end{aligned} \tag{2.13}$$

$$\lambda^*(\epsilon) = P^T(\epsilon)c_b(\epsilon) \tag{2.14}$$

where

$$P(\epsilon) \triangleq A_b^{-1}(\epsilon) \;;\; A_b(\epsilon) \triangleq \{A_j(\epsilon) \;,\; j \in J_0\}$$
$$c_b(\epsilon) \triangleq (c_j(\epsilon) \;,\; j \in J_0)$$

Note that $A_b(\epsilon)$, $c_b(\epsilon)$ include columns (elements) with the same indices as A_b, c_b determining the reduced solution.

Thus, by giving the structure of primal and dual solutions in the form (2.13) and (2.14), the *hypothesis of optimal basic set invariance* is realized. Let us prove that this hypothesis is valid for sufficiently small ϵ. To that end, substitute (2.13) and (2.14) in the optimality conditions. Evidently, it is sufficient to check only the validity of the inequalities:

$$P(\epsilon)b(\epsilon) \geq 0 \tag{2.15}$$
$$A_{\bar{b}}^T(\epsilon)P^T(\epsilon)c_b(\epsilon) \geq c_{\bar{b}}(\epsilon) \tag{2.16}$$

Call to mind that, by (2.5) and (2.7),

$$P(0)b(0) > 0 \tag{2.17}$$
$$A_{\bar{b}}^T(0)P^T(0)c_b(0) > c_{\bar{b}}(0) \tag{2.18}$$

If the functions $A(\epsilon)$; $b(\epsilon)$; $c(\epsilon)$ are sufficiently smooth, then for small ϵ, (2.17) and (2.18) imply (2.15) and (2.16), i.e., the hypothesis is true.

Let us formulate this result more rigorously.

Theorem 2.1.

Let $A(\epsilon)$; $b(\epsilon)$; $c(\epsilon)$ be continuous functions for $\epsilon \in [0\;;\epsilon']$ where $\epsilon' > 0$. Let the solution to the reduced program (2.2) be unique and nondegenerate. Then there exists a positive $\epsilon'' \leq \epsilon'$ such that, for all $\epsilon \in [0\;;\epsilon'']$, the optimal basic set of the perturbed program (2.1) is invariant, and its unique solution is given by the formula (2.13).

Corollary 1. Moreover, let $A(\epsilon)$; $b(\epsilon)$ be differentiable up to and including order m. Then there exists an ϵ'' such that, for $\epsilon \in [0\;;\epsilon'']$, the solution $x^*(\epsilon)$ may be written in the form

$$x^*(\epsilon) = x_0^* + \epsilon x^{(1)} + \cdots + \epsilon^m x^{(m)*} + o(\epsilon^{m+1}) \tag{2.19}$$

where the vectors $x^{(k)}$ can be calculated recursively. If in addition $A(\epsilon) = A_0 + \epsilon A_1$; $b(\epsilon) = b_0 + \epsilon b_1$ then

$$x^{(1)} = -P(0)(A_1 x_0^* - b_1) \ ; \ x^{(k+1)} = -P(0)A_1 x^{(k)}$$
$$k = 1,2,\ldots,m+1 \tag{2.20}$$

Proof of (2.19) is a consequence of the application of the implicit function theorem to the equation

$$A_b(\epsilon)\, x_b(\epsilon) = b(\epsilon)$$

Formula (2.20) can be obtained by substituting (2.10) into this equation.

Corollary 2. Under the conditions of the theorem, the optimal objective value of the perturbed program is given by the explicit formula

$$F^*(\epsilon) = c_b^T(\epsilon)P(\epsilon)b(\epsilon) \ , \ 0 \le \epsilon \le \epsilon'' \tag{2.21}$$

If in addition

$$A(\epsilon) = A_0 + \epsilon A_1 \ ; \ b(\epsilon) = b_0 + \epsilon b_1 \ ; \ c(\epsilon) = c_b + \epsilon c_1 \tag{2.22}$$

then the asymptotic form

$$F^*(\epsilon) = F_0^* + \epsilon F_1^* + O(\epsilon^2) \ , \ 0 \le \epsilon \le \epsilon'' \tag{2.23}$$

can be used where

$$F_1^* = c_1^T x_0^* + b_1^T \lambda^* - (\lambda^*)^T A_1 x_0^* \tag{2.24}$$

Proof. The formula (2.21) is implied by (2.13). The expansion (2.23) is easily obtained if one notes that

$$\begin{aligned} P(\epsilon) &\triangleq A_b^{-1}(\epsilon) = A_{b0}^{-1}[I + \epsilon A_{b1}A_{b0}^{-1}] = \\ &= P(0)[I - \epsilon A_{b1}P(0)] + O(\epsilon^2) \Rightarrow \\ P(\epsilon) &= P(\epsilon) - \epsilon P(0)A_{b1}P(0) + O(\epsilon^2) \end{aligned} \tag{2.25}$$

and substitutes (2.23) into (2.21).

Thus, we have a complete result for the case of uniqueness and non-degeneracy. We now show that, under some additional conditions, a result of similar simplicity can be obtained for the nonuniqueness case, which is of special relevance to the perturbation theory.

Theorem 2.2.

Let (2.22) be true. Let Θ^* be the solutions set to the reduced program and let Λ^* be the unique dual solution. If there exists the unique solution x^{**} to the auxiliary program

$$\max\left[(c_1 - A_1^T\lambda^*)^T x \,/\, x\in\Theta^*\right] \triangleq J_1^* \tag{2.26}$$

and x^{**} is simultaneously a nondegenerate basic solution to the reduced program, then the asymptotic presentation (2.23) is valid with

$$F_1^* = J_1^* + b_1^T\lambda^* \tag{2.27}$$

Moreover, for $0 < \epsilon < \epsilon''$, there exist solutions $x^*(\epsilon)$, $\lambda^*(\epsilon)$ to both the perturbed and dual perturbed programs, which may be represented in the form (2.13) and (2.14) where J_0 is the optimal basic set of the reduced program corresponding to x^{**}. Hence the expansion (2.19) is also valid.

Proof. By virtue of the duality theorem, $x \in \Theta^*$ is equivalent to the fact that x satisfies the conditions

$$A_0x = b_0 \;,\; x \geq 0 \;,\; x^T(A_0^T\lambda^* - c_0) = 0 \tag{2.28}$$

Since $x^{**} \in \Theta^*$ and is a nondegenerate basic solution, we have

$$\begin{aligned}
&x^{**} = (x_b^{**},0) \;,\; x_b^{**} = P(0)b > 0 \;,\\
&P(0) = A_b^{-1} \;,\; A_b \triangleq \{A_{0j} \;,\; j\in J_0\} \;,\\
&\Delta_j(0) \triangleq \lambda^{*T}A_{0j} - c_{0j} \;,\; \lambda^* = P(0)c_b \;,\\
&\Delta_j(0) = 0 \;,\; j \in J_0 \;;\; \Delta_j(0) \geq 0 \;,\; j \,\bar{\in}\, J_0
\end{aligned} \tag{2.29}$$

Let

$$\omega \triangleq \{j/\Delta_j(0) = 0\}$$

The conditions (2.28) and (2.29) imply that

$$\Theta^* = \Theta^\omega \triangleq \{x/\sum_{j\in\omega} A_{0j}x_j = b_0 \;,\; x_j \geq 0 \;,\; j \in \omega \;,\; x_j = 0 \;,\; j \,\bar{\in}\, \omega\} \tag{2.30}$$

In fact, it is obvious that $Ax = b$, $x \geq 0$ for $x\in \Theta^\omega$ and

$$x \in \Theta^{\omega} \Rightarrow \sum_{j \in \omega} x_j \Delta_j(0) + \sum_{j \in \omega} x_j \Delta_j(0) = 0$$

$$\Leftrightarrow x^T(A_0^T \lambda^* - c_0) = 0 \Rightarrow x \in \Theta^*$$

$$x \in \Theta^* \Rightarrow \sum_{j=1}^{n} x_j(\lambda^{*T} A_{0j} - c_{0j}) = 0$$

$$\Rightarrow \sum_{j \bar{\in} \omega} x_j \Delta_j(0) = 0 \Rightarrow x_j = 0 \ , \ j \bar{\in} \omega \Rightarrow x \in \Theta^{\omega}$$

Hence the solution to the auxiliary program is equivalent to the solution to a modified program

$$\max \{ \sum_{j \in \omega} \gamma_j x_j / \sum_{j \in \omega} A_{0j} x_j = b_0 \ , \ x_j \geq 0 \ , \ j \in \omega \} \tag{2.31}$$

where $\gamma_j \triangleq c_{1j} - \lambda^{*T} A_{1j}$ if we take into account that $x_j^{**} = 0, \ j \bar{\in} \omega$.

The uniqueness and nondegeneracy of x^{**} as the solution to the auxiliary program imply that the solution to (2.31) is also unique and nondegenerate, and so we have

$$\Delta_j^a = 0 \ , \ j \in J_0 \ ; \ \Delta_j^a > 0 \ , \ j \bar{\in} J_0 \ , \ j \in \omega \tag{2.32}$$

where

$$\Delta_j^a \triangleq \lambda_a^T A_{0j} - \gamma_j \ , \ \lambda_a^T = P^T(0)\gamma_b \ , \ \gamma_b \triangleq \{\gamma_j \ , \ j \in J_0\}$$

Let us check that $x^*(\epsilon)$, $\lambda^*(\epsilon)$ satisfy the optimality conditions (2.10)–(2.12) if they have the form (2.13) and (2.14) and ϵ is small enough [note that, by continuity, $P(\epsilon)$ exists for small ϵ]. It is sufficient to check (2.15) and (2.16).

By (2.25) we know that

$$x_b^*(\epsilon) \triangleq P(\epsilon)b(\epsilon) = x_b^{**} - \epsilon[P(0)A_{b1}x_b^{**} - P(0)b_1] + O(\epsilon^2) \tag{2.33}$$

$$\lambda^*(\epsilon) \triangleq P^T(\epsilon)c_b(\epsilon) = \lambda^* - \epsilon[P^T(0)A_{b1}^T \lambda^* - P^T(0)c_{b1}] + O(\epsilon^2) \tag{2.34}$$

The condition (2.15) is satisfied for small $\epsilon > 0$ since $x_b^{**} > 0$. It remains to show (2.16). Using (2.34) we obtain

$$\Delta_j(\epsilon) \triangleq A_j^T(\epsilon)P^T(\epsilon)c_b(\epsilon) - c_j(\epsilon)$$

$$= A_{0j}^T P^T(0)c_{0b} - c_{0j}$$

$$+ \epsilon[A_{1j}^T\lambda^* - A_{0j}^T(P^T(0)A_{b1}^T\lambda * - P^T(0)c_{b1}) - c_{1j}] + O(\epsilon^2)$$
$$= \Delta_j(0) + \epsilon\Delta_j^a + O(\epsilon^2) \;\; , \;\; j \bar{\in} J_0$$

and for small $\epsilon > 0$ (2.16) is valid since

$$\Delta_j(0) > 0 \;\; , \;\; j \bar{\in} \omega\backslash J_0 \Rightarrow \Delta_j(\epsilon) > 0 \;\; , \;\; j \bar{\in} \omega\backslash J_0$$
$$\Delta_j(0) = 0 \;\; , \;\; \Delta_j^a > 0 \;\; , \;\; j \in \omega\backslash J_0 \Rightarrow \Delta_j(\epsilon) > 0 \;\; , \;\; j \in \omega\backslash J_0$$

The proof is complete since the validity of (2.23) and (2.27) and the expansion (2.19) follows from Theorem 2.1.

Let us clarify the result in geometric terms (*Figure 3*). Let the perturbed program solution $x^*(\epsilon)$ be unique for $\epsilon \in (0,\epsilon'')$ but let the reduced program solution have the polyhedral solution set Θ^*. One can sketch the trajectory described by $x^*(\epsilon)$ as ϵ varies. The theorem proved above allow us to find a vertex of Θ^* which is the limiting point of the trajectory for $\epsilon \to 0$. This vertex was shown above to be the solution x^{**} to the auxiliary program. Moreover, one can see from (2.20) that the trajectory is smooth for some interval of ϵ. Thus, the hypothesis of optimal basic set invariance is also true under the nonuniqueness condition, but one needs the auxiliary program to define this set, which, in turn, determines the vertex of Θ^* and the corresponding vertex of $\Theta(\epsilon)$.

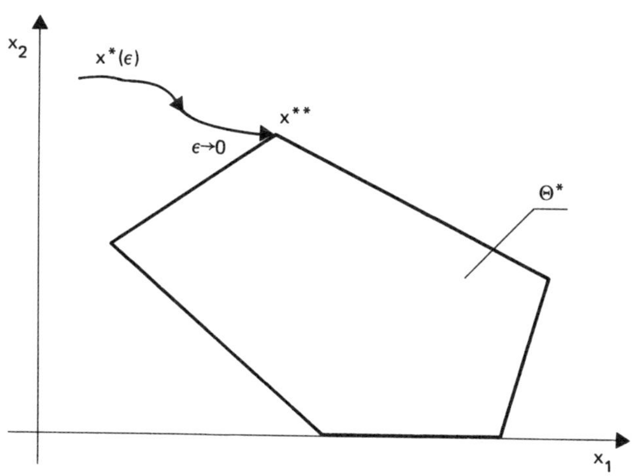

Figure 3.

The situation where the reduced program solution x^* is unique but degenerate, i.e., there is a set Λ of solutions to the dual program (2.3), is also of interest.

Theorem 2.3.

Suppose there exists a unique solution λ^{**} to the dual auxiliary program

$$\min \{(b_1 - A_1 x^*)^T \lambda / \lambda \in \Lambda\} \triangleq \varphi_d^* \tag{2.35}$$

and that λ^{**} is simultaneously a nondegenerate basic solution to the dual reduced program. Then the representation (2.23) is valid with

$$F_1^* = \varphi_d^* + c_1^T x^* \tag{2.36}$$

Other results of Theorem 2.2 are valid, too, but the basic set used in (2.13) must be determined by the solution to (2.35).

Proof. This is a dual analog of the proof of Theorem 2.2, and so we give a briefer presentation.

The condition $\lambda \in \Lambda$ is equivalent to the fact that

$$A_0^T \lambda \geq c_0 \ , \ x^{*T}(A_0^T \lambda - c_0) = 0$$

Since $\lambda^{**} \in \Lambda$ and λ^{**} is a nondegenerate basic solution, we have that

$$\lambda^{**} = P(0) c_b(0) \ , \ A_j^T \lambda^{**} > c_j \ , \ j \bar{\in} J_0$$
$$\Leftrightarrow \Delta_j(0) \triangleq A_j^T \lambda^{**} - c_j > 0 \ , \ j \bar{\in} J_0$$

Let $\omega_d \triangleq \{j / x_j^* = 0\}$. Then one can show that

$$\Lambda = \Lambda^\omega \triangleq \{\lambda / A_j^T \lambda \geq c_j \ , \ j \in \omega_d \ , \ A_j^T \lambda = c_j \ , \ j \bar{\in} \omega_d\}$$

Hence the program (2.35) is equivalent to the modified one

$$\min \{\gamma_d^T \lambda / \lambda \in \Lambda^\omega\} \ , \ \gamma_d \triangleq b_1 - A_1 x^* \tag{2.37}$$

The uniqueness and nondegeneracy of λ^{**} as the solution to that program imply that

$$[P(0)\gamma_d]_j < 0 \ , \ j \in \omega_d \cap J_0$$

Let us suppose that solutions to the perturbed and dual perturbed programs have the form (2.13) and (2.14). Then

$$x_b^*(\epsilon) \triangleq P(\epsilon)b(\epsilon) = x_b^* - \epsilon P(0)[A_{b1}x_b^* - b_1] + O(\epsilon^2)$$
$$\lambda^*(\epsilon) \triangleq P^T(\epsilon)c_b(\epsilon) = \lambda^{**} - \epsilon P^T(0)[A_{b1}\lambda^{**} - c_{b1}] + O(\epsilon^2)$$

Since

$$\Delta_j(\epsilon) = 0 \ , \ j \in J_0 \ ; \ \Delta_j(\epsilon) = \Delta_j(0) + O(\epsilon) \ , \ \Delta_j(0) > 0 \ , \ j \bar{\in} J_0$$
$$x_j^*(\epsilon) = 0 \ , \ j \bar{\in} J_0 \ ; \ x_j^*(\epsilon) = x_j^* - \epsilon[P(0)\gamma_d]_j + O(\epsilon^2) \ , \ j \in J_0$$
$$x_j^* > 0 \ , \ j \bar{\in} \omega_d \ , \ [P(0)\gamma_d]_j < 0 \ , \ j \in \omega_d \cap J_0$$

we obtain

$$\Delta_j(\epsilon) \geq 0 \ , \ x_j(\epsilon) \geq 0 \ , \ \Delta_j(\epsilon)x_j(\epsilon) = 0 \ , \ j = 1,\ldots,n \tag{2.38}$$

for small enough ϵ. Thus, the hypothesis is valid. The representation (2.23) and (2.36) is valid due to Theorem 2.1. QED.

Before we go on to consider more general cases, we shall outline some algorithmic aspects.

One can present the application of the perturbation method to linear programs as follows:

(1) Solve the reduced program and its dual.
(2) If the solutions to both programs are unique then the optimal basic set J_0 determines the perturbed program basic matrix that is optimal for small ϵ.
(3) Find (precisely or approximately) $x_b(\epsilon)$ as a solution to the basic equation

$$A_b(\epsilon)x = b(\epsilon)$$

(4) If the reduced program solution is nonunique and nondegenerate, then, after the determination of its optimal set Θ^*, solve the auxiliary program (2.26).
(5) If its solution is unique, then it gives the set J_0 and the matrix $A_b(\epsilon)$ simultaneously. After than, one can pass to (3).
(6) If the solution to the reduced program is unique but degenerate then one needs to determine the optimal set Λ of the dual program and to solve the dual auxiliary program (2.35).
(7) If the solution to (2.35) is unique then the corresponding basis gives J_0 and $A_b(\epsilon)$ and one can pass to (3) again.

The success of practical application of the procedure depends on the simplicity of the description of Θ^* (or Λ). A number of important problems occurring where this description is very simple will be shown in Chapter 2. Here we make some general remarks only.

As shown above, the obtained results can be treated in various ways.

(a) One can consider the problems to be one of parametric programming, i.e., one needs to state a formula determining the solution as a function of the parameter ϵ over some interval of its values. What precedes implies that if the length of this interval is sufficiently small, the formula (2.13) is true. Note that the perturbation method gives the precise result here (it is the specific feature of linear programs!), but that it is simpler to use the asymptotic expansion (2.19) and (2.20).

(b) The problem can be treated in the spirit of "sensitivity" theory. The theorems proved above give conditions under which the solution (or one of the solutions) changes in proportion to small changes in the parameter and permit one to find the coefficient of proportionality.

Both approaches mentioned consider, as an original program, the reduced one ($\epsilon = 0$). Consider now the same formal results from the point of view of the perturbation method. Here, from the very beginning, one needs to solve a program that can be seen as a perturbed one for some value ϵ_0 of the parameter ϵ, and so the reduced program ($\epsilon = 0$) is simpler than original one. The goal is to solve the original problem, and the construction of the function $x^*(\epsilon)$ is only necessary to extend the solution from $\epsilon = 0$ up to ϵ_0. Hence, it is essential to know whether this extension is possible. All the theorems show that there exists some finite internal where the results are true, but it is not known *a priori* whether ϵ_0 belongs to that interval. Therefore, one needs to check fulfillment of the optimality condition for $\epsilon = \epsilon_0$.

If the condition

$$x_b(\epsilon_0) \triangleq A_b^{-1}(\epsilon_0) b(\epsilon_0) \geq 0 \tag{2.39}$$

fails then the procedure is not directly applicable (it is possible that one needs to modify the scheme of reduced program separation).

If only the condition

$$\Delta_j(\epsilon_0) \geq 0 \ , \quad j \bar{\in} J_0 \tag{2.40}$$

is violated, then the basis obtained by the perturbation method can be used as an initial approximation for other computational procedures, e.g., for the simplex method.

It is also of interest to explain the connection between the simplex method as directly applied to the original program, and the two-stage procedure of the perturbation method, i.e., the solution of the reduced program and, after that, solution of the auxiliary program. This connection becomes clearer if one supposes that the determination of the reduced program solution set Θ^* is realized simultaneously with the solution of the auxiliary program (by the simplex method, too). Let us consider in detail the nondegeneracy case. We suppose that the application of the simplex procedure to the reduced program gives, at step t,

$$\Delta_j^{(t)} \geq 0 \qquad j \in J_0^{(t)} \tag{2.41}$$

Then the corresponding basic solution $x^{(t)}$ is an optimal one for the reduced program

$$\begin{aligned} &x^{(t)} = x^* = (P \cdot b\ , 0)\ ;\ \Delta_j^{(t)} = \Delta_j = A_{0j}^T \lambda^{(t)} - c_{0j}\ , \\ &\lambda^{(t)} = \lambda^* = P^T C_0 \end{aligned} \tag{2.42}$$

where $P^{(t)} = A_b^{-1}$ and the index b refers to an optimal basic set $J_0^t = J_0$.

If x^* is a nonunique solution, then there exists a J_1 such that

$$\Delta_j = 0\ ,\ j \in J_1 \cup J_0 \triangleq \omega$$

To choose the best basis, one needs to solve the auxiliary program (2.26) but, to that end, one can use the modified program (2.31), which is simpler than (2.26). Such a possibility follows from the proof of Theorem 2.2.

Let us compute the column estimates $\Delta_{1j}^{(t)}$ corresponding to the basic matrix A_{0b} in the program (2.31). By definition we have

$$\begin{aligned} &\Delta_{1j}^{(t)} = A_{0j}^T \lambda_a^{(t)} - \gamma_j \\ &\lambda_a^{(t)} \triangleq (P^{(t)})^T \gamma_b\ ,\ \gamma_b \triangleq (\gamma_j\ , j \in J_0) \end{aligned} \tag{2.43}$$

If all $\Delta_{1j} \geq 0$, $j \in J_1$ then x^* is the modified auxiliary program solution. Otherwise, one can find k such that

$$\Delta_{1k} \triangleq \min_{j \in J_1} \Delta_{1j} < 0$$

and introduce the column A_{0k} into the basis. As a result, the reduced program objective is unchanged, but the auxiliary program objective increases. It is not hard to prove that this procedure converges to the auxiliary program solution in a finite number of steps.

On the other hand, this procedure coincides with the simplex method applied to the perturbed program

$$\max \{(c_0 + \epsilon c_1)^T x \,/\, (A_0 + \epsilon A_1)x = b_0 + \epsilon b_1 \ , \ x \geq 0\} \tag{2.44}$$

if, at each step, one uses only two terms in the power expansion of $\lambda^{(t)}(\epsilon)$ to compute the column estimates. In fact, $\lambda^{(t)}(\epsilon)$ is given by the formula

$$\lambda^{(t)}(\epsilon) = [P^{(t)}(\epsilon)]^T c^{(t)}(\epsilon) \tag{2.45}$$

where

$$c^{(t)}(\epsilon) \triangleq [c_j(\epsilon) \ , \ j \in J_0^{(t)}] \tag{2.46}$$

Using the expansion of the (2.25)-type, we have that

$$P^{(t)}(\epsilon) = P^{(t)} - \epsilon P^{(t)} A_{b1}^{(t)} P^{(t)} + O(\epsilon^2) \tag{2.47}$$

and, hence,

$$\lambda^{(t)}(\epsilon) = \lambda_0^{(t)} + \epsilon \lambda_1^{(t)} + O(\epsilon^2) \tag{2.48}$$

where

$$\lambda_0^{(t)} \triangleq [P^{(t)}]^T c_0^{(t)} \ , \ \lambda_1^{(t)} \triangleq [P^{(t)}]^T \gamma^{(t)}$$

Moreover, at several initial steps of this "asymptotic simplex", one may compute $\lambda_0^{(t)}$ only because the sign of $\Delta_j^{(t)}(\epsilon)$ is determined by the main term if the latter is not equal to zero. These initial steps coincide with the first stage of the two-stage procedure of the perturbation method.

In conclusion, let us describe some possibilities of the perturbation method in complicated situations not covered by the basic procedure. Recall that this procedure gives the answer only if the solution to the reduced program or to the dual program is unique and that, moreover, it is necessary to obtain the unique solution to the auxiliary program (or to the dual auxiliary one). However, some useful results can be shown in the more general case, too.

Theorem 2.4.

Let the reduced program solution set Θ^* and the solution set Λ to the dual program be nonempty and bounded. Then

(a) The limiting solution set (for $\epsilon \to 0$) of the perturbed program is contained in the solution set of the auxiliary program, which has the following general form

$$\max_{x \in \Theta^*} \{c_1^T x + \min_{\lambda \in \Lambda} \lambda^T (b_1 - A_1 x)\} \triangleq F_1^* \tag{2.49}$$

(b) There exists ϵ' such that for $\epsilon \in [0\ ,\ \epsilon']$ the representation

$$F^*(\epsilon) = F^*(0) + \epsilon F_1^* + O(\epsilon^2) \tag{2.50}$$

is valid; and

(c) Any solution, x^{**}, to the auxiliary program is ϵ-feasible and ϵ^2-optimal, i.e., there exists a bounded vector $y(\epsilon)$ such that $x^{**} + \epsilon y(\epsilon)$ is feasible for the perturbed program and

$$F^*(\epsilon) = c^T(\epsilon)[x^{**} + \epsilon y(\epsilon)] + O(\epsilon^2)$$

If, moreover, the reduced program is nondegenerate, then the auxiliary program is linear and any optimal basic set of this program generates a basic feasible solution $x^{**}(\epsilon)$ to the perturbed program such that

$$F^*(\epsilon) = c^T(\epsilon)\, x^{**}(\epsilon) + O(\epsilon^2)$$

Proof of Theorem 2.4 is completely contained in the proof of the basic Theorem 4.8 in Section 1.4 relating to a more general class of problems. One should note only that condition (4.43) of Theorem 4.8 follows here from the fact that Θ^* and Λ are nonempty and bounded [21].

Note also that Theorem 2.4 does not cover all the possible situations in linear programs.

Counter example. Let the perturbed program have the following form

$$\max x_2$$

such that

$$x_1 + x_2 = 1 \ ; \ x_1 \geq 0$$
$$2(1 + \epsilon)x_1 + 2(1 + 2\epsilon)x_2 = 2(1 + \epsilon) \ ; \ x_2 \geq 0$$

Its solution is unique for $\epsilon > 0$ and does not depend on ϵ

$$x_1^*(\epsilon) = 1 \;\; ; \;\; x_2^*(\epsilon) = 0$$

At the same time, the solution to the reduced program

$$\max x_2$$

such that

$$x_1 + x_2 = 1 \;\; ; \;\; x_1 \geq 0$$
$$2x_1 + 2x_2 = 2 \;\; ; \;\; x_2 \geq 0$$

is $x_1* = 0$; $x_2^* = 1$ and bears no similarity to the perturbed solution. The origin of this effect is clear. The perturbed program restrictions imply the equation

$$\epsilon(x_1 + 2x_2 - 1) = 0$$

For any $\epsilon > 0$, this is equivalent to the restrictions

$$x_1 + 2x_2 = 1$$

which vanishes for $\epsilon = 0$. Hence the feasible sets of the perturbed problem and the reduced problem are not similar.

Formally, this is clear from the convergence of the dual perturbed program solution

$$\lambda_1^*(\epsilon) = -\frac{1+\epsilon}{\epsilon} \;\; , \;\; \lambda_2^*(\epsilon) = \frac{1}{2\epsilon}$$

to infinity as $\epsilon \to 0$. The conditions of Theorem 2.4 are violated, the solution set

$$\Lambda = \{\lambda \,/\, \lambda_1 + 2\lambda_2 = 1\}$$

of the dual reduced program

$$\min \{\lambda_1 + 2\lambda_2 / \lambda_1 + 2\lambda_2 \geq 0 \;\; , \;\; \lambda_1 + 2\lambda_1 \geq 1\}$$

being unbounded.

One can check that the conditions of Theorem 2.3 are also not fulfilled. The example establishes the condition of boundedness shown in Theorem 2.4 as essential for the stability of linear program solutions. *Singularly perturbed programs* where the stability conditions are violated are discussed in Chapter 3 with some practical applications.

1.3. Nonlinear programs: perturbations in objective functions

It is natural to begin the presentation of perturbation theory for nonlinear programs from the simplest case where perturbations are in objective functions only, i.e., the perturbed programs can be written in the form

$$\max \{f(x, \epsilon) / x \in \Theta\} \triangleq F^*(\epsilon) \tag{3.1}$$

where Θ is a nonempty closed domain in E^n, in particular, E^n itself. [In the last case, one calls (3.1) an unconstrained maximization problem.]

Let an arbitrary solution to (3.1) be $x^*(\epsilon)$. Let the solution set be $\Theta^*(\epsilon)$, and let the optimal value of the objective be $F^*(\epsilon)$. Associate with (3.1) the reduced program

$$\max \{f(x, 0) / x \in \Theta\} \triangleq F^*(0) \tag{3.2}$$

denoting

$$f(x, 0) \triangleq f_0(x) \quad , \quad f_1(x, \epsilon) \triangleq f(x, \epsilon) - f_0(x)$$

For simplicity, it is supposed in what follows that $f(x, \epsilon)$ is linear[†] in ϵ so that

$$f_1(x, \epsilon) = \epsilon f_1(x)$$

Generally speaking, it is impossible here to give as detailed a description of the solution behavior as was done for linear programs. Therefore, it is of interest to state some simple results which are true for any ϵ.

[†] The following results are essentially unchanged if $f(x,\epsilon)$ is assumed differentiable in ϵ such that $f_1(x,\epsilon) = \epsilon f_1(x) + o(\epsilon)$.

Theorem 3.1.

Let there exist a solution, x^*, to the reduced program and let an estimate f_1^* be established such that

$$f_1^* \geq \sup \{f_1(x) \,/\, x \in \Theta\} \tag{3.3}$$

Then

$$F^*(\epsilon) - f(x^*, \epsilon) \triangleq f(x^*(\epsilon)\,,\, \epsilon) - f(x^*, \epsilon) \leq \epsilon[f_1^* - f_1(x^*)] \tag{3.4}$$

Proof is concluded in the following:

$$f(x^*(\epsilon)\,,\, \epsilon) - f(x^*, \epsilon) \triangleq f_0(x^*(\epsilon)) - f_0(x^*) + \epsilon\{f_1(x^*(\epsilon)) - f_1(x^*)\} \leq$$
$$\leq \epsilon\{f_1(x^*(\epsilon)) - f_1(x^*)\} \leq \epsilon[f_1^* - f_1(x^*)] \quad \text{QED.}$$

By virtue of Theorem 3.1, one can use the reduced program solution instead of the optimal one, losses being limited by an amount of order ϵ if, of course, perturbations are bounded over the feasible set.

Note that the nonexistence of the reduced program solution does not imply that the perturbed program solution also does not exist. This is clear from the elementary example:

$$\max \{x - \epsilon x^2\} = \frac{1}{4\epsilon} \qquad x^*(\epsilon) = \frac{1}{2\epsilon}$$

At the same time, the reduced program has no solution. There is no correspondence here between the perturbed program and the reduced one, the perturbation is not bounded above.

Let us find both some simple estimates of neighborhood of solutions to the perturbed program and the solution to the reduced program, which are valid for all ϵ.

Theorem 3.2.

Let Θ be convex and let $f_0(x)$ be strongly concave:

$$f_0(x^*) - f_0(x) \geq \rho \|x - x^*\|^2 \tag{3.5}$$

where ρ is a known scalar. Then

$$\|x^*(\epsilon) - x^*\| \leq \sqrt{\frac{\epsilon}{\rho}}\sqrt{f_1^* - f_1(x^*)} \tag{3.6}$$

where f_1^* is defined by (3.3).

If, moreover, the function $f_1(x)$ satisfies a Lipschitz condition

$$\|f_1(x^{(1)}) - f_1(x^{(2)})\| \leq k\|x^{(1)} - x^{(2)}\| \tag{3.7}$$

for any $x^{(1)}$, $x^{(2)}$ belonging to Θ, then

$$\|x^*(\epsilon) - x^*\| \leq \epsilon\frac{k}{\rho} \tag{3.8}$$

Proof. Since

$$f_0(x^*) - f_0(x^*(\epsilon)) \leq \epsilon[f_1(x^*(\epsilon)) - f_1(x^*)] \leq \epsilon[f_1^* - f_1(x^*)] \tag{3.9}$$

we have that

$$\rho\|x^*(\epsilon) - x^*\|^2 \leq \epsilon[f_1^* - f_1(x^*)]$$

which implies (3.6).

The estimate (3.8) follows from the inequalities (3.9) and (3.7):

$$\begin{aligned}&\rho\|x^*(\epsilon) - x^*\|^2 \leq f_0(x^*) - f_0(x^*(\epsilon)) \leq \\ &\leq \epsilon[f_1(x^*(\epsilon)) - f_1(x^*)] \leq \epsilon k\|x^*(\epsilon) - x^*\| \qquad \text{QED.}\end{aligned}$$

Example. Let the problem be stated in the form

$$\max\{\sum_{i=1}^{n}(-\frac{1}{2}a_i x_i^2 + b_i x_i) + f_1(x)\}\ ,\ x = (x_i\ ,\ i = 1,\dots,n)$$

where $a_i > 0$ and $f_1(x)$ satisfies (3.7). This is equivalent to the perturbed problem

$$\max\{f_0(x) + \epsilon f_1(x)\} \tag{3.10}$$

if

$$f_0(x) = \sum_i(-\frac{1}{2}a_i x_i^2 + b_i x_i)\ ,\ \epsilon = 1$$

The reduced solution is $x^* = (x_i^*)$,

$$x_i^* = \frac{b_i}{a_i} \quad , \quad i = 1,\ldots,n$$

and $f_0(x)$ satisfies a condition of type (3.5) since

$$f_0(x) - f_0(x^*) \geq \|x - x^*\|^2 \frac{1}{2} \min_i a_i$$

Thus, one can use the inequality (3.8) to estimate the solution to the original problem, $x^*(1)$:

$$\sum_i [x_i^*(1) - \frac{b_i}{a_i}]^2 \leq \left[\frac{2k}{\min a_i}\right]^2$$

The assumptions of Theorem 3.2 are very restrictive. In many cases they are not satisfied, or it is difficult to test for them. Nevertheless, one still needs to know if the perturbed program solution is close to the reduced one for small ϵ. The following result will aid us in this.

Theorem 3.3.

Let the functions $f_0(x)$, $f_1(x)$ be continuous over Θ and let the perturbed program solution set $\Theta^*(\epsilon)$ be bounded for all sufficiently small ϵ, i.e., there exists $\hat{\epsilon} > 0$ such that

$$\Theta^*(\epsilon) \subset \bar{\Theta} \quad , \quad 0 < \epsilon \leq \hat{\epsilon} \tag{3.11}$$

where $\bar{\Theta}$ is closed and bounded. Let the reduced program solution x^* be unique. Then for any solution $x^*(\epsilon) \in \Theta^*(\epsilon)$

$$\lim_{\epsilon \to 0} x^*(\epsilon) = x^* \tag{3.12}$$

Proof. Suppose (3.12) is false. By (3.11), this is equivalent to the existence of a subsequence $\{\epsilon'\}$ such that

$$\lim_{\epsilon' \to 0} x^*(\epsilon) = \tilde{x} \neq x^* \tag{3.13}$$

By definition

$$f_0(x^*(\epsilon')) + \epsilon' f_1(x^*(\epsilon')) \geq f_0(x^*) + \epsilon' f_1(x^*)$$

Taking the limit, we obtain

$$f_0(\tilde{x}) \geq f_0(x^*)$$

Since Θ is closed, $\tilde{x} \in \Theta$. This means that $\tilde{x} = x^*$ and so contradicts (3.13). QED.

Remark. The boundedness of $\Theta^*(\epsilon)$ required by the theorem conditions is trivially true if the feasible set Θ is bounded. It is not hard to prove also that (3.11) is valid for any sufficiently small $\hat{\epsilon}$, if and only if (a) $f_1^* < \infty$ and (b) there does not exist a sequence $\{x_i\} \in \Theta$ such that

$$\lim_{i \to \infty} f_0(x_i) = F^*(0) \quad , \quad \lim_{i \to \infty} \|x_i\| = \infty \tag{3.14}$$

It may seem that the restriction (b) is not necessary, but we have a counter example.

Counter example.

$$\max \{f_0(x) + \epsilon f_1(x) \,/\, x \geq 0 \quad , \quad x \in E^1\}$$

the functions $f_0(x)$, $f_1(x)$ satisfying the following conditions

$$f_0(x) \leq c \quad , \quad f_0(x^*) = c \quad , \quad x^* \leq 2\pi \quad , \quad \lim_{x \to \infty} f_0(x) = c$$

$$f_1(x) = \begin{cases} 0 & , \quad 0 \leq x \leq 2\pi \\ \sin x & , \quad x \geq 2\pi \end{cases}$$

By definition, the reduced solution is x^*, but the perturbed problem may have no solution for any $\epsilon > 0$. It only has the maximizing sequence $\{x_i\}$

$$x_i = \frac{\pi}{2} + 2\pi i \quad , \quad i = 0,1,\cdots$$

so that

$$\sup \{f_0(x) + \epsilon f_1(x) \,/\, x \geq 0\} = c + \epsilon$$

The reason for this is the violation of requirement (3.14).

If a zero-order approximation given by the reduced program solution is not satisfactory, then, under some conditions, it is possible to construct approximations of higher order.

Let us restrict our attention to the case where x^* belongs to the interior of the feasible set. This hypothesis is trivially satisfied for unconstrained optimization problems (the general situation is considered below, in Section 1.4).

Theorem 3.4.

Let the conditions of Theorem 3.3 be fulfilled. Let the functions $f_0(x)$, $f_1(x)$ be l-times differentiable ($l \geq 1$) at the point x^* belonging to the interior of the set Θ. Let the matrix $Q \triangleq \nabla\nabla f_0(x^*)$ be nonsingular

$$|Q| \neq 0 \tag{3.15}$$

Then one can chose $\epsilon' > 0$ such that, for $\epsilon \in [0\ ,\ \epsilon']$, there exists the unique solution, $x^*(\epsilon)$, to the perturbed problem. This solution can be written in the form

$$x^*(\epsilon) = x^* + \epsilon x^{(l)} + \cdots + \epsilon^l x^{(l)} + o(\epsilon^l) \triangleq \bar{x}^{(l)} + o(\epsilon^l) \tag{3.16}$$

If $\bar{x}^{(l)}$ is used instead of $x^*(\epsilon)$ then the objective approaches the optimal value to within $o(\epsilon^{l+1})$.

Proof. By Theorem 3.3, there exists small enough $\hat{\epsilon}$ such that $x^*(\epsilon)$ belongs to the interior of Θ if $0 \geq \epsilon \leq \hat{\epsilon}$. Hence, the stationary condition

$$\nabla_x f(x^*(\epsilon)\ ,\ \epsilon) = 0 \tag{3.17}$$

is satisfied. One can interpret this condition as an implicit representation of the function $x^*(\epsilon)$. Recall that

$$\nabla f_0(x^*) \triangleq \nabla_x f(x^*\ ,\ 0) = 0 \tag{3.18}$$

Since the Jacobian differs from zero at the point ($x = x^*$, $\epsilon = 0$), then by virtue of the implicit function theorem, there exists the unique and l times continuously differentiable function $x^*(\epsilon)$, i.e., (3.16) is valid. The coefficients $x^{(i)}$, $i = 1,\ldots,l$, can be found by substituting (3.16) in (3.17) and writing the coefficients for different powers of ϵ. In particular, we have

$$x^{(1)} = -\,Q^{-1}\nabla f_1(x^*) \tag{3.19}$$

The last result of the theorem can be obtained by a direct estimation of the error

$$\begin{aligned} &f(x^*(\epsilon)\ ,\epsilon) - f(\overline{x}^l\ ,\epsilon) = \\ &= f_0(\overline{x}^l + o(\epsilon^l)) - f_0(\overline{x}^l) + \epsilon[f_1(\overline{x}^l + o(\epsilon^l)) - f_1(\overline{x}^l)] = \\ &= (\nabla f_0(\overline{x}^l))^T o(\epsilon^l) + o(\epsilon^{l+1}) = o(\epsilon^{l+1}) \qquad \text{QED.} \end{aligned}$$

Example. Let $f_1(x)$ in (3.10) be twice differentiable. Then, according to (3.16) and (3.19)

$$x_i^*(\epsilon) = \frac{b_i}{a_i} + \epsilon\frac{1}{a_i}\frac{\partial f_1}{\partial x_i} + \epsilon^2\frac{1}{a_i}\sum_{j=1}^{n}\frac{1}{a_j}\cdot\frac{\partial f_1}{\partial x_i \partial x_j}\cdot\frac{\partial f_1}{\partial x_j} + o(\epsilon^2)$$

where all derivatives are calculated at the point

$$x^* = (\frac{b_i}{a_i}\ ,\ i = 1,2,\ldots,n)$$

Remark. Note that the expansion (3.16) is not valid for any case.

Counter example. Let the perturbed program have the form

$$\max\ \{-x^{k_1} + \epsilon x^{k_2} / 0 \le x \le 1\}$$

where k_1, k_2 are integers, $k_1 > k_2 > 0$. Its solution,

$$x^*(\epsilon) = \left(\frac{\epsilon k_2}{k_1}\right)^{1/(k_1-k_2)}$$

converges under $\epsilon \to 0$ to the reduced one, $x^* = 0$, as ϵ^l, l being any fractional number.

The previous results relating to the proximity of perturbed and reduced program solutions are restricted by the condition that x^* is unique.

Let, however, the set Θ^* contain a collection of reduced program solutions. First of all, it is of importance to select from this collection a solution which is asymptotically (for small ϵ) close to the perturbed one. A heuristic approach suggests the use of the auxiliary program

$$\max\ \{f_1(x) / x \in \Theta^*\} \triangleq F_1^* \tag{3.20}$$

which allows an improvement in resolution of the analysis.

The following general result is true.

Theorem 3.5.

Let all the conditions of Theorem 3.3 be fulfilled except the condition of the uniqueness of the reduced program solution. Then

(a) The set Θ^{**} of limiting points of $\Theta^*(\epsilon)$-element trajectories for $\epsilon \to 0$ is contained in the solution set Θ_1^* of the auxiliary program,

$$\Theta^{**} \subset \Theta_1^* \tag{3.21}$$

(b) The optimal value, $F^*(\epsilon)$, of the perturbed program objective may be written in the form

$$F^*(\epsilon) = F^*(0) + \epsilon F_1^* + o(\epsilon) \tag{3.22}$$

where F_0^*, F_1^* are the optimal objective values of the reduced and auxiliary programs respectively.

Proof. As in the proof of Theorem 3.1, one can obtain

$$f_1(x^*(\epsilon)) \geq f_1(x^*) \tag{3.23}$$

where x^* is an arbitrary solution to the reduced program. Let $\tilde{x}$ be a partial limit, i.e., there exists a sequence $\{\epsilon'\}$ such that

$$\lim_{\epsilon' \to 0} x^*(\epsilon') \triangleq \tilde{x} \overline{\in} \Theta_1^* \tag{3.24}$$

As in Theorem 3.3, it is not hard to prove that $\tilde{x} \in \Theta^*$. Taking the limit in (3.23) along $\{\epsilon'\}$, we obtain

$$f_1(\tilde{x}) \geq f_1(x^*) \quad , \ \forall x^* \in \Theta^*$$

This contradicts (3.24) and the contradiction proves (3.21). In turn, (3.21) and (3.9) imply that

$$\lim_{\epsilon \to 0} f_1(x^*(\epsilon)) = f_1(x^*) \triangleq F_1^*$$

$$0 \le f_0(x^*) - f_0(x^*(\epsilon)) \le \epsilon[f_1(x^*(\epsilon)) - f_1(x^*)] = o(\epsilon) \Rightarrow$$
$$f_0(x^*(\epsilon)) = f_0(x^*) + o(\epsilon) \triangleq F^*_{(0)} + o(\epsilon)$$

We have, finally, that

$$F^*(\epsilon) \triangleq f_0(x^*(\epsilon)) + \epsilon f_1(x^*(\epsilon)) = F^*_{(0)} + \epsilon F^*_1 + o(\epsilon) \qquad \text{QED.}$$

Remark. Results of the type (3.21) have appeared in papers devoted to the penalty function method (see, e.g., [81]). Recall that, in accordance with this method, the solution to the program

$$\max \{f(x) \,/\, g_l(x) \le 0 \;\;,\;\; l = 1,\ldots,m\}$$

must be replaced by the solution to a sequence of unconstrained programs

$$\max \{f(x) + \frac{1}{\epsilon}\psi[g(x)]\}$$

where $\epsilon \to 0$ and where $\psi[g(x)]$ is a continuous function such that

$$\psi(g) = \begin{cases} = 0 \;\;, & \text{if } g_l \le 0 \text{ for all } \;\; l \\ < 0 \;\;, & \text{if } g_l > 0 \text{ for some } \; l \end{cases}$$

Evidently, the unconstrained problem is equivalent, for $\epsilon \to 0$, to the perturbed one

$$\max \{\psi[g(x)] + \epsilon f(x)\}$$

The corresponding reduced problem has the form

$$\max \psi[g(x)]$$

and the auxiliary one is equivalent to the original constrained maximization problem. Thus, results relating to the convergence of the penalty function method are a particular case of Theorem 3.5.

It is also clear that the penalty function scheme can be thought of as a "counter example" for the perturbation method since, generally speaking, the two-stage approach is more difficult here than a direct one.

Let us give a simple corollary to Theorem 3.5.

Corollary. Let the conditions of Theorem 3.5 be fulfilled and, moreover, let the solution x^{**} to the auxiliary program (3.20) be unique. Then for any solution, $x^*(\epsilon)$, to the perturbed program

$$\lim_{\epsilon \to 0} x^*(\epsilon) = x^{**}$$

Proof of the corollary is left as an exercise.

Remark. Recall that, for linear programs,

$$x^*(\epsilon) = x^{**} \quad , \quad 0 < \epsilon \leq \hat{\epsilon}$$

if x^{**} is the unique solution to the auxiliary program. This result follows directly from (3.23) since the linear program has a basic solution in Θ^* but there are no basic solutions in a finite neighborhood of x^{**} except x^{**} itself.

The results proved above permit us to find a zero-order approximation by the standard scheme: solve the reduced program and, if it is necessary, the auxiliary one.

To improve the approximation in the case of nonuniqueness, one can use the representation (3.16) the auxiliary program solution x^{**} being used instead of an arbitrary reduced solution x^*. We shall postpone the proof of this result until Section 1.5.

In conclusion, we consider some applications to problems of suboptimal decomposition and aggregation.

It is evident that the results of this section give a rigorous basis for the heuristic suppositions mentioned in the previous section if optimization problems have the form

$$\max \{f_0(z(x)) + \epsilon f_1(x) \,/\, x \in X\} \tag{3.25}$$

However, it is of interest to consider the situation where the explicit representation of type (3.25) is absent and where the objective function $f(x)$ is given in an arbitrary form, in particular, where it is determined algorithmically.

It is well-known (see, e.g., [112]) that the "valley problem" is the main difficulty in numerical optimization. The appearance of a "valley" in the objective surface results in a slow, "zigzag" motion if one uses gradient methods for the optimum search. The test-function of Rosenbrock

$$f(x) = (x_1^2 - x_2)^2 + 0.01(1 - x_1^2) \tag{3.26}$$

is a standard example.

The minimum search process comes quickly to "the bottom of the valley", which coincides here with the parabola $x_2 = x_1^2$ but, after that, has a zigzag character oscillating around the bottom and moving very slowly to the minimum point $x_1^* = x_2^* = 1$.

The Rosenbrock function may be written in the form (3.25), if

$$f_0 = z^2 \quad , \quad z = x_1^2 - x_2$$
$$f_1(x) = 1 - x_1^2 \quad , \quad \epsilon = 0.01$$

and where X is a plane (x_1 , x_2). In fact, one can hypothesize that any function having the "valley-type" surface is representable in the form (3.25). At least, the converse is true, i.e., if a function has such a structure, then its surface has "valleys" since a change in the surface level is determined by the perturbation term only when the point moves along the curves (generally, along the hypersurfaces) given by the equations

$$z(x) = z^* = \text{constant}$$

and any displacements from these curves result in growth of the function cost.

If the structure is given explicitly, then application of the perturbation method is the best option. The reduced problem solution generates the "valley" equation $z(x) = z^*$ and the solution to the auxiliary one approximates the minimum point on "the valley bottom" with high accuracy.

However, as was shown above, this structure may be unknown. Therefore one needs to use some approximation scheme. Apparently, a local approximation using information about the function behavior in a neighborhood of the "valley" is most effective.

Let us describe a possible realization of this idea, restricting ourselves to unconstrained optimization problems (some results are shown in Section 2.2 for linear programs as well).

We consider the following iterative algorithm, which, beginning from an arbitrary $x^{(0)}$ gives a sequence of points $x^{(t)}$ by the rule:

(1) Estimate $\nabla f(x^{(t)})$.

(2) If $|\nabla f(x^{(t)})| > \epsilon$ where ϵ is some given small value, pass to the point $x^{(t+1)}$ using the steepest descent formula.

(3) Otherwise, estimate the matrix $\nabla\nabla f(x^{(t)}) \triangleq Q^{(t)}$.

(4) Calculate the largest eigenvalues of $Q^{(t)}$ up to $0 < \lambda_{k+1}^{(t)} \leq \epsilon$ and the corresponding eigenvectors $c_k^{(t)}$, to obtain the local approximation of the objective

$$f(x) = f(x^{(t)}) + \sum_{i=1}^{k} \lambda_i^{(t)} z_i^2 + O(\epsilon) \tag{3.27}$$

$$z_i = (c_i^{(t)})^T(x - x^{(t)}) \tag{3.28}$$

[In (3.27) $O(\epsilon)$ contains both linear and quadratic terms corresponding to small eigenvalues.]

(5) Apply the perturbation method to minimize (3.27).

Obtaining straightforward solution

$$z = z_j^* = 0 \qquad i = 1,\ldots,k$$

to the reduced program

$$\min \{\sum \lambda_i^{(b)} z_i^2\}$$

solve the auxiliary one.

This program is equivalent to the direct minimization of $f(x)$ itself, but under the additional restrictions

$$(c_i^{(t)})^T(x - x^{(t)}) = 0 \ , \ i = 1,\ldots,k \tag{3.29}$$

which can be used to exclude k variables.

The solution gives $x^{(t+1)}$.

Of course, this algorithm is applicable to constrained problems, too, if they are previously transformed to the unconstrained form by a penalty function or augmented Lagrangian method [145].

1.4. Necessary and sufficient conditions for an extremum. Quasiconvex and quasilinear programs

This section is devoted to an exposition of a number of fundamental nonlinear programming theorems and also to an investigation of program with linear or convex reduced structure. Interest in such problems is stimulated by two circumstances. In the first place, the results for the convex case are simpler and more complete. In the second, these results are highly applicable to the use of linear (convex) models, which, while traditional in mathematical economics, do not always lead to reliable results, and the transition to nonlinear (nonconvex) problems sharply complicates the optimization procedures. The perturbation method permits one to answer the questions:

(1) Whether the linear (convex) model is correct, i.e., whether the solution to the program with a nonlinear (nonconvex) small perturbation remains in a small neighborhood of solutions to the reduced linear (convex) program.

(2) Whether it is possible to calculate corrections to the reduced solution without solving the whole nonlinear problem.
(3) If the reduced model is not correct — in particular, if it admits solutions that are not close to the solution to the nonlinear (nonconvex) program — then how does one find those that are close?

This approach enables one to use the classical linear convex analysis and takes into account different nonlinear effects.

A basis for the consideration of quasiconvex and, in general, nonlinear problems is given by the theory of necessary and sufficient conditions for an extremum. We shall detail some of this theory mainly following W. Zangwill [28].

Definition 1. A vector $\bar{\lambda} \triangleq \{\bar{\lambda}_l\}$, $l = 1,\ldots,m$ is called a Lagrange multipliers vector† for the program

$$\max \{f(x) \,/\, g_l(x) \geq 0 \ , \ l = 1,\ldots,m\} \tag{4.1}$$

at the feasible point $\bar{x}$ if it satisfies the conditions

$$\nabla f(\bar{x}) = \sum_{l=1}^{m} \bar{\lambda}_l \nabla g_l(\bar{x}) \ , \ \bar{\lambda} \geq 0 \ , \ \bar{\lambda}^T f(\bar{x}) = 0 \tag{4.2}$$

The function $f(x)$ and the vector-function $g(x) \triangleq \{g_l(x)\}$, $l = 1,\ldots,m$ are here and later on assumed to be continuously differentiable.

Definition 2. The set

$$D(\bar{x}) = \{y \,/\, \exists \mu_0 \ , \ g(\bar{x} + \mu y) \leq 0 \ , \ 0 < \mu \leq \mu_0\} \tag{4.3}$$

is called the cone of feasible directions at the admissible point $\bar{x}$.

It is almost obvious that

$$Cl\, D(\bar{x}) \subset \hat{D}(\bar{x}) \tag{4.4}$$

where $Cl\, D(\bar{x})$ is the closure of $D(\bar{x})$, and $\hat{D}(x)$ is the polyhedral set defined by the gradients of the binding constraints:

$$\hat{D}(\bar{x}) = \{y \,/\, (\nabla g_l(\bar{x}))^T y \leq 0 \ , \ l \in \Omega(\bar{x})\} \tag{4.5}$$

† Sometimes, for the sake of simplicity, we shall omit the word "multipliers".

where $\Omega(\bar{x})$ is the set of indices of binding constraints

$$l \in \Omega(\bar{x}) \Leftrightarrow g_l(\bar{x}) = 0$$

The converse inclusion to (4.4), generally speaking, does not hold and although the corresponding counter-examples are rather artificial [28], it proves necessary to introduce the following definition.

Definition 3. The program (4.1) is said to satisfy regularity conditions if, together with (4.4), the converse inclusion holds for all solution points, i.e.,

$$ClD(x) = \hat{D}(x) \quad , \quad x \in \Theta^* \tag{4.6}$$

where Θ^* is the set of solutions to (4.1). It is supposed that $\Theta^* \neq \phi$.

Theorem 4.1. [28]

Let the functions $f(x)$, $g(x)$ be continuously differentiable and the program (4.1) satisfy the regularity conditions. Then, for each solution $x \in \Theta^*$, one can construct a Lagrange multipliers vector $\lambda(x)$.

Note that, although the theorem is formulated for the nonconvex problem, it is based on convex analysis — namely, the main part of the proof uses the separability of the convex feasible directions set $\hat{D}(x)$ and the open halfspace

$$\{y \, / \, (\nabla f(x))^T y > 0\}$$

For convex programs the existence of a Lagrange vector appears to be sufficient condition of optimality.

Theorem 4.2. [28]

Let the functions $f(x)$, $g(x)$ be continuously differentiable and let $-f(x)$, $g_l(x)$, $l = 1,\ldots,m$ be convex. Then

(1) If, at some feasible point $\bar{x}$, there exists a Lagrange vector $\lambda(\bar{x})$, then this point $\bar{x}$ is the solution to (4.1) and $\{\bar{x} \, , \, \lambda(\bar{x})\}$ is a saddle point of the Lagrange function

$$L(x \, , \lambda) \triangleq f(x) - \lambda^T g(x)$$

(2) If λ is a Lagrange vector for one of the solutions, it is also a multipliers vector for any other solution; provided that the regularity conditions are fulfilled, the whole set Λ of Lagrange vectors is not empty and is determined as the solutions set of the dual program[†]

$$\Lambda = \text{Arg min}\ \{L^*(\lambda) \,/\, \lambda \geq 0\}\ ,\ L^*(\lambda) = \max_x L(x\,,\lambda) \tag{4.7}$$

The proof of the theorem is elementary. On the one hand, the definition of the saddle point:

$$L(x\,,\hat{\lambda}) \leq L(\hat{x}\,,\hat{\lambda}) \leq L(\hat{x}\,,\lambda)\ ,\ \lambda \geq 0\ ,\ \hat{\lambda} \geq 0 \tag{4.8}$$

is used. On the other hand, the fact that if the derivative of the concave function equals zero at some point, this point ensures the maximum:

$$\nabla_x L(\bar{x}\,,\bar{\lambda}) \triangleq \nabla f(\bar{x}) - \sum_{l=1}^{m} \bar{\lambda}_l \nabla g_l(\bar{x}) = 0 \Rightarrow$$
$$\Rightarrow f(\bar{x}) = L(\bar{x}\,,\bar{\lambda}) \geq L(x\,,\bar{\lambda}) \geq f(x)$$

is taken into account, where the latter inequality is valid for all feasible x.

Being sufficiently general, the regularity conditions introduced in Definition 3 are difficult to check. Therefore, other hypotheses — also named regularity conditions — are often used.

Theorem 4.3.

Let the solution set of the program (4.1) be nonempty. The regularity conditions will be fulfilled if one of the following hypotheses hold:

(1) (4.1) is a linear program. This point of the theorem is equivalent to the known duality theorem of linear programming: the solvability of the prime problem yields the solvability of the dual one and vice versa.

(2) (4.1) is a convex program [that is, $-f(x)$, $g(x)$ are convex functions], and the Slater conditions are satisfied

$$\exists\bar{x}\ ,\ g_l(\bar{x}) < 0\ ,\ l = 1,\ldots,m \tag{4.9}$$

[†] If $f(x)$, $g_l(x)$, $l = 1,\ldots,m$, are linear, it is equivalent to the dual problem of linear programming.

(3) At each solution point the gradients of binding constraints are linearly independent.

The proof of Theorem 4.3 is so simple that in [28] it is offered as an exercise.

For our purposes it is more convenient to use a slightly different definition of regularity, the conditions of which were introduced in [152].

Definition 4. The program (4.1) is termed regular, if to each solution point $x \in \Theta^*$ one can assign a vector $y(x)$ such that

$$(\nabla g_l(x))^T y(x) < 0 \quad , \quad l \in \Omega(x) \tag{4.10}$$

The condition stated in Definition 4, as well as the suppositions of Theorem 4.3, stipulate the validity of (4.6) since

$$y \in \hat{D}(x) \Rightarrow \exists \mu_0 \;, \; y + \mu y(x) \in D(x) \;, \; 0 < \mu \le \mu_0 \Rightarrow$$
$$\Rightarrow y \in Cl\, D(x) \Rightarrow \hat{D}(x) \subset Cl\, D(x) \Rightarrow \hat{D}(x) = Cl\, D(x)$$

At the same time it is more general than hypotheses (2) or (3) in the sense that their fulfillment yields (4.10). It is easy to show that, under condition (2), one can take

$$y(x) = \bar{x} - x$$

since

$$(\nabla g_l(x))^T(\bar{x} - x) \le g_l(\bar{x}) - g_l(x) < 0 \quad , \quad l \in \Omega(x)$$

Under condition (3), any solution to the system

$$(\nabla g_l(x))^T y(x) = -L \quad , \quad L > 0 \quad , \quad l \in \Omega(x)$$

can be taken as $y(x)$. The solvability of this system is guaranteed by the linear independence of the gradients of binding constraints. Note also that if $-f(x)$, $g(x)$ are convex and (4.10) is satisfied, then there exists an interior point of the feasible domain, i.e., the Slater conditions are fulfilled.

In Section 1.2 we considered a simple example showing that, to ensure the proximity of the solutions to the perturbed and reduced problems, it is essential to assume that the Lagrange vectors set of the reduced problem is bounded. Now we shall establish that this condition is equivalent to the regularity condition in the sense of Definition 4.

Theorem 4.4. [153]

Let x be a solution to problem (4.1). Then the corresponding set of Lagrange vectors is nonempty and bounded if and only if there exists a vector $y(x)$ satisfying (4.10).

Proof. Let the set of Lagrange vectors corresponding to x be nonempty and be bounded. Then, obviously,

$$\exists L_l \geq 0 \ , \ l \in \Omega(x) \ , \ \sum_{l \in \Omega(x)} L_l = 1 \ , \ \sum_{l \in \Omega(x)} L_l \nabla g_l(x) = 0 \tag{4.11}$$

i.e., the zero point does not belong to the polyhedron

$$Z \triangleq \{Z \,/\, Z = \sum_{l \in \Omega(x)} L_l \nabla g_l(x) \ , \ \sum_{l \in \Omega(x)} L_l = 1 \ , \ L_l \geq 0\} \tag{4.12}$$

Hence, by virtue of the separability theorem, there exists a vector $p(m \times 1)$ such that

$$\|p\| = 1 \ , \ p^T(\sum_{l \in \Omega(x)} L_l \nabla g_l(x)) < 0 \ , \ L_l \geq 0 \ , \ \sum_{l \in \Omega(x)} L_l = 1$$

Denoting $p = y(x)$ and equating to unity each of the L_l successively, we obtain (4.10). Conversely, assume that there exists $y(x)$ satisfying (4.10). In view of Theorem 4.1, the set of Lagrange vectors is nonempty at x. We now show that it is also bounded. Suppose the contrary. Then

$$\exists \bar{L}_l \geq 0 \ , \ \sum_{l \in \Omega(x)} \bar{L}_l = 1 \ , \ \sum_{l \in \Omega(x)} \bar{L}_l \nabla g_l(x) = 0 \Rightarrow \tag{4.13}$$
$$\Rightarrow \sum_{l \in \Omega(x)} \bar{L}_l (\nabla g_l(x))^T y(x) = 0$$

The former contradicts (4.10). QED.

Let us proceed to the investigation of the relationship between the reduced and perturbed programs, which we shall write in the form

$$F^*(0) \triangleq \max \{f_0(x) \,/\, g_0(x) \leq 0\} \tag{4.14}$$

$$F^*(\epsilon) \triangleq \max \{f_0(x) + \epsilon f_1(x) \,/\, g_0(x) + \epsilon g_1(x) \leq 0\} \tag{4.15}$$

Later on, for the sake of convenience, we shall denote by $\Theta(0)$ and $\Theta(\epsilon)$ the admissible sets of the reduced and perturbed problems, respectively, and by $\Theta^*(0)$, $\Theta^*(\epsilon)$ the corresponding solutions sets.

First of all, we shall formulate a theorem establishing an asymptotic proximity of the solutions to (4.14) and (4.15) independent of linearity and convexity properties.

Theorem 4.5.

Let the functions $f_0(x)$, $g_0(x)$ be continuously differentiable and let $f_1(x)$, $g_1(x)$ be continuous. Let the solution x^* to the reduced problem exist and be unique, this program itself satisfying the regularity conditions of Definition 4. Suppose also that there exists $\hat{\epsilon} > 0$ such that, for $0 < \epsilon \leq \hat{\epsilon}$, the solutions set to the perturbed program is nonempty and bounded:

$$\phi \neq \Theta^*(\epsilon) \subset \bar{\Theta} \tag{4.16}$$

where $\bar{\Theta}$ is a compact set in R^n. Then there exists the limit

$$\lim_{\epsilon \to 0} x^*(\epsilon) = x^* \ , \ \forall\, x^*(\epsilon) \in \Theta^*(\epsilon) \tag{4.17}$$

i.e., an arbitrary solution to the perturbed program tends, with $\epsilon \to 0$, to the reduced program solution.

Proof. Assume that (4.17) does not hold. Then, by virtue of (4.16), one can choose a subsequence $\{\epsilon'\}$ such that

$$\lim_{\epsilon' \to 0} x^*(\epsilon') = \tilde{x} \neq x^*$$

Since the functions $g_0(x)$, $g_1(x)$ are continuous, we have

$$\tilde{x} \in \Theta(0) \Rightarrow f_0(\bar{x}) < f_0(x^*) \tag{4.18}$$

On the other hand, by (4.10) and in view of the fact that the functions $g_l(x)$ are continuously differentiable, we may write for all sufficiently small ϵ:

$$x^* + \epsilon y(x^*) \in \Theta(\epsilon) \Rightarrow f_0(x^* + \epsilon y(x^*)) + \epsilon f_1(x^* + \epsilon y(x^*)) \leq$$
$$\leq f_0(x^*(\epsilon)) + \epsilon f_1(x^*(\epsilon))$$

Taking the limit along the subsequence $\{\epsilon'\}$ in the last inequality leads to

$$f_0(x^*) \leq f_0(\tilde{x})$$

which contradicts (4.18) and so proves the theorem.

Corollary. Under the conditions of the theorem if the function $f_0(x)$ is strongly concave and the functions $g_{0l}(x)$, $l = 1,\dots,m$ are convex, the solution $x^*(\epsilon)$ to the perturbed program is unique and continuous as a function of ϵ with $0 \le \epsilon \le \bar{\epsilon}$, $\bar{\epsilon} > 0$.

Remark. Theorem 4.5 gives sufficiently general conditions for the reduced program (4.14) to be correct. The omission of the assumptions that the solution set is bounded or that the regularity conditions are fulfilled may, as examples show, lead to solutions or Lagrange vectors tending to ∞ with $\epsilon \to 0$ (a similar singular situation is considered for linear programs in Chapter 3) and, hence, disturb the correspondence between the perturbed and reduced programs. Note also that if the functions $-f_0(x)$, $g_{0l}(x)$ are convex, the assumption of their differentiability may be replaced by one of their continuity. In this case subgradients $-\partial f_0(x)$, $\partial g_{0l}(x)$ should be used instead of $-\nabla f_0(x)$, $\nabla g_{0l}(x)$ [114].

Once the asymptotic neighborhood is established, it is natural to try to improve the zero-order approximation.

Theorem 4.6.

Let the conditions of Theorem 4.5 be fulfilled. Let us suppose also that the following conditions are satisfied:

(1) The function $f_0(x)$ is strongly concave and the functions $g_{0l}(x)$, $l = 1,\dots,m$ are convex.

(2) The function $f_0(x)$, $g_0(x)$, $f_1(x)$, $g_1(x)$ are $m + 1$ times ($m \ge 1$) continuously differentiable.

(3) The gradients of the binding constraints $\nabla g_{0l}(x^*)$, $l \in \Omega(x^*)$ are linearly independent at the solution point x^* (i.e., regularity conditions stronger than (4.10) are fulfilled).

Then for sufficiently small ϵ the solution $x^*(\epsilon)$ to the perturbed problem and the corresponding Lagrange vector $\lambda^*(\epsilon)$ are unique, continuous as functions of ϵ, and admit the expansions

$$x^*(\epsilon) = x^* + \epsilon x^{1*} + \cdots + \epsilon^m x^{m*} + o(\epsilon^m) \tag{4.18'}$$

$$\lambda^*(\epsilon) = \lambda^* + \epsilon \lambda^{1*} + \cdots + \epsilon^m \lambda^{m*} + o(\epsilon^m) \tag{4.18''}$$

where λ^* is the Lagrange vector of the reduced program, which, by the postulated regularity conditions, is also unique.

Proof. Since (a) $\nabla g_l(x^*)$, $l \in \Omega(x^*)$ are linearly independent, (b) there exists the limit (4.17), and (c) the indices set of binding [at the point $x^*(\epsilon)$] constraints of the perturbed problem is contained in $\Omega(x^*)$, one can conclude that the gradients of these constraints are also linearly independent. It follows that there exists a unique Lagrange vector for the perturbed problem $\lambda^*(\epsilon)$, which is determined by the equations

$$\nabla f_0(x^*(\epsilon)) + \epsilon \nabla f_1(x^*(\epsilon)) = \sum_{l=1}^{m} \lambda_l^*(\epsilon)(\nabla g_{0l}(x^*(\epsilon)) + \epsilon \nabla g_{1l}(x^*(\epsilon))) \tag{4.19}$$

$$(\lambda^*(\epsilon))^T(g_0(x^*(\epsilon)) + \epsilon g_1(x^*(\epsilon)) = 0 \ , \ \lambda^*(\epsilon) \geq 0 \tag{4.20}$$

By (4.19) and the corollary of Theorem 4.5 the function $\lambda^*(\epsilon)$ is continuous and

$$\lim_{\epsilon \to 0} \lambda^*(\epsilon) = \lambda^* \tag{4.21}$$

Denote by $\omega(\epsilon)$ the set of indices l such that $\lambda_l^*(\epsilon) > 0$ and by ω the set of indices l such that $\lambda_l^* > 0$. It is easy to show that, for ϵ small enough,

$$\omega \subset \omega(\epsilon) \subset \Omega(x^*) \tag{4.22}$$

Thus, the point-set mapping $\omega(\epsilon)$ can take only a finite number of values, which we shall denote by $\omega_1, \omega_2, \ldots, \omega_q$. That is, for each ϵ, $\omega(\epsilon)$ equals either ω_1 or $\omega_2 \cdots$ or ω_q.

Using the introduced notation we can rewrite the system (4.19) and (4.20) in the form

$$\begin{cases} \nabla f_0(x^*(\epsilon)) + \epsilon \nabla f_1(x^*(\epsilon)) = \displaystyle\sum_{l \in \omega(\epsilon)} \lambda_l^*(\epsilon)(\nabla g_{0l}(x^*(\epsilon)) + \epsilon \nabla g_{1l}(x^*(\epsilon))) \\ g_{0l}(x^*(\epsilon)) + \epsilon g_{1l}(x^*(\epsilon)) = 0 \ , \ l \in \omega(\epsilon) \end{cases} \tag{4.23}$$

Assume that $\omega(\epsilon) = \omega_p$ $(1 \leq p \leq q)$ and determine the Jacobian of the system (4.23) with respect to $x \triangleq \{x_r\}$, $r = 1,\ldots,n$, λ_l, $l \in \omega_p$ at the point $x = x^*$, $\lambda = \lambda^*$, $\epsilon = 0$.

This is obviously equal to the determinant of the matrix

$$\nabla \nabla_p L_0(x^*, \lambda^*) \triangleq \begin{bmatrix} \nabla_x \nabla_x L_0(x^*, \lambda^*) & \nabla_p \tilde{g}_0(x^*) \\ \nabla_p \tilde{g}_0(x^*)^T & 0 \end{bmatrix} \tag{4.24}$$

where $\nabla \nabla_p L_0(x^*, \lambda^*)$ is the matrix of second derivatives of the reduced Lagrange function

$$L_0(x,\lambda) \triangleq f_0(x) - \lambda^T g_0(x)$$

with respect to $x \triangleq \{x_r\}$, $r = 1,\ldots,n$, λ_l, $l \in \omega_p$ calculated at the point $x = x^*$, $\lambda = \lambda^*$, and $\nabla_p \tilde{g}_0(x^*)$ is the matrix with columns $\nabla g_{0l}(x^*)$, $l \in \omega_p$.

Now show that

$$\det \nabla \nabla_p L_0(x^*,\lambda^*) \neq 0 \tag{4.25}$$

Suppose that it is not true. Then there exist y, z such that

$$\begin{aligned} &\nabla_x \nabla_x L_0(x^*, \lambda^*) y + \nabla_p \tilde{g}_0(x^*) z = 0 \ , \\ &(\nabla_p \tilde{g}_0(x^*))^T y = 0 \ , \ \|y\| + \|z\| = 1 \end{aligned} \tag{4.26}$$

In view of the strong concavity of $f_0(x)$ and the convexity of $g_0(x)$, the matrix $\nabla_x \nabla_x L_0(x^*, \lambda^*)$ is negative definite. Hence, by (4.26) we have that

$$\begin{aligned} &y^T \nabla_x \nabla_x L_0(x^*, \lambda^*) y + y^T \nabla_p \tilde{g}_0(x^*) z = 0 \Rightarrow y^T \nabla_x \nabla_x L_0(x^*, \lambda^*) y = 0 \Rightarrow \\ &\Rightarrow y = 0 \Rightarrow \nabla_p \tilde{g}_0(x^*) z = 0 \Rightarrow z = 0 \end{aligned} \tag{4.27}$$

since the columns of $\nabla_p \tilde{g}_0(x^*)$ are linearly independent. However, $\|y\| + \|z\| = 1$. This contradiction proves (4.25).

Thus, if $\omega(\epsilon) = \omega_p$ then the system (4.23) satisfies the conditions of the implicit function theorem in some neighborhood of the point $\{x = x^*, \lambda = \lambda^*, \epsilon = 0\}$. Hence, the expansions

$$\begin{aligned} x^*(\epsilon) &= x^* + \epsilon x_p^{1*} + \cdots + \epsilon^m x_p^{m*} + o(\epsilon^m) \\ \lambda^*(\epsilon) &= \lambda^* + \epsilon \lambda_p^{1*} + \cdots + \epsilon^m \lambda_p^{m*} + o(\epsilon^m) \end{aligned} \tag{4.28}$$

are valid. As has been established earlier, the functions $x^*(\epsilon)$, $\lambda^*(\epsilon)$, are continuous over the right neighborhood of zero. This is only possible if the coefficients of ϵ^s $(0 \le s \le m)$ coincide for different $p = 1,2,\ldots,q$, — i.e.,

$$\begin{aligned} &x_1^{1*} = x_2^{1*} = \cdots = x_q^{1*} \triangleq x^{1*}, \cdots, x_1^{m*} = x_2^{m*} = \cdots = x_q^{m*} \triangleq x^{m*} \ , \\ &\lambda_1^{1*} = \lambda_2^{1*} = \cdots = \lambda_q^{1*} \triangleq \lambda^{1*}, \cdots, \lambda_1^{m*} = \lambda_2^{m*} = \cdots = \lambda_q^{m*} \triangleq \lambda^{m*} \end{aligned}$$

which is equivalent to the result of the theorem. QED.

Substitute (4.18′) into the objective function of the perturbed program assuming that the coefficients of ϵ powers are not known

$$\begin{aligned} F^*(\epsilon) = \ & f_0(x^*) + \epsilon[(\nabla f_0(x^*))^T x^1 + f_1(x^*)] + \\ & + \epsilon^2[\frac{1}{2}(x^1)^T \nabla\nabla f_0(x^*) x^1 + (\nabla f_1(x^*))^T x^1 + \\ & + (\nabla f_0(x^*))^T x^2] + \cdots \end{aligned} \tag{4.29}$$

It is not difficult to see that the vectors of the coefficients x^{1*}, x^{2*}, $\cdots$, x^{m*} must attain the maxima (in certain domains) of the coefficients of ϵ powers in the expansion (4.29). This allows the construction of algorithms for their determination. Consider, for example, how to find x^{1*}. First of all it is obvious that the vector x^{1*} must be a solution to the program

$$\max \{(\nabla f_0(x^*))^T x^1 \,/\, (\nabla g_{0l}(x^*)^T x^1 \leq -g_{1l}(x^*) \;,\; l \in \Omega(x^*)\} \tag{4.30}$$

This, however, is not sufficient to detect x^{1*} since (4.30) has a set of solutions described by the equations and inequalities

$$\begin{aligned} &(\nabla g_{0l}(x^*))^T x^1 = -g_{1l}(x^*) \;,\; l \in \omega \\ &(\nabla g_{0l}(x^*))^T x^1 \leq -g_{1l}(x^*) \;,\; l \in \Omega(x^*) \backslash \omega \end{aligned} \tag{4.31}$$

Denote by Θ_1^* the solutions set to (4.31) and by $\Omega_1(x^1)$ the index set corresponding to the binding constraints in (4.31). One can state that x^{1*}, x^{2*} constitute a solution to the program

$$\begin{aligned} \max \; &\{(x^1)^T \nabla \nabla f_0(x^*) x^1 + (\nabla f_1(x^*))^T x^1 + \\ &+ (\nabla f_0(x^*))^T x^2 \,/\, x^1 \in \Theta_1^* \;,\; (\nabla g_{0l}(x^*))^T x_2 \leq \\ &\leq -(x^1)^T \nabla \nabla g_{0l}(x^*) x^1 - (\nabla g_{1l}(x^*))^T x^1 \;,\; l \in \Omega_1(x^1)\} \end{aligned} \tag{4.32}$$

and thus attain the maximum of the coefficient of ϵ^2 terms in (4.29). Obviously,

$$\begin{aligned} \max \; &\{(\nabla f_0(x^*))^T x^2 \,/\, (\nabla g_{0l}(x^*))^T x^2 \leq \\ &\leq -(x^1))^T \nabla \nabla g_{0l}(x^*) x^1 - (\nabla g_{1l}(x^*))^T x^1 \;,\; l \in \Omega_1(x^1)\} = \\ &= (\lambda^*)^T [-(x^1)^T \nabla \nabla g_{0l}(x^*) x^1 - (\nabla g_{1l}(x^*))^T x^1] \end{aligned} \tag{4.33}$$

and, hence, the program (4.32) can be rewritten in the form

$$\begin{aligned} \max \; &\{(x^1)^T \nabla_x \nabla_x L_0(x^* \,,\, \lambda^*) x^1 + (\nabla f_1(x^*) - \\ &- \sum_{l \in \omega} \lambda_l^* \nabla g_{1l}(x^*))^T x^1 \,/\, x^1 \in \Theta_1^*\} \end{aligned} \tag{4.34}$$

This is a quadratic program. Since the matrix $\nabla_x \nabla_x L_0(x^*,\lambda^*)$ is negative definite, its solution is unique and consequently coincides with x^{1*}, the corresponding Lagrange vector being equal to λ^{1*}. Similarly, the coefficients of ϵ^s to higher powers can be found. Unlike (4.34), the problems arising here will be linear owing to the fact that the coefficients of ϵ^s, $s > 2$ can only be linear in x^{s-1}.

Remark. Theorem 4.6 and the procedure based thereon remain valid for the nonconvex case as well. Instead of the convexity of the functions $-f_0(x)$, $g_0(x)$, it is sufficient that the matrix $\nabla_x \nabla_x L_0(x^*,\lambda^*)$ be positive or negative definite; this leads to (4.20). Note that this matrix will be positive definite if, for example, $f_0(x)$ is strongly convex and $g_0(x)$ are linear. In that special case $x^*(\epsilon)$ is the solution of the quasilinear system

$$g_{0l}(x) + \epsilon g_{1l}(x) = 0 \ , \ l \in \Omega(x^*)$$

Let us consider another simpler procedure for constructing the corrections. It is based on Theorem 4.7.

Theorem 4.7.

Let the conditions of Theorem 4.6 be fulfilled excepting the convexity hypothesis (1). Let us also assume that (a) the program (4.14) satisfies the conditions of strict complementarity slackness [81], i.e.,

$$\lambda_l^* > 0 \ , \ l \in \Omega(x^*) \ ; \ \lambda_l^* = 0 \ , \ l \,\bar{\in}\, \Omega(x^*) \tag{4.35}$$

and (b)

$$\det \{\nabla \nabla L_0(x^*, \lambda^*)\} \neq 0 \tag{4.36}$$

where $\nabla \nabla L_0(x^*, \lambda^*)$ is the matrix of second derivatives of the reduced Lagrange functions with respect to $x \triangleq \{x_r\}$, $r = 1,\ldots,n$, λ_l, $l \in \Omega(x^*)$. Then the perturbed solution $x^*(\epsilon)$ and the associated Lagrange vector $\lambda^*(\epsilon)$ admit the expansions (4.18′) and (4.18″). Moreover,

$$\left\| \begin{matrix} x^{i*} \\ \lambda^{i*} \end{matrix} \right\| = \{\nabla \nabla L_0(x^*, \lambda^*)\}^{-1} \left\| \begin{matrix} \varphi_i(x^*, x^{1*}, \ldots, x^{i-1*}, \lambda^*, \lambda^{1*}, \ldots, \lambda^{i-1*}) \\ \psi_i(x^*, x^{1*}, \ldots, x^{i-1*}, \lambda^*, \lambda^{1*}, \ldots, \lambda^{i-1*}) \end{matrix} \right\| \tag{4.37}$$

where a way to construct $\varphi_i(\cdot)$, $\psi_i(\cdot)$ is shown below.

The proof is similar to that of Theorem 4.6 with the unique difference that, by virtue of (4.21), under the conditions of Theorem 4.7

$$\omega = \omega(\epsilon) = \Omega(x^*)$$

This allows us to rewrite (4.19) in the form

$$\begin{cases} \nabla f_0(x^*(\epsilon)) + \epsilon \nabla f_1(x^*(\epsilon)) = \sum\limits_{l \in \Omega(x^*)} \lambda_l^*(\epsilon)(\nabla g_{0l}(x^*(\epsilon)) + \epsilon \nabla g_{1l}(x^*(\epsilon))) \ , \\ g_{0l}(x^*(\epsilon)) + \epsilon g_{1l}(x^*(\epsilon)) = 0 \ , \ l \in \Omega(x^*) \end{cases} \tag{4.38}$$

The Jacobian of the system is equal to the determinant of the matrix $\nabla\nabla L_0(x^*, \lambda^*)$. Hence it is nonzero. It follows that the expansions (4.18′) and (4.18″) are valid, the explicit expression for the functions $\varphi_i(\cdot)$, $\psi_i(\cdot)$ being obtained by substituting (4.18′) and (4.18″) into (4.38) and expanding the corresponding functions as asymptotic series in ϵ.

In view of Theorem 4.7, the perturbation method procedure may in this case be described as follows:

(1) Find the solution x^* to the reduced problem.
(2) Substitute the expansions (4.18′) and (4.18″) into (4.38).
(3) Determine λ^* from the system

$$\nabla f_0(x^*) = \sum_{l\in\omega} \lambda_l^* \nabla g_{0l}(x^*)$$

(4) Solve the consecutive linear systems beginning with the system

$$\begin{aligned} &\nabla_x\nabla_x L_0(x^*, \lambda^*)x^1 + \nabla\tilde{g}_0(x^*)\lambda^1 = -\nabla f_1(x^*) + \sum_l \lambda_l^*\nabla g_{1l}(x^*) \ , \\ &(\nabla\tilde{g}_0(x^*))^T x^1 = -\tilde{g}_1(x^*) \end{aligned} \tag{4.39}$$

where $\nabla\tilde{g}_0(x^*)$ is a matrix with columns $\nabla g_{0l}(x^*)$, $l \in \Omega(x^*)$, $\tilde{g}_1(x^*)$ is the subvector of $g_1(x^*)$ corresponding to the components $l \in \Omega(x^*)$, and $\nabla_x\nabla_x L_0(x^*, \lambda^*)$ is the matrix of second derivatives of $L_0(x, \lambda)$ with respect to x.

Remark. The strict complementarity slackness conditions are checked in the reduced program. The facts that the matrix $\nabla_x\nabla_x L_0(x^*, \lambda^*)$ is nonsingular and that the gradients of the binding constraints are linearly independent are examined simultaneously when solving the system (4.39) since the matrix requiring inversion has the form

$$\begin{bmatrix} \nabla_x\nabla_x L_0(x^*, \lambda^*) & \nabla\tilde{g}_0(x^*) \\ (\nabla\tilde{g}_0(x^*))^T & 0 \end{bmatrix} \triangleq \nabla\nabla L_0(x^*, \lambda^*)$$

As an example, let us consider a quasilinear program

$$\max\ \{c_0^T x + \epsilon f_1(x) \,/\, A_0 x + \epsilon g_1(x) \le b\} \tag{4.40}$$

For the conditions of the theorem to be satisfied here, it is sufficient to suppose that both the reduced linear program and the dual program have unique solutions. The precise solution to the perturbed program is completely determined by the second subsystem in (4.38), written as

$$\bar{A}x^*(\epsilon) + \epsilon\bar{g}_1(x^*(\epsilon)) = \bar{b} \tag{4.41}$$

where $\bar{A}$ is the optimal basic matrix of the reduced program

$$\det \bar{A} \neq 0$$

and $\bar{g}_1(\cdot)$, $\bar{b}$ are the corresponding subvectors of $g_1(\cdot)$, b. Upon substitution of (4.18′) into (4.41) we obtain

$$x^* = \bar{A}^{-1}\bar{b} \ , \ x^{1*} = -\bar{A}^{-1}\bar{g}_1(x^*) \ , \ x^{2*} = -A^{-1}(\nabla\,\bar{g}_1(x^*)x^{1*})$$

Omit the assumption that the solution is unique. In the same way as was done in Section 1.2, introduce the auxiliary program

$$F_1^* = \max_{x \in \Theta^*(0)} \{f_1(x) + \min\{-\sum_1^m \lambda_l g_{1l}(x) \,/\, \lambda \in \Lambda\}\} \tag{4.42}$$

where $\Theta^*(o)$ is the set of solutions to the reduced program, and Λ is the set of its Lagrange vectors. The relationship among the perturbed, reduced, and auxiliary programs is established by the following theorem, which in the linear case coincides with Theorem 2.4.

Theorem 4.8.

Let

(1) $f_0(x)$, $f_1(x)$, $g_0(x)$, $g_1(x)$ be continuously differentiable.
(2) $-f_0(x)$, $g_{0l}(x)$, $l = 1,\ldots,m$ be convex.
(3) There exist a positive number ϵ_0 and a compact set $\bar{\Theta} \subset R^n$ such that, for $0 \leq \epsilon \leq \epsilon_0$,

$$\phi \neq \Theta^*(\epsilon) \subset \bar{\Theta} \tag{4.43}$$

(4) The set Λ of Lagrange vectors of the reduced program (4.14) be not empty and bounded.

Then

(1) The set Θ^{**} of limit points of solutions to the perturbed program (obtained with $\epsilon \to 0$) is contained in the set of solutions to the auxiliary program

$$\lim_{\epsilon\to 0} \Theta^*(\epsilon) \triangleq \Theta^{**} \subset \operatorname{Arg} \max_{x\in\Theta^*(0)} \{f_1(x) + \min \{-\lambda^T g_1(x) \,/\, \lambda \in \Lambda\}\} \tag{4.44}$$

(2) The optimal objective value of the perturbed program is written in the form

$$F^*(\epsilon) = F^*(0) + \epsilon F_1^* + o(\epsilon) \tag{4.45}$$

where F_1^* is given by (4.42).

(3) For each solution x^* to the auxiliary program there is a vector $y(x^*,\epsilon)$ such that

$$\bar{x}(\epsilon) = x^* + \epsilon y(x^*, \epsilon) \in \Theta(\epsilon) \quad , \quad \|y(x^*, \epsilon)\| \le M = \text{const}$$

the value of the objective function for the feasible solution $\bar{x}(\epsilon)$ being, to within $o(\epsilon)$, equal to the optimal perturbed value, i.e.,

$$f_0(\bar{x}(\epsilon)) + \epsilon f_1(\bar{x}(\epsilon)) = F^*(\epsilon) + o(\epsilon) \tag{4.46}$$

Proof. Let us verify the inequality

$$F^*(\epsilon) \ge F^*(0) + \epsilon F_1^* + o(\epsilon) \tag{4.47}$$

Let x^* be a solution to the auxiliary program and let x_1^* be a solution to the program

$$\begin{aligned} \max \ \{(\nabla f_0(x^*))^T x_1 \,/\, (\nabla g_{0l}(x^*))^T x_1 \le \\ \le -g_{1l}(x^*) \ , \ l \in \Omega(x^*)\} \triangleq \varphi_1^* \end{aligned} \tag{4.48}$$

By virtue of the duality

$$\begin{aligned} \varphi_1^* = \min \{- \sum_{l\in\Omega(x^*)} \lambda_l g_{1l}(x^*) \,/ \sum_{l\in\Omega(x^*)} \lambda_l \nabla g_{0l}(x^*) = \nabla f_0(x^*) \\ \lambda_l \ge 0 \ , \ l \in \Omega(x^*)\} \end{aligned}$$

but the latter is equivalent to

$$\min \{-\sum_{l=1}^{m} \lambda_l g_{1l}(x^*) \,/\, \lambda \in \Lambda\}$$

since, by definition (4.2),

$$\lambda \in \Lambda \Rightarrow \lambda_l = 0 \;,\; l \bar{\in} \Omega(x^*) \;,\; \sum_{l=1}^{m} \lambda_l \nabla g_{0l}(x^*) = \nabla f_0(x^*)$$

By virtue of Theorem 4.4, one can find $y(x^*)$ satisfying the relation

$$(\nabla g_{0l}(x^*))^T y(x^*) < 0 \;,\; l \in \Omega(x^*)$$

Hence, there exists a positive number ρ such that, for all sufficiently small ϵ,[†]

$$\bar{x}(\epsilon) = x^* + \epsilon x_1^* + \epsilon^2 \rho y(x^*) \in \Theta(\epsilon) \Rightarrow \tag{4.49}$$

$$\begin{aligned} &F^*(\epsilon) \geq f_0(\bar{x}(\epsilon)) + \epsilon f_1(\bar{x}(\epsilon)) = f_0(x^*) + \\ &+ \epsilon[(\nabla f_0(x^*))^T x_1^* + f_1(x^*)] + o(\epsilon) = \\ &f_0(x^*) + \epsilon[f_1(x^*) + \min\{-\lambda^T g_1(x^*) \,/\, \lambda \in \Lambda\}] + o(\epsilon) \triangleq \\ &\triangleq F^*(0) + \epsilon F_1^* + o(\epsilon) \end{aligned} \tag{4.50}$$

Let us obtain the reverse inequality. Assume that $\bar{x}^*$ is a partial limit of a perturbed solution, i.e.,

$$\begin{aligned} &\exists \{\epsilon'\} \;,\; x^*(\epsilon') \in \Theta^*(\epsilon') \\ &\lim_{\epsilon' \to 0} x^*(\epsilon') = \bar{x}^* \end{aligned} \tag{4.51}$$

Similarly as for Theorem 4.5, it is easy to show that

$$\bar{x}^* \in \Theta^*(0) \tag{4.52}$$

† For simplicity, we assume here that $f_0(x)$, $g_0(x)$ are twice continuously differentiable.

We then have

$$F^*(\epsilon') = f_0(x^*(\epsilon')) + \epsilon' f_1(x^*(\epsilon')) \le \min_{\lambda\in\Lambda} \{f_0(x^*(\epsilon')) + \\ + \epsilon' f_1(x^*(\epsilon')) - \sum_{l=1}^{m} \lambda_l (g_{0l}(x^*(\epsilon')) + \epsilon g_{1l}(x^*(\epsilon')))\} \tag{4.53}$$

However, the reduced Lagrange function is concave in x. Hence,

$$\begin{aligned} &\nabla_x L_0(\bar{x}^*, \lambda) = 0 \ , \ \lambda \in \Lambda \Rightarrow \\ &f_0(\bar{x}^*) = L_0(\bar{x}^*, \lambda) \ge L_0(x , \lambda) \ , \ \lambda \in \Lambda \ , \ \forall x \Rightarrow \\ &F^*(\epsilon') \le \{f_0(\bar{x}^*) + \epsilon'[f_1(x^*(\epsilon')) + \min_{\lambda\in\Lambda} \sum_{l=1}^{m} (-\lambda_l) g_{1l}(x^*(\epsilon'))] = \\ &= F^*(0) + \epsilon'[f_1(\bar{x}^*) + \min_{\lambda\in\Lambda} \sum_{l=1}^{m} (-\lambda_l) g_{1l}(\bar{x}^*)] + o(\epsilon') \end{aligned} \tag{4.54}$$

The conjunction of (4.54) and (4.47) yields

$$f_1(\bar{x}^*) + \min_{\lambda\in\Lambda} \sum_{l=1}^{m} (-\lambda_l) g_{l1}(\bar{x}^*) = F_1^* \tag{4.54'}$$

i.e., $\bar{x}^*$ is the solution to the perturbed program. The second and third statements of the theorem now follow from (4.47), (4.54), (4.49), and (4.50). QED.

The inclusion (4.44) naturally implies existence of the limit

$$\lim_{\epsilon\to 0} x^*(\epsilon) = x^* \ , \ \forall x^*(\epsilon) \in \Theta^*(\epsilon) \tag{4.55}$$

if the solution to the auxiliary program is unique.

1.5. Perturbations in nonconvex programs

Let us consider a nonlinear perturbed program

$$F^*(\epsilon) \triangleq \max \{f_0(x) + \epsilon f_1(x) \,/\, g_0(x) + \epsilon g_1(x) \le 0\} \tag{5.1}$$

and the corresponding reduced program

$$F^*(0) \triangleq \max \{f_0(x) \,/\, g_0(x) \le 0\} \tag{5.2}$$

We shall suppose here that (5.2) is not necessarily a convex program.

Let (5.2) be solvable and let its solutions set be $\Theta^*(0)$. Some basic results concerning more general cases are presented in [154] and [155]. For simplicity, we introduce the main constraint that the vectors $\nabla g_{0l}(x)$, $l \in \Omega(x)$ are linearly independent at each point $x \in \Theta^*(0)$ where, as above,

$$\Omega(x) \triangleq \{l \,/\, g_{0l}(x) = 0\}$$

By Definition 1 (Section 1.4) $\lambda(x)$ is the Lagrange multipliers vector at the point $x \in \Theta^*(0)$ if

$$\nabla f(x) = \sum_{l=1}^{m} \lambda_l(x) \nabla g_{0l}(x) \;\; ; \;\; \lambda_l(x) \geq 0 \;\; , \;\; \lambda_l(x) g_{0l}(x) = 0 \tag{5.3}$$

The linear independence of $\nabla g_{0l}(x)$, $l \in \Omega(x)$ implies the regularity condition (4.10) (discussed on pp. 40–41) and thus provides the existence and boundedness of the set of Lagrange vectors (Theorem 4.4). Moreover, it is evident that under the linear independence condition there is only one vector $\lambda(x)$ satisfying (5.3) for each $x \in \Theta^*(0)$.

Let us write the auxiliary program in the form

$$F_1^* \triangleq \max \, \{f_1(x) - \lambda^T(x) g_1(x) \,/\, x \in \Theta^*(0)\} \tag{5.4}$$

Theorem 5.1.

Let there exist $\hat{\epsilon} > 0$ such that

$$\Theta^*(\epsilon) \subset \bar{\Theta} \tag{5.5}$$

over $0 \leq \epsilon \leq \hat{\epsilon}$, where $\bar{\Theta}$ is a compact set in R^n.

Let the functions $f_0(x)$, $g_0(x)$, $f_1(x)$, $g_1(x)$ be continuous and continuously differentiable in some vicinity of $\bar{\Theta}$. Let $\nabla g_{0l}(x)$, $l \in \Omega(x)$, $x \in \Theta^*(0)$ be linear-independent.

Then

(1) The set Θ^{**} of limit points of solutions to the perturbed program (obtained with $\epsilon \to 0$) is contained in the set of solutions to the auxiliary program

$$\Theta^{**} \subset \text{Arg max}\{f_1(x) - \lambda^T(x) g_1(x) \,/\, x \in \Theta^*(0)\} \tag{5.6}$$

(2) The optimal objective value of the perturbed program can be written in the form

$$F^*(\epsilon) = F^*(0) + \epsilon F_1^* + o(\epsilon) \tag{5.7}$$

where F_1^* is given by (5.4).

(3) For each solution x^* to (5.4) there exists a vector $y(x^*, \epsilon)$ such that

$$\bar{x}(\epsilon) \triangleq x^* + \epsilon y(x^*, \epsilon) \in \Theta(\epsilon) \;\; , \;\; \|y(x^*, \epsilon)\| \leq M = \text{const}$$

$$f_0(\bar{x}(\epsilon)) + \epsilon f_1(\bar{x}(\epsilon)) = F^*(\epsilon) + o(\epsilon)$$

Proof is based on the following lemmata.

Lemma 5.1. Let the conditions of the Theorem 5.1 be fulfilled. Then

(a) The point-set mapping $\Theta^*(\epsilon)$ is uper semicontinuous on $0 \leq \epsilon \leq \hat{\epsilon}$, $\hat{\epsilon} > 0$ and, in particular,

$$\Theta^{**} \subset \Theta^*(0) \tag{5.8}$$

(b) The vectors $\nabla(g_{0l}(x) + \epsilon g_{1l}(x))$, $l \in \Omega_\epsilon(x)$ are linearly independent if $x \in \Theta^*(\epsilon)$, $\Omega_\epsilon(x) \triangleq \{l / g_{0l}(x) + \epsilon g_{1l}(x) = 0\}$ and $0 \leq \epsilon \leq \hat{\epsilon}$, $\hat{\epsilon} > 0$.

Proof of Lemma 5.1. Let (5.8) be false. Then, taking into account (5.5), we have the existence of a subsequence $\{\epsilon'\}$, convergent to zero, such that

$$\lim_{\epsilon \to 0, \epsilon \in \{\epsilon'\}} x^*(\epsilon) = \tilde{x} \bar{\in} \Theta^*(0) \tag{5.9}$$

where $x^*(\epsilon) \in \Theta^*(\epsilon)$ for $\epsilon \in \{\epsilon'\}$. Similarly to (4.18)

$$\tilde{x} \in \Theta(0) \tag{5.10}$$

Let $\bar{x} \in \Theta^*(0)$. As in the proof of Theorem 4.5, one can state that there exists $y(\bar{x})$ such that

$$\bar{x} + \epsilon y(\bar{x}) \in \Theta(\epsilon) \Rightarrow f_0(\bar{x} + \epsilon y(\bar{x})) + \epsilon f_1(\bar{x} + \epsilon y(\bar{x})) \leq f_0(x^*(\epsilon)) + \epsilon f_1(x^*(\epsilon))$$

for all $\epsilon \in \{\epsilon'\}$ and $0 < \epsilon \leq \epsilon'$.

Taking the limit gives

$$f_0(\bar{x}) \leq f_0(\tilde{x})$$

This contradicts (5.9) and (5.10), and so (5.8) is valid.

Let result (b) of the lemma be false. Then there exists a subsequence $\{\epsilon'\}$ convergent to zero such that the gradients mentioned in (b) are linearly dependent for $x = x^*(\epsilon) \in \Theta^*(\epsilon)$, $\epsilon \in \{\epsilon'\}$.

Let $\hat{x}$ be a partial limit of $x^*(\epsilon)$, $\epsilon \in \{\epsilon'\}$, then the hypothesis implies that the gradients $\nabla g_{0l}(x)$ are linearly dependent at $x = \hat{x}$. However, as follows from (5.8), $\hat{x} \in \Theta^*(0)$, i.e., we obtain a contradiction to the condition of Theorem 5.1, where linear-independence of the gradients in the reduced program was assumed.

Let $\epsilon_1 \in [0, \hat{\epsilon}]$. Then, bearing in mind result (b), one can easily state that

$$\Theta^{**}(\epsilon_1) \subset \Theta^*(\epsilon_1) \tag{5.11}$$

where $\Theta^{**}(\epsilon_1)$ is the set of limit points of $\Theta^*(\epsilon)$ obtained when $\epsilon \to \epsilon_1$. [This can be done analogously to the proof of (5.8).] The inclusion (5.11) implies the required upper semicontinuity. QED.

Lemma 5.2. Let the conditions of Theorem 5.1 be fulfilled. Let $\{\epsilon'\} = \{\epsilon_1, \epsilon_2, \cdots\}$, $\epsilon_{l+1} > \epsilon_l > 0$ be a subsequence of the values ϵ convergent to zero and let there exist

$$\lim_{\epsilon \to 0, \epsilon \in \{\epsilon'\}} x^*(\epsilon) = x^* \in \Theta^*(0) \tag{5.12}$$

Then

$$x^*(\epsilon) = x(\epsilon) + \epsilon x_1(\epsilon) \;\; , \;\; \|x_1(\epsilon)\| \leq M = \text{const} \;\; , \;\; \epsilon \in \{\epsilon'\} \tag{5.13}$$

where

$$x(\epsilon) \in \Theta(0) \;\; , \;\; \epsilon \in \{\epsilon'\} \tag{5.14}$$

and $x(\epsilon)$ satisfies the additional conditions

$$g_l(x(\epsilon)) = 0 \;\; , \;\; l \in \omega(x^*) \tag{5.15}$$

$$\omega(x^*) = \{l \,/\, \lambda_l(x^*) > 0\} \tag{5.16}$$

Proof of Lemma 5.2. By (5.12) we have

$$\begin{aligned} &g_{0l}(x^*(\epsilon)) + \epsilon g_{1l}(x^*(\epsilon)) = \mu_l(\epsilon) \leq 0 \; , \; \mu_l(\epsilon) \to 0 \; , \; l \in \Omega(x^*)) \; , \; \epsilon \to 0 \; , \; \epsilon \in \{\epsilon'\} \; ; \\ &g_{0l}(x^*(\epsilon)) + \epsilon g_{1l}(x^*(\epsilon)) \leq \delta_l < 0 \;\; , \;\; \delta_l = \text{const} \;\; , \;\; l \,\overline{\in}\, \Omega(x^*) \;\; , \;\; \epsilon \in \{\epsilon'\} \end{aligned} \tag{5.17}$$

where

$$\Omega(x^*) \triangleq \{l \,/\, g_{0l}(x^*) = 0\}$$

Without loss of generality, one can assume that

$$\Omega(x^*) = \Omega_1 \cup \Omega_2 \;\; , \;\; \Omega_1 \cap \Omega_2 = \phi$$

where

$$\begin{aligned} &l \in \Omega_1 \Leftrightarrow \lim_{\epsilon \to 0, \epsilon \in \{\epsilon'\}} \frac{\mu_l(\epsilon)}{\epsilon} = \mu_l \leq 0 \;\; ; \\ &l \in \Omega_2 \Leftrightarrow \lim_{\epsilon \to 0, \epsilon \in \{\epsilon'\}} \frac{\epsilon}{\mu_l(\epsilon)} = 0 \end{aligned} \tag{5.18}$$

Let

$$\tilde{x}(\epsilon) = \frac{x^*(\epsilon_l)(\epsilon_{l+1} - \epsilon) + (\epsilon - \epsilon_l)x^*(\epsilon_{l+1})}{\epsilon_{l+1} - \epsilon_l}$$

$$\tilde{x}(\epsilon) = \tilde{x}(-\epsilon)$$

where ϵ_l, ϵ_{l+1} are consecutive elements of $\{\epsilon'\}$. Note that $\tilde{x}(\epsilon)$ is, by definition, continuous. Let us construct vectors z_l, $l = 1,2,\ldots,n-|\Omega_1|$, forming a basis for the subspace that is orthogonal to the subspace spanned by the vectors $\nabla g_{0l}(x^*)$, $l \in \Omega_1$. As follows from the implicit function theorem, the system

$$g_{0l}(\tilde{x}(\epsilon) + y) = 0 \ , \ l \in \Omega_1 \ ; \ z_l^T y = 0 \ , \ l = 1,2,\ldots,n-|\Omega_1| \tag{5.19}$$

determines a function $y(\epsilon)$, $y(0) = 0$, which is continuous in a neighborhood of the point $\{\epsilon = 0 \ , \ y = 0\}$.

Let us show that

$$\|y(\epsilon)\| \leq M\epsilon \ , \ M = \text{const} \ , \ \epsilon \in \{\epsilon'\} \tag{5.20}$$

for all ϵ small enough.

If one assumes that (5.20) is false then a subsequence $\{\epsilon''\} \subset \{\epsilon'\}$ exists such that

$$\lim_{\substack{\epsilon \to 0 \\ \epsilon \in \{\epsilon''\}}} \frac{\epsilon}{\|y(\epsilon)\|} = 0 \ , \quad \lim_{\substack{\epsilon \to 0 \\ \epsilon \in \{\epsilon''\}}} \frac{y(\epsilon)}{\|y(\epsilon)\|} = \rho \ , \ \|\rho\| = 1 \tag{5.21}$$

Let us use the Lagrange formulae in (5.19), taking into account (5.17) and (5.18). We obtain

$$[\nabla g_{0l}(x^*(\epsilon) + \Theta_l(\epsilon) y(\epsilon))]^T y(\epsilon) = \epsilon g_{1l}(x^*(\epsilon)) - \mu_l(\epsilon) \ , \ \epsilon \in \{\epsilon''\} \ ,$$
$$l \in \Omega_1 \ ; \ z_l^T y(\epsilon) = 0 \ , \ l = 1,2,\ldots,n-|\Omega_1|$$

Dividing the latter equation by $\|y(\epsilon)\|$ and letting ϵ tend toward zero, $\epsilon \in \{\epsilon''\}$, we have

$$\left.\begin{array}{l} (\nabla g_{0l}(x^*))^T \rho = 0 \ , \ l \in \Omega_1 \ ; \\ z_l^T \rho = 0 \ , \ l = 1,2,\ldots,n-|\Omega| \end{array}\right\} \Rightarrow \rho = 0$$

This contradicts (5.21) and proves (5.20).

Let

$$x_1(\epsilon) \triangleq -\frac{y(\epsilon)}{\epsilon} \ , \ x(\epsilon) \triangleq x^*(\epsilon) + y(\epsilon) \triangleq x^*(\epsilon) - \epsilon x_1(\epsilon)$$

By (5.20) and (5.19) we obtain

$$\|x_1(\epsilon)\| \leq M \ , \ g_{0l}(x(\epsilon)) = 0 \ , \ l \in \Omega_1 \tag{5.22}$$

and in addition, by virtue of (5.17) and (5.18) for $\epsilon \in \{\epsilon'\}$

$$\begin{aligned} g_{0l}(x(\epsilon)) &= g_{0l}(x^*(\epsilon)) + O(\epsilon) = \mu_l(\epsilon) + O(\epsilon) \leq 0 \ , \ l \in \Omega_2 \ ; \\ g_{0l}(x(\epsilon)) &= g_{0l}(x^*(\epsilon)) + O(\epsilon) \leq \delta_l + O(\epsilon) < 0 \ , \ l \bar{\in} \Omega(x^*) \end{aligned} \tag{5.23}$$

The estimates (5.22) and (5.23) imply (5.13) and (5.14). Moreover, if

$$\omega(x^*) \subset \Omega_1 \tag{5.24}$$

then (5.15) is also true.

To prove the inclusion (5.24), let us note first that (5.12) implies existence of the limit

$$\lim_{\epsilon \to 0, \epsilon \in \{\epsilon'\}} \lambda_\epsilon(x^*(\epsilon)) = \lambda(x^*) \tag{5.24'}$$

where $\lambda_\epsilon(x^*(\epsilon))$ is the Lagrange vector of the perturbed program corresponding to $x^*(\epsilon)$.

In fact, by definition of the Lagrange vector,

$$\begin{aligned} &\nabla f_0(x^*(\epsilon)) + \epsilon \nabla f_1(x^*(\epsilon)) = \sum_{l \in \Omega(x^*)} \lambda_{\epsilon l}(x^*(\epsilon))[\nabla g_{0l}(x^*(\epsilon) + \epsilon \nabla g_{1l}(x^*(\epsilon))] \ ; \\ &\lambda_{\epsilon l}(x^*(\epsilon)) \geq 0 \end{aligned} \tag{5.25}$$

Let us suppose that there exists a subsequence $\{\epsilon''\}$ such that the partial limit of $\lambda_\epsilon(x^*(\epsilon))$ is not equal to $\lambda(x^*)$, i.e.,[†]

$$\lim_{\epsilon \to 0, \epsilon \in \{\epsilon''\}} \lambda_\epsilon(x^*(\epsilon)) = \tilde{\lambda} \neq \lambda(x^*)$$

Let $\epsilon \to 0$ in (5.25). Then

$$\nabla f_0(x^*) = \sum_{l \in \Omega(x^*)} \tilde{\lambda}_l \nabla g_{0l}(x^*) \ , \ \tilde{\lambda}_l \geq 0$$

Thus, we have the existence of two different Lagrange vectors at the point x^* which, as was noted above, cannot be true.

[†] Note that $\lambda_\epsilon(x^*(\epsilon))$ remains uniformly bounded when $\epsilon \to 0$ since, if this were false, dividing (5.25) by $||\lambda_\epsilon(x^*(\epsilon))||$ and letting ϵ tend toward zero would give

$$0 = \sum_{l \in \Omega(x^*)} \rho_l \nabla g_{0l}(x^*) \ , \ \rho_l = \lim_{\epsilon \to 0} \frac{\lambda_{\epsilon l}(x^*(\epsilon))}{||\lambda_\epsilon(x^*(\epsilon))||}$$

which contradicts the linear-independence.

From (5.24′) and (5.16) it follows that, for all sufficiently small ϵ,

$$\begin{aligned} &\lambda_{\epsilon l}(x^*(\epsilon)) > 0 \ , \ l \in \omega(x^*) \Rightarrow \\ &g_{0l}(x^*(\epsilon)) + \epsilon g_{1l}(x^*(\epsilon)) = 0 \ , \ l \in \omega(x^*) \end{aligned} \tag{5.25'}$$

This means that $\mu_l(\epsilon)$, $l \in \omega(x^*)$ defined in (5.17) are equal to zero for all $\epsilon \in \{\epsilon'\}$ small enough. Returning to (5.18), one can see that (5.24) is true. QED.

Proof of Theorem 5.1. Let x^* be a partial limit of $x^*(\epsilon) \in \Theta^*(\epsilon)$, i.e., there exists a subsequence $\epsilon \in \{\epsilon'\}$ such that the relation (5.12) is true. Statement (1) of the theorem will be proved if one verifies that

$$x^* \in \text{Arg max } \{f_1(x) - \lambda^T(x) g_1(x) \,/\, x \in \Theta^*(0)\} \tag{5.26}$$

Using Lemma 5.2, we can write

$$\begin{aligned} F^*(\epsilon) &\triangleq f_0(x^*(\epsilon)) + \epsilon f_1(x^*(\epsilon)) = \\ &= f_0(x(\epsilon)) + \epsilon[(\nabla f_0(x(\epsilon)))^T x_1(\epsilon) + f_1(x(\epsilon))] + o(\epsilon) \ , \ \epsilon \in \{\epsilon'\} \end{aligned} \tag{5.27}$$

Since the linear-independence of the gradients of the binding constraints implies the regularity condition (4.10) and since the convexity properties were not used to obtain the inequality (4.47), the latter remains true in the case under consideration. By virtue of (5.27) it can be rewritten in the form

$$\begin{aligned} &f_0(x(\epsilon)) + \epsilon[(\nabla f_0(x(\epsilon))^T x_1(\epsilon) + f_1(x(\epsilon))] + o(\epsilon) \geq \\ &\geq F^*(0) + \epsilon F_1^* \ , \ \epsilon \in \{\epsilon'\} \end{aligned} \tag{5.28}$$

Owing to (5.28) and the inequality

$$f_0(x(\epsilon)) \leq F^*(0)$$

we have

$$(\nabla f_0(x(\epsilon)))^T x_1(\epsilon) + f_1(x(\epsilon)) + \nu(\epsilon) \geq F_1^* \ , \quad \nu(\epsilon) \underset{\epsilon \to 0 \ , \ \epsilon \in \{\epsilon'\}}{\longrightarrow} 0 \tag{5.29}$$

Let us now obtain the inequalities, which are opposite to (5.28) and (5.29). Substituting (5.13) into (5.25′) and taking into account (5.15), we have for $l \in \omega(x^*)$

$$(\nabla g_{0l}(x(\epsilon)))^T x_1(\epsilon) + g_{1l}(x(\epsilon)) + \nu_l(\epsilon) = 0 \ , \quad \nu_l(\epsilon) \underset{\epsilon \to 0 \ , \ \epsilon \in \{\epsilon'\}}{\longrightarrow} 0 \tag{5.30}$$

Via (5.12) and (5.13)

$$\lim_{\epsilon \to 0, \epsilon \in \{\epsilon'\}} x(\epsilon) = x^*$$

Hence the relation (5.30) can be rewritten as

$$(\nabla g_{0l}(x^*))^T x_1(\epsilon) + g_{1l}(x^*) + \tilde{\nu}_l(\epsilon) = 0 \ , \quad \tilde{\nu}_l(\epsilon) \underset{\epsilon\to 0, \epsilon\in\{\epsilon'\}}{\to} 0$$

and so

$$\begin{aligned}(\nabla f_0(x(\epsilon))^T x_1(\epsilon) + f_1(x(\epsilon)) &\leq (\nabla f_0(x^*))^T x_1(\epsilon) + f_1(x^*) + \tilde{\nu}(\epsilon) \leq \\ &\leq \max\{(\nabla f_0(x^*))^T x_1 \,/\, (\nabla g_{0l}(x^*))^T x_1 + g_{1l}(x^*) + \tilde{\nu}_l(\epsilon) \ , \ e \in \omega(x^*)\} + \\ &+ f_1(x^*) + \tilde{\nu}(\epsilon) \quad \text{where} \quad \tilde{\nu}_l(\epsilon) \ , \ \tilde{\nu}(\epsilon) \underset{\epsilon\to 0 \ , \ \epsilon\in\{\epsilon'\}}{\to} 0\end{aligned}$$

Transforming the right-hand side by means of the duality, we obtain

$$\begin{aligned}&(\nabla f_0(x(\epsilon)))^T x_1(\epsilon) + f_1(x(\epsilon)) \leq \\ &\leq -\sum_l \lambda_l(x^*) g_{1l}(x^*) + f_1(x^*) + \tilde{\tilde{\nu}}(\epsilon) \ , \quad \tilde{\tilde{\nu}}(\epsilon) \underset{\epsilon\to 0 \ , \ \epsilon\in\{\epsilon'\}}{\to} 0 \ ;\end{aligned} \tag{5.31}$$

$$f_0(x(\epsilon)) + \epsilon[(\nabla f_0(x(\epsilon)))^T x_1(\epsilon) + f_1(x(\epsilon))] \leq F^*(0) + \epsilon F_1^* + o(\epsilon) \tag{5.32}$$

The conjunction of (5.29) and (5.31) gives

$$f_1(x^*) - \sum_l \lambda_l(x^*) g_{1l}(x^*) = F_1^*$$

which shows the validity of (5.26) and so proves the first statement of the theorem.

The conjunction of (5.32), (5.28), and (5.27) shows that

$$\lim_{\epsilon\to 0, \epsilon\in\{\epsilon'\}} \varphi(\epsilon) = 0$$

where $\varphi(\epsilon) \triangleq \epsilon^{-1} | F^*(\epsilon) - F^*(0) - \epsilon F_1^* |$.

To prove (5.7) we must contradict the hypothesis that there exists another sequence $\{\bar{\epsilon}'\}$ such that

$$\varliminf_{\epsilon\to 0} \varphi(\epsilon) = m > 0 \ , \ \epsilon \in \{\bar{\epsilon}'\}$$

One can select from $\{\bar{\epsilon}'\}$ a subsequence $\{\bar{\epsilon}''\}$ convergent to zero such that

$$\lim_{\epsilon\to 0, \epsilon\in\{\epsilon'\}} x^*(\epsilon) = \bar{x} \in \Theta^*(0)$$

However, the consideration applied earlier to $\{\epsilon'\}$ in connection with $\{\bar{\epsilon}''\}$ implies that

$$\lim_{\epsilon \to 0, \epsilon \in \{\bar{\epsilon}''\}} \varphi(\epsilon) = 0 \ , \ \epsilon \in \{\bar{\epsilon}''\}$$

We have the contradiction that proves (5.7).

Result (3) of the theorem follows from (5.7) by the scheme used in the proof of Theorem 4.8. QED

Corollary 1. Let the conditions of Theorem 5.1 be fulfilled and let the solution x^* to the auxiliary program be unique. Then, for any solution $X^*(\epsilon)$ to the perturbed program, there exists the limit

$$\lim_{\epsilon \to 0} x^*(\epsilon) = x^* = \arg\max \{f_1(x) - \lambda^T(x) g_1(x) \,/\, x \in \Theta^*(0)\} \tag{5.33}$$

Proof of the corollary is based on Theorem 5.1 and Lemma 5.1 and is straightforward.

Theorem 5.2.

Let the conditions of Theorem 5.1 be fulfilled and let the solution x^* to the auxiliary program be unique. Let us assume also that the functions $f_0(x)$, $g_0(x)$, $f_1(x)$, $g_1(x)$ are m times ($m \geq 1$) continuously differentiable and, moreover, that:

(a) The program (5.2) satisfies the conditions of strict complementarity slackness at the point x^* (4.35).

(b) The determinant

$$\det \{\nabla \nabla L_0(x^*, \lambda(x^*))\} \neq 0 \tag{5.34}$$

where the derivatives of the reduced Lagrange function are calculated at the point $x = x^*$, $\lambda = \lambda(x^*)$ with respect to $x = \{x_r\}$, $r = 1,\ldots,n$; λ_l, $l \in \Omega(x^*)$.

Then the solution to the perturbed program $x^*(\epsilon)$ and the corresponding Lagrange vector $\lambda^*(\epsilon)$ are unique and admit the expansions (4.18′) and (4.18″).

Proof. The single thing that distinguishes Theorem 5.2 from Theorem 4.7 is that the solution to the reduced program is not assumed to be unique. The existence of the limit (5.33) implies that this difference is not critical and the proof of the Theorem 4.7 holds true. QED.

As was shown by the example in Section 1.3, the expansions (4.18′) and (4.18″) are not generally true even when the solution to the reduced program is unique. Consequently, it is reasonable to estimate the difference between the perturbed and the auxiliary solutions, i.e., to estimate the convergence rate in (5.33).

Let us consider the following construction. Let

$$\delta_1(\epsilon) \triangleq \min \{\|x^*(\epsilon) - x\| \,/\, x \in \Theta^*(0)\} \triangleq \|x^*(\epsilon) - x_0(\epsilon)\| \ , \ x_0(\epsilon) \in \Theta^*(0) \tag{5.35}$$

By definition one can write

$$x^*(\epsilon) = x_0(\epsilon) + \delta(\epsilon) x_1(\epsilon) \tag{5.36}$$

where

$$\delta(\epsilon) = \max(\epsilon \, , \, \delta_1(\epsilon)) \tag{5.37}$$

$$x_1(\epsilon) \triangleq \frac{x^*(\epsilon) - x_0(\epsilon)}{\delta(\epsilon)} \Rightarrow \|x_1(\epsilon)\| = \frac{\delta_1(\epsilon)}{\delta(\epsilon)} \leq 1 \tag{5.38}$$

Let us suppose that an estimate of $\delta(\epsilon)$ is known (later an example of such an estimate for the programs with linear reduced structure will be given). Then to evaluate $\|x^*(\epsilon) - x^*\|$ one must evaluate $\|x_0(\epsilon) - x^*\|$.

Lemma 5.3. Let the conditions of Theorem 5.1 be fulfilled and the solution x^* to the auxiliary problem be unique. Let the functions $f_0(x)$, $g_0(x)$ be twice continuously differentiable and let there exist a neighborhood $R(x^*)$ of the point x^* such that $\lambda(x) = \lambda(x^*) \triangleq \lambda^*$ for all $x \in \Theta^*(0) \cap R(x^*)$. This condition is satisfied if (5.2) is a convex program since in this case $\lambda(x) = \lambda^*$ for all $x \in \Theta^*(0)$.

Then

$$\Phi(x^*) - \Phi(x_0(\epsilon)) = O(\delta(\epsilon)) \quad , \quad \Phi(x) \triangleq f_1(x) - \lambda^T(x) g_1(x) \tag{5.39}$$

and the expansion (5.7) takes a more precise form

$$F^*(\epsilon) = F^*(0) + \epsilon F_1^* + O(\epsilon\delta(\epsilon)) \tag{5.40}$$

Remark. The estimate (5.39) permits us to evaluate $\|x_0(\epsilon) - x^*\|$ and so to evaluate $\|x^*(\epsilon) - x^*\|$. For example, if $\Phi(x)$ is strongly convex and the set $\Theta^*(0)$ is convex then [see formulae (3.5)]

$$\begin{aligned} \|x_0(\epsilon) - x^*\| &= O(\sqrt{\Phi(x^*) - \Phi(x_0(\epsilon))}) = O(\delta^{1/2}(\epsilon)) \\ \|x^*(\epsilon) - x^*\| &= O(\delta^{1/2}(\epsilon)) \end{aligned} \tag{5.41}$$

Proof of Lemma 5.3. Let $\{\epsilon'\}$ be an arbitrary subsequence, tending toward zero. By virtue of corollary 1 there exists the limit (5.12) and, hence, the relations (5.13)–(5.15) are valid. Substituting (5.13) into the objective function of the perturbed program and taking into account the fact that $f_0(x)$ is twice continuously differentiable we have, for $\epsilon \in \{\epsilon'\}$,

$$F^*(\epsilon) = f_0(x(\epsilon)) + \epsilon[(\nabla f_0(x(\epsilon)))^T x_1(\epsilon) + f_1(x(\epsilon))] + O(\epsilon^2) \tag{5.42}$$

On the other hand, the smoothness of $f_0(x)$ permits one to clarify the inequality (4.47)

$$F^*(\epsilon) \geq F^*(0) + \epsilon F_1^* + O(\epsilon^2)$$

which, together with (5.42), leads (as in the proof of Theorem 5.1) to

$$\begin{aligned} & f_0(x(\epsilon)) + \epsilon[(\nabla f_0(x(\epsilon)))^T x_1(\epsilon) + f_1(x(\epsilon))] + O(\epsilon^2) \geq \\ & \geq F^*(0) + \epsilon F_1^* + O(\epsilon) \quad , \quad \epsilon \in \{\epsilon'\} \end{aligned} \tag{5.43}$$

$$(\nabla f_0(x(\epsilon)))^T x_1(\epsilon) + f_1(x(\epsilon)) + O(\epsilon) \geq F_1^* + O(\epsilon) \quad , \quad \epsilon \in \{\epsilon'\} \tag{5.44}$$

Similarly, given the fact that $g_0(x)$ is twice continuously differentiable, we obtain for $\epsilon \in \{\epsilon'\}$

$$(\nabla g_{0l}(x(\epsilon))^T x_1(\epsilon) + g_{1l}(x(\epsilon)) + O(\epsilon) = 0 \ , \ l \in \omega(x^*) \tag{5.45}$$

Let us substitute (5.36) into (5.44) and (5.45). We have for $\epsilon \in \{\epsilon'\}$

$$\begin{aligned} &(\nabla f_0(x_0(\epsilon)))^T x_1(\epsilon) + f_1(x_0(\epsilon)) \geq F_1^* + O(\delta(\epsilon)) \ ; \\ &(\nabla g_{0l}(x_0(\epsilon)))^T x_1(\epsilon) + g_{1l}(x_0(\epsilon)) + O(\delta(\epsilon)) \ , \ l \in \omega(x^*) \end{aligned} \tag{5.46}$$

The latter relations verify the inequality

$$\begin{aligned} &(\nabla f_0(x_0(\epsilon)))^T x_1(\epsilon) + f_1(x_0(\epsilon)) \leq \\ &\leq f_1(x_0(\epsilon)) + \max_{x_1} \{(\nabla f(x_0(\epsilon)))^T x_1 \,/\, (\nabla g_{0l}(x_0(\epsilon)))^T x_1 + \\ &+ g_{1l}(x_0(\epsilon)) + O(\delta(\epsilon)) = 0 \ , \ l \in \omega(x^*)\} \end{aligned} \tag{5.47}$$

Since $x_0(\epsilon) \in \Theta^*(0)$, and, as it follows from (5.38), $\lim_{\epsilon \to 0} x_0(\epsilon) = x^*$, then for ϵ small enough ($\epsilon \in \{\epsilon'\}$),

$$\begin{aligned} &x_0(\epsilon) \in \Theta^*(0) \cap R(x^*) \Rightarrow \lambda(x_0(\epsilon)) = \lambda(x^*) \triangleq \lambda^* \Rightarrow \\ &\nabla f_0(x_0(\epsilon)) = \sum_l \lambda_l(x^*) \nabla g_{0l}(x_0(\epsilon)) \Rightarrow \\ &\max_{\epsilon \to 0} \{(\nabla f_0(x_0(\epsilon)))^T x_1 \,/\, (\nabla g_{0l}(x_0(\epsilon)))^T x_1 + g_{1l}(x_0(\epsilon)) + O(\delta(\epsilon)) = 0 \ , \\ &\qquad l \in \omega(x^*)\} = -\sum_l \lambda_l(x^*) g_{1l}(x_0(\epsilon)) + O(\delta(\epsilon)) \end{aligned}$$

Thus, (5.47) can be rewritten in the form

$$\begin{aligned} &f_1(x_0(\epsilon)) + (\nabla f_0(x_0(\epsilon)))^T x_1(\epsilon) \leq f_1(x_0(\epsilon)) - \sum_l \lambda_l(x^*) g_{1l}(x_0(\epsilon)) + \\ &+ O(\delta(\epsilon)) \triangleq \Phi_1(x_0(\epsilon)) + O(\delta(\epsilon)) \leq F_1^* + O(\delta(\epsilon)) \ , \ \epsilon \in \{\epsilon'\} \end{aligned}$$

The latter together with (5.46) give the relations

$$\begin{aligned} &\Phi_1(x_0(\epsilon)) = F_1^* + O(\delta(\epsilon)) \triangleq \Phi(x^*) + O(\delta(\epsilon)) \Rightarrow \\ &f_0(x_0(\epsilon)) + \epsilon[(\nabla f_0(x_0(\epsilon)))^T x_1(\epsilon) + f_1(x_0(\epsilon))] + O(\epsilon^2) \geq \\ &\geq F^*(0) + \epsilon F_1^* + O(\epsilon\delta(\epsilon)) \ , \ \epsilon \in \{\epsilon'\} \end{aligned}$$

which, if we take into account (5.42) and (5.43) and the fact that $\{\epsilon'\}$ is an arbitrary subsequence, completes the proof. QED.

Let us show how to evaluate $\delta(\epsilon)$ in the case where the perturbed program is of the form

$$F^*(\epsilon) \triangleq \max \{c_0^T x + \epsilon f_1(x) \,/\, Ax + \epsilon g_1(x) \leq b\} \tag{5.48}$$

and the reduced one is linear

$$F^*(0) \triangleq \max \{c_0^T x \,/\, Ax \leq b\} \tag{5.49}$$

Using the notation of Theorem 5.1 [see (5.16) , (5.17), and (5.25′)] we can write

$$\begin{aligned} &A_l x^*(\epsilon) + \epsilon g_{1l}(x^*(\epsilon)) = b_l \ , \ l \in \omega(x^*) \ ; \\ &A_l x^*(\epsilon) + \epsilon g_{1l}(x^*(\epsilon)) \leq b_l \ , \ l \in \Omega(x^*)\backslash\omega(x^*) \ ; \\ &A_l x^*(\epsilon) + \epsilon g_{1l}(x^*(\epsilon)) \leq b_l - \delta \ , \ \delta > 0 \ , \ l \bar{\in} \Omega(x^*) \end{aligned} \tag{5.50}$$

where A_l are the rows of A_0, $g_{1l}(x)$ are the components of the vector function $g_1(x)$, $\omega(x^*)$ is the set of indices corresponding to positive components of $\lambda(x^*)$. The rows A_l, $l \in \Omega(x^*)$ are assumed linearly independent. Hence, there exists a vector $x_1(\epsilon)$ such that

$$A_l x_1(\epsilon) = -g_{1l}(x^*(\epsilon)) \ , \ l \in \Omega(x^*) \ , \ \|x_1(\epsilon)\| \leq M = \text{const}$$

If we put

$$x^*(\epsilon) = x_0(\epsilon) + \epsilon x_1(\epsilon) \tag{5.51}$$

and substitute (5.51) into (5.50), it takes the form

$$\begin{aligned} &A_l x_0(\epsilon) = b_l \ , \ l \in \omega(x^*) \\ &A_l x_0(\epsilon) \leq b_l \ , \ l \in \Omega(x^*)\backslash\omega(x^*) \\ &A_l x_0(\epsilon) \leq b_l - \delta + O(\epsilon) \ , \ l \in \Omega(x^*) \end{aligned} \tag{5.52}$$

The latter is equivalent to the inclusion

$$x_0(\epsilon) \in \Theta^*(0) \tag{5.53}$$

The relations (5.51) and (5.53) mean that

$$\delta(\epsilon) = O(\epsilon)$$

Thus, Lemma 5.3 gives rise to the following theorem.

Theorem 5.3.

Let the conditions of the Theorem 5.1 be fulfilled for the problems (5.48) and (5.49). (The linearity of the reduced problem implies the fulfillment of a part of these conditions immediately.) Let the solution to the auxiliary problem be unique. Then

$$\Phi(x^*) - \Phi(x_0(\epsilon)) = O(\epsilon) \ , \ x_0(\epsilon) \in \Theta^*(0) \ , \ x^*(\epsilon) = x_0(\epsilon) + O(\epsilon)$$

The optimal value of the perturbed problem admits the expansion

$$F_*(\epsilon) = F^*(0) + \epsilon F_1^* + O(\epsilon^2)$$

Let us note in conclusion that, although throughout this section we considered problems with inequality constraints only, all the results can be obtained if there are some equality constraints.

CHAPTER 2

Approximate Decomposition and Aggregation for Finite Dimensional Deterministic Problems

The effectiveness of the perturbation method is determined by the structural peculiarities of the original problem. The main goal of this chapter is to demonstrate the variety of such structures and to indicate the possibility of reaching some meaningful conclusions by means of the formal apparatus described in the previous chapter.

At the same time some additional mathematical results are given here with special reference to nonlinear problems where the exactness of approximation is not estimated in the general case. The content of the material corresponds to the chapter title. The only problems considered here are those where the perturbation method involves approximate decomposition and aggregation procedures. Consideration of these problems was stimulated by practical demands in the analysis of economic and industrial complexes. Some of the suggested schemes were described earlier in the book [72] at the heuristic level.

This chapter begins with a discussion of how to optimize the activity of a system that includes a number of weakly connected subsystems. It is shown (Section 2.1) that the application of the perturbation method effectively implies a decomposition scheme. It is a trivial but rather interesting point that a two-level coordination procedure results if the hypothesis of strongly restricted coordinating efforts is adopted. This enables one to find an explicit form of the suboptimal coordinating policy.

Linear programs that have properties similar to those of direct aggregation problems are considered in Section 2.2. These programs are interpreted as a mathematical model of an enterprise engaged in a number of similar activities. The perturbation method reduces the original problem to that of an aggregated program and elementary disaggregation procedures. Further, a situation is analyzed where there are resource constraints with similar properties. The application of the perturbation method reduces the effective number of constraints in the program to be solved, and can yield a significant reduction in computational effort. The final part of Section 2.2 details some elementary procedures for reducing the number of variables and constraints. It is recommended here that "strong differences" and "bottleneck" effects be exploited.

Section 2.3 is devoted to the important practical class of nonlinear programs introduced by Dantzig and Wolfe in connection with problems of optimal control and scheduling of chemical plants. Here one must distribute resources between a number of processing units, taking into account the possibility of adjusting the characteristics of each unit by means of a technological control system. This difficult problem is simplified if one adopts the realistic hypothesis that there is relatively little freedom to change the input–output coefficients of the processing units.

Using this hypothesis and the formal procedure presented in Section 1.5, one can construct a two-level decomposition scheme where, at the first stage, one must solve a linear program corresponding to the "pure" distribution problem and, after that, solve a number of independent problems of the optimal adjustment of each unit. A typical structure in the chemical industry includes a basic unit, which produces half-finished goods, and a group of finishing units, which is why this case is considered in detail. An algorithm is given to construct the suboptimal control law. Finally, some estimates are presented for the nonlinear generalization of a weakly dynamic control problem.

The subsequent sections (2.4 and 2.5) have the common feature that they deal with the same model of input–output analysis widely used for planning at both national and industrial plant levels. It is for these problems that the aggregation approach is most important and, therefore, it has been the subject of intense research effort. For the sake of simplicity, the presentation is divided into two parts. Initially, a standard problem of how to find the total outputs of the different "industries" where the final consumption vector is given is described under various hypotheses about the structure of the input–output matrix. Then some optimization problems are considered under similar assumptions. The basic assumption is that the success of heuristic aggregation schemes used in practice is determined, as a rule, by the aggregation of industries with similar input–output coefficients. The perturbation method gives a theoretical basis for such heuristic schemes and shows a simple way to improve the approximation and to assess its accuracy.

Section 2.6 is devoted to large transportation problems where intraregional shipping is essentially cheaper than interregional one.

The perturbation method gives an easily comprehended procedure for hierarchic scheduling. From the mathematical point of view, this program is also of interest since the uniqueness of Lagrange multipliers for the reduced program does not occur here.

The concluding section (2.8) is devoted to a practical control problem where the state variables are constrained. By using a weak inertia hypothesis, the control law is constructed explicitly and its optimality proved (in the standard local sense).

2.1. Perturbed decomposable structures and two-level planning

Programs of a directly decomposable structure rarely occur in practice. However, a typical situation involves weak connections between the subsystems. Such problems may be represented in the form

$$\max \{F_0[z_l(x_l) \ , \ l = 1,\ldots,L] + \epsilon F_1(x)\} \tag{1.1}$$

under the conditions

$$g_l(x_l) + \epsilon g_{1l}(x) \leq 0 \tag{1.2}$$

where $z_l(x_l)$ are scalar functions, x_l are L_l-vectors (columns), $x = (x_l \ , \ l = 1,\ldots,L)$, g_l, g_{1l} are m_l-vector-functions, and ϵ is a small value characterizing the intensity of the subsystems' mutual dependence. The function $F_0(z)$, $z = (z_l)$, is assumed to be a monotone. Hence, the reduced problem has a directly decomposable structure, and its solution is equivalent to that of L independent problems

$$\max \{z_l(x_l) \ / \ g_l(x^l) \leq 0\} \tag{1.3}$$

It is this decomposition that implies the feasibility of the perturbation method.

The general algorithms were described in Chapter 1. Here we shall only demonstrate the useful properties of some particular problem types. Our attention will center mainly on an interpretation of formal procedures from the viewpoint of the planning coordination of complex systems. To simplify the consideration we shall describe only linear programs using the results of Section 1.2.

A. Let us consider a linear program of (1.1) and (1.2) type:

$$\max \{\sum_{l=1}^{L} \sum_{j\in\Omega_l} c_j x_j \ / \ (F + \epsilon B)x = b \ , \ x \geq 0\} \tag{1.4}$$

where $F = \text{diag}\,\{F_l\}$, $b = (b_l)$, $x = (x_j)$. We suppose that the conditions of Theorem 2.2 are fulfilled. Then one can construct the following procedure.

(1) Solve L subproblems of (1.3) type

$$\max \{ \sum_{j\in\Omega_l} c_j x_j \ / \ F_l x_l = b_l \ , \ x_l \geq 0\}$$
$$l = 1,\ldots,L \ , \ x_l = (x_j \ , \ j \in \Omega_l) \tag{1.5}$$

If the solutions x_l^*, $l = 1,\ldots,L$ are unique then the vector

$$x^* \triangleq (x_l^* \ , \ l = 1,\ldots,L) \tag{1.6}$$

is equal to the solution of (1.4) to within $O(\epsilon)$ (see Theorem 2.1). To improve this approximation one must fix the optimal basic sets σ_l in each problem and then proceed to step (3). If, on the other hand, there is nonuniqueness in any problem 1, one needs to describe the optimal solution set Θ_l^* and to fix the solution vector λ_l^* of the corresponding dual program, which we suppose to be unique.

(2) Solve

$$\min \{ \sum_{s=1}^{L} \lambda_s^{*T} B_{sl} x_l \,/\, x_l \in \Theta_l^* \} \ , \ l = 1,\ldots,L \tag{1.7}$$

where B_{sl} is the submatrix of B formed by rows corresponding to b_s and columns corresponding to x_l.

Supposing that all problems (1.7) have the unique nondegenerate solution x_l^*. One can adopt $x^* = (x_l^*)$ as a suboptimal [to within $O(\epsilon)$] solution to (1.4) (see Theorem 2.2).

To improve the accuracy one must fix the optimal basic sets σ_l in the problems (1.7) and pass on to step (3).

(3) Put

$$x_j = 0 \ , \ j \bar{\in} \bigcup_l \sigma_l \tag{1.8}$$

in the system

$$(F + \epsilon B)x = b$$

After that, it is reduced to the form of

$$(\tilde{F} + \epsilon \tilde{B})\tilde{x} = b \tag{1.9}$$

where $\tilde{x} = (\tilde{x}_l)$ is the vector formed by the basic components, $\tilde{x}_l = (x_j \ , \ j \in \sigma_l)$. If the system (1.9) has a nonnegative solution $\tilde{x}^*$ for the given ϵ, then this solution and (1.8) form the admissible solution of the original problem. ent can write $\tilde{x}_l^*$ in the series form

$$\tilde{x}_l^* = x_l^{(0)} + \epsilon x_l^{(1)} + \cdots \ , \ l = 1,\ldots,L \tag{1.10}$$

where

$$x_l^0 = x_l^* = \tilde{F}_l^{-1} b_l \ , \ x_l^{(t+1)} = -\tilde{F}_l^{-1} B_l x^{(t)} \ , \ x^{(t)} \triangleq (x_l^{(t)})$$

To check its optimality, one can solve the equation for the simplex multipliers

$$(\tilde{F} + \epsilon \tilde{B})^T \lambda = \tilde{c}$$

where $\tilde{c} \triangleq (c_j, j \in \bigcup_l \sigma_l)$, and after that verify the validity of the inequality $(F + \epsilon B)^T \lambda \geq c$, which will enable one to ensure that the fixed value of ϵ belongs to the segment given in the conditions of Theorem 2.1 and 2.2 (Chapter 1).

Let us give an economic interpretation of the program (1.4) and the algorithm described above. One can consider this program as a mathematical formalization of the problem of how to optimize the planning for an industrial concern of L enterprises. Each enterprise uses mainly its "own" resources (in preference) but needs (in small quantities) resources belonging to other enterprises to realize or perhaps improve its own technological activities. One can interpret the objective function as the overall income of all partners. (Of course, an additive objective function may be replaced by any function increasing with the growth of each partner's contribution.) Then the algorithm can be interpreted as the coordination procedure.

At the first stage, each enterprise solves the planning problem (programs its own activity), neglecting the requirements of "foreign" resources, which makes it possible to find either a suboptimal plan or the "shadow prices" λ_l^* of its own resources only. At the second stage, by using information about "prices" received from other partners, each enterprise can construct a suboptimal plan (but nonfeasible only!) and, moreover, can isolate the optimal set of its own activities.

The concluding stage may be interpreted in two different ways. It is possible either to balance the consumption of all resources simultaneously, gathering all information about optimal activities in the "coordinating center", or to realize the sequential correction of the resource distribution by means of an exchange procedure. The first approach corresponds to a direct solution of equation (1.9). The second approach corresponds to the sequential construction of expansions (1.10). Actually, this formal procedure is equivalent to the following steps: starting from his initial plan x_l^0, each participant determines his additional needs $\epsilon B_{sl} x_l^0$, $s = 1,\ldots,L$. After that he passes these values to the other participants who modify their plans, taking into account the necessary loss of a part $\epsilon B_l x^{(0)}$ of their resources, and so on.

Note that, this procedure has an infinite number of steps since, at any step, the correction values are calculated using the approximate description $\tilde{F}_l$ of optimal activities. It should be kept in mind that such a "decentralization" is conditional and presupposes the coordination of the actions of all participants.

B. Let us consider the model where the coordinating element is presented explicitly. The formal description of the class of problems of interest here may be written in the form

$$\max \sum_l c_l^T x_l \tag{1.11}$$

such that

$$A_l x_l + \epsilon B_l x = b_l + u_l \ , \ \sum_l u_l \leq \epsilon b_0$$
$$x_l \in X_l \ , \ x = (x_l) \ , \ u_l \geq 0$$

where x_l, u_l are vectors of unknowns; c_l, b_0, A_l, B_l, b_l are constant matrices; and X_l are polyhedrons.

The economic interpretation of (1.11) is clear. As in the previous model, each enterprises l uses mainly its own resources b_l but can receive additional resources u_l from the "central unit" (the firm's administration). Our basic assumption is that the central unit may distribute relatively small quantities of resources only and that the potential to organize exchanges among the enterprises is also small. If this assumption is false then (1.11) has the standard block structure and only iterative decomposition procedures [48] are applicable.

Let us try to apply the perturbation method algorithm (see Section 1.2) to (1.11). It is not difficult to see that this cannot be done directly since the specific structure of the inequality makes the reduced problem irregular (at $\epsilon = 0$). However a simple substitution, $u_l = \epsilon v_l$ changes the situation, transforming (1.11) into

$$\begin{aligned} & A_l x_l + \epsilon B_l x = B_l + \epsilon v_l \ , \ \sum_l v_l \leq b_0 \\ & x_l \in X_l \ , \ v_l \geq 0 \end{aligned} \tag{1.12}$$

so that the reduced problem

$$\max \{ \sum_l c_l^T x_l / \ A_l x_l = b_l \ , \ x_l \in X_l \ , \ \sum_l v_l \leq b_0 \ , \ v_l \geq 0 \} \tag{1.13}$$

is generally regular, but its solution is nonunique.

In fact, even if all programs

$$\max \{ c_l^T x_l / \ A_l x_l = b_l \ , \ x_l \in X_l \} \tag{1.14}$$

have unique solutions x_l^* (it will be supposed for simplicity), then the variables v_l will not be determined.

If one needs to find its suboptimal values, then the auxiliary problem

$$\max \{\sum_l \lambda_l^{*T} v_l \,/\, \sum_l v_l \leq b_0 \;,\; v_l \geq 0\} \tag{1.15}$$

where λ_l^* are the optimal simplex multipliers [the solutions of the problems that are dual to (1.14)] must be solved. However, this problem is trivial: if components of λ_l^* are denoted by λ_{il} and components of b_0 by b_{i0}, then we have

$$v_{il}^* = \begin{cases} b_{i0} \;, & l = k_i \\ 0 \;, & l \neq k_i \end{cases} \;, \quad k_i \triangleq \arg\max_l \lambda_{il} \tag{1.16}$$

Thus, in the initial approximation, the resource is passed to the enterprise that uses it most efficiently.

Further, in accordance with the general scheme, one can solve the equation

$$\tilde{A}_l \tilde{x}_l + \epsilon \tilde{B}_l \tilde{x} = b_l + \epsilon v_l^* \tag{1.17}$$

where $\tilde{x}_l$ is a vector formed by the components x_j with indices belonging to the optimal basic set of (1.14), and where $\tilde{A}_l$, $\tilde{B}_l$ are the corresponding submatrices of matrices A_l, B_l. Once again, it is true that if (1.17) admits a unique solution under the given value of ϵ, then this solution is feasible for the original problem. The verification of its optimality is simple (it is not necessary to check if $B_l \geq 0$).

It is of interest to compare this scheme with the algorithm suggested in [48] to solve the program

$$\max \{\sum_l c_l^T x_l \,/\, \sum_l A_l x_l \leq b_0 \;,\; x_l \in X_l\} \tag{1.18}$$

If $A_l x_l \geq 0$ on X_l, this program may be transformed into

$$\max \{\sum_l c_l^T x_l \,/\, \sum_l w_l \leq b_0 \;,\; w_l \geq 0 \;,\; A_l x_l = w_l \;,\; x_l \in X_l\} \tag{1.18'}$$

This is similar to (1.12), but the right-hand sides of equations

$$A_l x_l = w_l$$

may change appreciably and not be close to b_l.

This "freedom" to redistribute resources makes the decomposition procedure essentially more difficult. The optimum may be achieved by an iterative procedure in which the first iteration is similar to the local scheme described above, but subsequent iterations are complicated since it is possible (and even necessary!) to obtain a distribution where the "shadow prices" λ_l^* are nonunique.

As for the scheme considered here, it is not a universal tool, but it is a very simple one. An assumption about the limited potential of the coordinating center is natural enough, considering the real conditions under which a production firm has to operate, when all the planning usually amounts to merely correcting those plans prepared by the enterprises on the basis of rigid norms.

Let us consider yet another model that imitates the coordination process more realistically [146]. We take into account that the objectives of the coordinating unit and those of the enterprises being coordinated do not coincide. Formally, the objective function of the coordinating unit is not representable as a monotonic functional of the "separate" objective functions. However, the coordinating unit can assume or even require that the real output x_l must not be lower than the fixed part of the level y_l given by the "center", i.e.,

$$x_l \geq (1 - k_l) y_l \tag{1.19}$$

where k_l are small values, which will be referred to as "independence coefficients".

Let an enterprise l elaborate its plan on the basis of a standard linear programming model

$$J(y_l, u_l) \triangleq \max \{c_l^T x_l \,/\, A_l x_l \leq b_l + u_l \ , \ x_l \geq (1 - k_l) y_l \geq 0\} \tag{1.20}$$

where u_l denotes the additional resources that can be directed to this enterprise from the central reserves,

Having released the planned assignment y_l, the firm will receive from the enterprises all the information about the resources $A_l y_l$ required to fulfill the assignment and so assumes the obligation to provide it with these resources. Thus,

$$A_l y_l \leq b_l + u_l \tag{1.21}$$

and the plan exhausts at least some resources completely. Therefore, central control actions are to order y_l and to distribute the limited volume of additional resources

$$\sum_l u_l \leq b_0 \ , \ u_l \geq 0 \tag{1.22}$$

These actions must be directed to maximize the central objective function

$$\sum_l d_l^T \bar{x}_l$$

where $\bar{x}_l$ is the real output of enterprise l.

In general, the planned level y_l is different from $\bar{x}_l$, but $\bar{x}_l$ depends on the plan as well as on the level of resources. Such a dependence was referred to in [64] as a plan-production function (PPF). The center can use the solution $\bar{x}(y_l, u_l)$ to problem (1.20) as a PPF, i.e., the center can adopt the hypothesis of the optimal behavior of enterprises.

Let us look for the optimal structure of central plan (y_l) and the real outputs $(\bar{x}_l)$ for various values of the "independence coefficients" k_l.

First of all, what is the function $\bar{x}_l(y_l, u_l)$? Rewriting (1.20) by introducing the variables

$$w_l \triangleq \frac{1}{k_l}[x_l - (1 - k_l)y_l]$$

we have

$$\begin{aligned} J(y_l,u_l) = {} & (1 - k_l)c_l^T y_l + \\ & + k_l \max \{c_l^T w_l \,/\, A_l w_l \le b_l + u_l + \frac{1 - k_l}{k_l} s_l \;,\; w_l \ge 0\} \end{aligned} \tag{1.23}$$

where

$$s_l \triangleq b_l + u_l - A_l y_l$$

Let us also take into account that, due to (1.21), $y_l \ge 0$ and $s_l \ge 0$ and, moreover, that some components s_{il}, $l \in I_l$ are equal to zero, as was supposed above. Obviously, there are small enough values of $\bar{k}_l$ for which, under $k_l \le \bar{k}_l$ only, the inequalities with $i \in I_l$ are limiting. Finally, we obtain the following expression for the PPF:

$$\bar{x}(y_l \,,\, u_l) = (1 - k_l)y_l + k_l \bar{w}_l(u_l) \;,\; k_l \le \bar{k}_l \tag{1.24}$$

where $\bar{w}_l(u_l)$ is the solution to

$$\max \{c_l^T w_l \,/\, \tilde{A}_l w_l \le \tilde{b}_l + \tilde{u}_l \;,\; w_l \ge 0\} \tag{1.25}$$

($\tilde{A}$, $\tilde{b}$, $\tilde{u}_l$ are submatrices of the corresponding matrices formed by the rows, A, b, u_l).

The interpretation of this result is obvious. For small "independence coefficient" k_l, the real output is a linear combination of the central planned level y_l and of the optimal (from the enterprise's point of view!) plan which is found by taking into account the limiting resources. Note also that $s_l = 0$, if the central plan exhausts all the resources and, hence, the result (1.24) is true for any k_l, $0 < k_l \leq 1$.

Let us further consider the problem to find (y_l), (u_l) for this simple case. We have

$$\max \sum_l d_l^T[(1 - k_l)y_l + k_l\bar{w}_l(u_l)]$$

such that

$$A_l y_l = b_l + u_l \ , \ y_l \geq 0 \ , \ \sum_l u_l \leq b_0 \ , \ u_l \geq 0 \ , \ l = 1,\ldots,L \tag{1.26}$$

This problem is similar to (1.11) and is reducible to a linear program. However, its solution may be difficult if the dimension is large. At the same time it is not hard to obtain an explicit result if the hypothesis of small distributing values is adopted.

Then the perturbation method enables one to establish that suboptimal planned levels are the solutions to the following problems

$$\max \{d_l^T y_l \,/\, A_l y_l = b_l \ , \ y_l \geq 0\} \tag{1.27}$$

and that the resource distribution is determined by the solution to

$$\max \{\sum_l((1 - k_l)\lambda_l^* + k_l\mu_l)^T u_l \,/\, \sum_l u_l \leq b_0 \ , \ u_l \geq 0\} \tag{1.28}$$

where λ_l^* are the unique optimal solutions to the programs that are dual to (1.27) and whose μ_l are the basic plans of the same problems corresponding, however, to the optimal bases of the problems (1.25).

The optimal distribution rule is obvious from (1.28): the entire additional resource i must be passed to the enterprise where its combined "shadow price" $(1 - k_l)\lambda_{lil} + k_l\mu_{il}$ is maximal.

It should be noted that, by virtue of (1.27), the suboptimal central policy is determined only by the objectives of the center itself, but the resource distribution depends on the enterprise's objectives, too; and the more "independent" the enterprise, the more the central unit must take into account its objectives.

If the distribution effects are absent, the real output is equal to

$$(1 - k_l)\bar{y}_l + k_l\bar{x}_l$$

where $\bar{y}_l$ is the planned level, and $\bar{x}_l$ is the "selfish" optimum of the enterprise.

2.2. Aggregation of activities

This section is devoted to some linear programs that are close to being directly aggregatable and decoposable in the sense of Section 1.1.

First, we consider programs that are representable in the form

$$\max \sum_{s=1}^{k} \sum_{j\in\Omega_s} (c_s + \epsilon\beta_{js})x_j \tag{2.1}$$

such that

$$\sum_{s=1}^{k} \sum_{j\in\Omega_s} (A_s + \epsilon B_{js})x_j \le b \ , \ x = (x_j) \ge 0 \tag{2.2}$$

where ϵ is a small parameter.

The variables x_j, $j = 1,\ldots,n$, can be interpreted as the activity levels of an industrial complex composed of many aggregates. A specific feature of the problem under consideration is the hypothesis that the vectors $A_s + \epsilon B_{js}$ and scalars $c_s + \epsilon\beta_{js}$, which can be interpreted as input–output coefficients and "prices" corresponding to the different activities, are similar if those activities belong to a group s, $j \in \Omega_s$. This arises if one attempts to schedule the complex taking into account the slightly different ways of producing one and the same product, or the difference between technologically similar products.

The application of standard computational procedures to programs of the structure (2.1) and (2.2) is often frustrated by difficulties of two kinds: (1) increasing the detail in the description of the complex results in a large number of variables, (2) if two or more similar columns appear in the basis, the program becomes ill-conditioned ([18]), which in turn may lead to error accumulation. To avoid these difficulties we suggest the following suboptimization scheme, which is based on the results of the perturbation method (see Theorem 2.2 and 2.4).

Let us introduce the variables z_s,

$$z_s = \sum_{j\in\Omega_s} x_j \tag{2.3}$$

Then (2.1) and (2.2) can be written in the form

$$\max \{ \sum_{s=1}^{k} c_s z_s + \epsilon \sum_{s=1}^{k} \sum_{j\in\Omega_s} \beta_{js}x_j \} \tag{2.4}$$

such that

$$\sum_{s=1}^{k} A_s z_s + \epsilon \sum_{s=1}^{k} \sum_{j\in\Omega_s} B_{js}x_j \le b \ , \ \sum_{j\in\Omega_s} x_j \ge z_s \ , \ x_j \ge 0 \tag{2.5}$$

If one puts $\epsilon = 0$, the *reduced* program arises

$$\max \{ \sum_{s=1}^{k} c_s z_s \,/\, (x\,,\,z) \in \Theta(0)\} \tag{2.6}$$

where

$$\Theta(0) \triangleq \{(x\,,\,z) \,/\, \sum_{s=1}^{k} A_s z_s \leq b \;,\; z_s = \sum_{j\in\Omega_s} x_j \;,\; x_j \geq 0 \;,\; j = 1,\ldots,n\} \tag{2.7}$$

Let us introduce the *aggregated* program

$$\max \{ \sum_{s=1}^{k} c_s z_s \,/\, \sum_{s=1}^{k} A_s z_s \leq b \;,\; z_s \geq 0\} \tag{2.8}$$

and suppose that it has the unique solution $z^* = (z_s^*)$. Then the reduced program has a bounded set $\Theta^*(0)$ of solutions

$$\Theta^*(0) = \{x \,/\, \sum_{j\in\Omega_s} x_j = z_s^* \;,\; s = 1,\ldots,k \;,\; x_j \geq 0 \;,\; j = 1,\ldots,n\} \tag{2.9}$$

Moreover, if

$$\sigma \triangleq \{s \,/\, z_s^* > 0\} \tag{2.10}$$

then

$$\Theta^*(0) = \prod_{s\in\sigma} \Theta_s(0) \tag{2.11}$$

where

$$\Theta_s(0) \triangleq \{x_j \,/\, \sum_{j\in\Omega_s} x_j = z_s^*\} \;,\; s \in \sigma \tag{2.12}$$

The reduced solution determines suboptimal levels of aggregated activities z_s. By virtue of Theorem 2.4 (see Chapter 1) any distribution of activity levels among the activities within the same aggregated group is admissible and suboptimal to within $O(\epsilon)$.

Let the program dual to the aggregated one have the unique solution λ^*. Then, according to Theorem 2.4, one can obtain a suboptimal solution (2.1) and (2.2) that is more precise.

This solution is found by means of the auxiliary program

$$\max\{\sum_{s\in\sigma}\sum_{j\in\Omega_s}(\beta_{js}-\lambda^{*T}B_{js})x_j \,/\, x\in\Theta_0^*\} \tag{2.13}$$

which, being separable, separates into $|\sigma|$ elementary programs

$$\max\{\sum_{j\in\Omega_s}(\beta_{js}-\lambda^{*T}B_{js})x_j \,/\, \sum_{j\in\Omega_s}x_j=z_s^* \;,\; x_j\geq 0 \;,\; j\in\Omega_s\} \;,\; s\in\sigma \tag{2.13'}$$

If

$$J_s \triangleq \arg\max_j (\beta_{js}-\lambda^{*T}B_{js}) \;,\; s\in\sigma \tag{2.14}$$

then one can write the solutions to these programs in the explicit form

$$\sum_{j\in J_s}x_j^*=z_s^* \;,\; x_j^*\geq 0 \;,\; j\in J_s \;,\; x_j^*=0 \;,\; j\bar{\in} J_s \tag{2.15}$$

i.e., the auxiliary program allows one to exclude activities that do not conflict, but in general the distribution between the activities in each group is nonunique.

By Theorem 2.4, the solutions (2.15) are suboptimal in (2.1) and (2.2) to within $O(\epsilon^2)$ and feasible to within $O(\epsilon)$.

If the solutions to all programs (2.13′) are unique, then only one effective activity is selected in each group. Let its index be j_s. Then we have

$$x_{j_s}^*=z_s^* \;,\; s\in\sigma \;,\; x_j^*=0 \;,\; j\neq j_s$$

By means of Theorem 2.2 one can in that case construct an optimal basic index set J_0 of the reduced program, which is the optimal index set for the perturbed program for all small enough ϵ. The set J_0 consists of all j_s, $s\in\sigma$ and indices of all inequalities in (2.8), which are strictly fulfilled at the point z^*.

Taking this fact into account, one can find all positive components of the locally (for ϵ small enough) optimal solution to the perturbed program as the solution to the linear system

$$\sum_{s\in\sigma} (\tilde{A}_s + \epsilon B_{j_s}) x_{j_s} = \tilde{b} \tag{2.16}$$

where $\tilde{A}_s$, $\tilde{B}_{j_s}$, $\tilde{b}$ are subvectors of the vectors A_s, b_{j_s}, b formed by components whose indices coincide with those of inequalities that are fulfilled as equality at the point z^*.

Thus, the complete algorithm of the perturbation method as applied to (2.1) and (2.2) can be written in the form

(1) Solve (2.8) and the dual to (2.8) to find z^*, λ^*.
(2) If λ^* is unique, then find the index sets J_s defined by (2.14).
(3) Find the suboptimal solutions in the form (2.15).
(4) If only one index j_s belongs to each J_s, then find the locally optimal solution by means of the system (2.16).

The problem of how to check whether the given ϵ belongs to the feasible region can be solved by standard methods. If, for this ϵ, the solution to (2.16) is nonnegative, then the result is at least feasible. If, in addition, there exists λ such that conditions

$$\lambda^T (A_s + \epsilon B_{j_s}) \geq c_s + \epsilon \beta_{j_s} \tag{2.17}$$

are true, then ϵ belongs to the segment mentioned in Theorems 2.2 and 2.4, and, therefore, the given perturbed problem has all the properties described above.

To illustrate the algorithm let us consider an elementary example.

Example

Let the original problem be to maximize

$$f = x_1 + 1.01x_2 + 2x_3 + 2.04x_4$$

subject to

$$\begin{aligned} x_1 + 1.05x_2 + 3x_3 + 2.9x_4 &\leq 6 \ ; \\ 2x_1 + 2.04x_2 + x_3 + 0.9x_4 &\leq 6 \ ; \quad x_i \geq 0 \ ; \quad i = 1,\ldots,4 \end{aligned}$$

One can present this in the form (2.8) by setting

c_1	c_2
1.	2.
A_1	A_2
1.	3.
2.	1.

β_{11}	β_{21}	β_{32}	β_{42}
0.	1.	0.	4.
B_{11}	B_{21}	B_{32}	B_{42}
0.	5.	0.	−10.
0.	2.	0.	−10.

$\epsilon = \epsilon_0 = 0.01$.

According to the algorithm, we have:

(1) The primal aggregated program is

$$\max \{z_1 + 2z_2 / \, z_1 + 3z_2 \leq 6 \; ; \; 2z_1 + z_2 \leq 6 \; ; \; z_1 \, , \, z_2 \geq 0\} \Rightarrow$$
$$\Rightarrow z_1^* = 2.4 \quad ; \quad z_2^* = 1.2$$

The dual one is

$$\min \{6\lambda_1 + 6\lambda_2 / \, \lambda_1 + 2\lambda_2 \geq 1 \; ; \; 3\lambda_1 + \lambda_2 \geq 2 \; ; \; \lambda_1 \, , \, \lambda_2 \geq 0\} \Rightarrow$$
$$\Rightarrow \lambda_1^* = 0.6 \; ; \; \lambda_2^* = 0.2$$

(2) $\max \{0 \; ; \; 1 - (3 + 0.4)\} = 0 \qquad j_1 = 1$

$\max \{0 \; ; \; 4 - (-6 - 2)\} = 12 \qquad j_2 = 4$

(3) $x_1^* = 2.4 \;\; ; \;\; x_2^* = 0 \;\; ; \;\; x_3^* = 0 \;\; ; \;\; x_4^* = 1.2$

(4) $(2.16) \Leftrightarrow x_1 + 2.9x_4 = 6 \;\; ; \;\; 2.0x_1 + 0.9x_4 = 6$ for $\epsilon = \epsilon_0 \Rightarrow x_1^* = 2.45$;
$x_2^* = 0 \;\; ; \;\; x_3^* = 0 \;\; ; \;\; x_4^* = 1.22$

The solution is feasible. To check its optimality one needs to find simplex multipliers satisfying the basic system

$$\begin{aligned} \lambda_1 + 2\lambda_2 &= 1 \\ 2.9\lambda_1 + 0.9\lambda_2 &= 2.04 \end{aligned} \quad \Leftrightarrow \quad \begin{aligned} \lambda_1^* &= 0.649 \\ \lambda_2^* &= 0.175 \end{aligned}$$

and to make sure that

$$\begin{aligned} 1.05\lambda_1^* + 2.02\lambda_2^* &\geq 1.01 \\ 3.0\lambda_1^* + \lambda_2^* &\geq 2.0 \end{aligned} \quad \Leftrightarrow \quad \begin{aligned} 1.03 &\geq 1.01 \\ 2.12 &\geq 2.0 \end{aligned}$$

Hence, $x^* = (2.45 \,;\, 0 \,,\, 0 \,;\, 1.22)$ is the optimal solution to the original problem.

The dual situation is of interest, too. It arises if there are some scarce resources with similar technological characteristics. The corresponding model has the following form:

$$\max \{\sum_j c_j x_j \,/\, (a_s + \epsilon b_{is})^T x \le b_s + \epsilon \beta_{is} \,,\, i \in \Omega_s \,,\, s = 1,\ldots,k \,,\, x \ge 0\} \quad (2.18)$$

where $a_s + \epsilon b_{is}$ is a vector with components reflecting the input of resource i, if this resource belongs to the group Ω_s of "similar" resources.

The dual of (2.18) is

$$\min \{\sum_s \sum_{i \in \Omega_s} (b_s + \epsilon \beta_{is}) y_i \,/\, \sum_s \sum_{i \in \Omega_s} (a_s + \epsilon b_{is}) y_i \ge c \;,\; y \ge 0\} \quad (2.19)$$

where $c = (c_j)$, $y = (y_i)$.

Since (2.19) is of the same type as (2.1) and (2.2), the following algorithm to isolate the basis that is optimal under sufficiently small ϵ can be used.

(1) Solve the problem

$$\min \{\sum_s b_s v_s \,/\, \sum_s a_{js} v_s - \eta_j = c_j \;,\; \eta_j \ge 0 \;,\; v_s \ge 0\} \quad (2.20)$$

and the dual problem

$$\max \{c^T z \,/\, \sum_j a_{sj} z_j \le b_s \;,\; z \ge 0\} \quad (2.21)$$

obtaining optimal solutions $(v^* \,,\, \eta^*)$ and $\bar{z} = (\bar{z}_j)$. (These solutions are assumed here to be unique.)

(2) Find

$$i_s = \arg \min_{i \in \Omega_s} \{\beta_{is} - (\bar{z})^T b_{is}\} \quad (2.22)$$

(The uniqueness is assumed again.)

(3) Since indices i_s determine the dual optimal basis, find the locally optimal solution to (2.18) by solving the system of equations

$$(a_s + \epsilon b_{i_s s})^T \tilde{x} = b_s + \epsilon \beta_{i_s s} \;,\; s \in \sigma \quad (2.23)$$

where σ is the optimal basic set of indices in (2.20), $\tilde{x} = (x_j\ ,\ j \in \sigma)$ and $x_j^{\text{opt}} = 0$ if j is the basic variable in (2.20).

One can prove the validity of the algorithm in a similar fashion to the proof of the algorithm to solve (2.18) if one uses Theorem 2.3 (Chapter 1). It is of interest to note that regularity conditions are not fulfilled in (2.19), but that the uniqueness in (2.22) saves our approach.

The application of the dual aggregation scheme can markedly reduce the computations, since (2.21) generally has substantially fewer constraints than (2.18). As an illustration, one can analyze the behavior of the solution to a simple perturbed program of (2.18) type.

Example.

The data are as follows:

	c_1	c_2		
	1	1		
a_1	2	3	b_1	1
a_2	3	2	b_2	1

b_{11}	-1	2	β_{11}	0
b_{21}	2	-1	β_{21}	1
b_{31}	1	1	β_{31}	20
b_{42}	-1	0	β_{42}	0
b_{52}	1	-1	β_{52}	-1

By the algorithm, we have

(1) $\min\{6(v_1+v_2)\,/\,2v_1+3v_2-\eta_1=1\,;3v_1+2v_2-\eta_2=1\,;\eta_1,\eta_2,v_1,v_2,\geq 0\} \Rightarrow$

$v_1^*=0.2\ ;\ v_2^*=0.3\ ;\ \eta_1^*=\eta_2^*=0.$

$\max\{z_1+z_2\,/\,2z_1+3z_2\leq 6\,;3z_1+2z_2\leq 6\,;z_1,z_2\geq 0\} \Rightarrow \bar{z}_1=\bar{z}_2=1.2$

(2) $\min\limits_{i\in\Omega_1}\{-1.2;-0.2;17.6\} = -1.2 \Rightarrow i_1 = 1$

$\min\limits_{i\in\Omega_2}\{1.2;-1\} = -1 \Rightarrow i_2 = 5$

(3) $$(2.23) \Leftrightarrow \begin{cases} (2-\epsilon)\tilde{x}_1 + (3+2\epsilon)\tilde{x}_2 = 6 \\ (3+\epsilon)\tilde{x}_1 + (2-\epsilon)\tilde{x}_2 = 6-\epsilon \end{cases}$$

$$\tilde{x}_1 = \frac{6+15\epsilon-2\epsilon^2}{5+13\epsilon+\epsilon^2}\ ;\ \tilde{x}_2 = \frac{6+14\epsilon-\epsilon^2}{5+13\epsilon+\epsilon^2}\ ;$$

ϵ	0	0.1	0.5	1.0
$\tilde{x}_1$	1.2	1.185	1.106	1.0
$\tilde{x}_2$	1.2	1.171	1.085	1.0

It is not hard to check that the optimality of $\tilde{x}$ is guaranteed for all values of ϵ shown above. Note that, for $\epsilon = 1$, perturbations are very strong and the original problem has the form:

$$\max\{x_1 + x_2 \,/\, x_1 + 5x_2 \leq 6\,;4x_1 + 2x_2 \leq 7\,;3x_1 + 4x_2 \leq 26\,;2x_1 + 2x_2 \leq 6\,;\\ 4x_1 + x_2 \leq 5\,;x_j > 0\}$$

where it is hard to anticipate any possibility of aggregation. Of course, we do not mean to say that the perturbation method is always so successful. However, there is a special case for which this is so. The third inequality in our example has a right-hand side, which is quite different from that of those within the first group, but this does not matter. Indeed, this inequality is not essential since the difference is a large positive one!

Further we shall consider some simple approaches with which to decrease the program dimension. These approaches use the ideas of "strong differences" and "bottlenecks", which are complementary in a definite sense, to the idea of "similarity" used above. If "bottlenecks" and "strong differences" arise, one can select inessential constraints and ineffective activities by simple estimates.

Let us consider a class of programs with linear constraints

$$\begin{aligned} &\max\{\varphi(x)\,/\,x \in \Theta\} \\ &\Theta = \{x\,/\,Ax \leq b\,,\,x \geq 0\} \end{aligned} \tag{2.24}$$

Theorem 2.1.

Let $A = (a_{ij}) \geq 0$, $b = (b_i) > 0$. If

$$\sum_j a_{ij} \min_{\substack{l \\ (a_{lj}>0)}} \frac{b_l}{a_{lj}} \leq b_i \tag{2.25}$$

or

$$\frac{a_{ij}}{b_i} \leq \frac{a_{kj}}{b_k} \tag{2.26}$$

for all j and any k, then the inequality

$$\sum_i a_{ij} x_j \leq b_i$$

is not essential to describing Θ.

The proof is elementary. Moreover, analogous properties occur, too, if

$$\Theta = \{x \,/\, A^I x \leq b^I \,,\, A^{II} x \geq b^{II} \,,\, x \geq 0\}$$

where $A^I \,,\, A^{II} \geq 0 \,;\, b^I \,,\, b^{II} > 0$.

Since the tests (2.25) and (2.26) are based on the properties of the feasible set Θ only, one can use it with any objective function. The following theorem provides a stronger test of inessentiality.

Theorem 2.2.

Let

$$\begin{aligned} \bar{\varphi}_i &\triangleq \max \{\varphi(x) \,/\, a_i^T x \leq b_i \,,\, x \geq 0\} \quad, \quad i = 1,\ldots,m \\ \underline{\varphi}_i &\triangleq \min \{\varphi(x) \,/\, a_i^T x \geq b_i \,,\, x \geq 0\} \quad, \quad i = 1,\ldots,m \end{aligned} \tag{2.27}$$

If

$$\underline{\varphi}_i > \min_k \bar{\varphi}_k \tag{2.28}$$

then the constraint $a_i^T x \leq b_i$ is inessential.

If φ is a monotonically increasing function (at least over the set $a_i^T x = b_i$), then the condition (2.28) can be replaced by

$$\underline{\varphi}_i \geq \min_k \bar{\varphi}_k \tag{2.28$'$}$$

Proof. Let us suppose the converse, i.e., that

$$a_i^T \bar{x} = b_i$$

where $\bar{x}$ is the solution to (2.24). Then

$$T_i \triangleq \Theta \bigcap \{x \,/\, a_i^T x = b_i\} \neq \emptyset$$

but

$$\underline{\varphi}_i \leq \min\,\{\varphi(x) \,/\, x \in T_i\} \geq \max\,\{\varphi(x) \,/\, x \in T_i\} \leq$$
$$\leq \max\,\{\varphi(x) \,/\, x \in \Theta\} \triangleq \varphi_{\max}$$

Hence, we have

$$\underline{\varphi}_i \leq \varphi_{\max}$$

However, by the theorem condition

$$\underline{\varphi}_i > \min_k \bar{\varphi}_k \geq \varphi_{\max}$$

The contradiction proves the first desired result.

Note that the constraint may be essential only if T_i contains more than one point. But then the monotonicity of $\varphi(x)$ yields

$$\min\,\{\varphi(x) \,/\, x \in T_i\} < \max\,\{\varphi(x) \,/\, x \in T_i\}$$

i.e., $\underline{\varphi}_i < \varphi_{\max}$ and this contradicts (2.28′).

Of course, the use of the test (2.28) is rational only if the computation of the estimates (2.27) is elementary, e.g., if $\varphi(x)$ is separable.

Theorems 2.1 and 2.2 provide ways to discard those resource constraints that are certain not to be "bottlenecks" for activity scheduling.

If $\varphi(x) = c^T x$ then the duality theory provides analogous tests to omit ineffective activities. The following result is given without proof.

Theorem 2.3.

Let $A \geq 0$; $c > 0$; $b > 0$ and let at least one of the following conditions be true:

(a) For all i and any k

$$\frac{c_i}{a_{ij}} \leq \frac{c_k}{a_{ik}} \tag{2.29}$$

(b) $\bar{\psi}_j \leq \max_k \underline{\psi}_k$ (2.30)

where

$$\bar{\psi}_j = \max \{b^T\Lambda / A_j^T\Lambda \leq c_j \; ; \; \Lambda \geq 0\} = c_j \min \{x_j / A_j x_j \geq b \; ; \; x_j \geq 0\}$$
$$\underline{\psi}_j = \min \{b^T\Lambda / A_j^T\Lambda \geq c_j \; ; \; \Lambda \geq 0\} = c_j \max \{x_j / A_j x_j \leq b \; ; \; x_j \geq 0\}$$

Then the optimal value of the program

$$\max \{c^T x / Ax \leq b \; ; \; x \geq 0\} \tag{2.31}$$

is preserved if one assumes $x_j = 0$.

Note that to compute the estimates $\bar{\varphi}_i$, $\underline{\varphi}_i$ (or $\underline{\psi}_j$, $\bar{\psi}_j$) in the linear case is no harder than sorting out the data sets. In conclusion, we shall summarize practical recommendations for reducing the linear program size. Apparently, it is expedient to select essential constraints by tests following from Theorems 2.1 and 2.2 and, after that, to remove ineffective activities by tests from Theorem 2.3. The number of resources and activities having strongly distinct properties must decrease after such procedures. Therefore, one can pass to the stage of aggregation of similar activities, providing its preliminary classification.

If there is no *a priori* information about a potential similarity, one can use some formal classification scheme. A number of schemes is known for classification of vector objects (see, e.g., [93] and [147]), but it is advisable only to use the simplest ones.

To be precise, we describe a very simple procedure, with no pretensions to its being the most effective one.

Let the original problem be a linear program

$$\max \{\sum_s c_j x_j / \sum_j A_j x_j = b \; ; \; x_j \geq 0\}$$

As a preliminary, normalize all vectors A_j that are essential to bring the quasicollinear vectors closer together. If ρ_j is the norm of (c_j , A_j), then normalizing yields

$$\max \{\sum_j \bar{c}_j \bar{x}_j / \sum_j \bar{A}_j \bar{x}_j = b \; ; \; \bar{x}_j \geq 0\} \tag{2.32}$$

where $\bar{x}_j = \rho_j x_j$, $\bar{c}_j = c_j / \rho_j$, $\bar{A}_j = A_j / \rho_j$.

Let A be any normed vector, and let P be a group of normed vectors, $A = (A_l , l = 1,\ldots,N)$. One can introduce the cost

$$T(A\ ,\ P) = \frac{1}{N} \sum_{l=1}^{N} A_l^T A$$

as an estimate of the closeness of A to P.

Let A_1 be an arbitrary vector included in the description of (2.32) and $P_1 = A_1$. Order all other vectors in accordance with the following algorithm. Let P_i be a set of the ordered vectors at a step i. Then

$$P_{i+1} = P_i \cup A_{i+1}$$
$$i + 1 = \arg\max \{T(A_s\ ,\ P_i) \,/\, A_s \,\overline{\subset}\, P_i\}\ ,\ T_{i+1} \triangleq T(A_{i+1}\ ,\ P_i)$$

The sequence T_i has an evident property: if the set of all A_i separates into groups of similar vectors, then "jumps" appear in passing from one group to another. Selecting the $k - 1$ largest values in the sequence $\{\delta_i\}$

$$\delta_i \triangleq \frac{T_i - T_{i+1}}{T_i}$$

one can detect the $k - 1$ largest "jumps".

Vectors whose indices belong to the interval between two sequential jumps, including the beginning and the end, are united in one group $\Omega_s\ ,\ s = 1,\ldots,k$.

Let us note that the values $c_j\ ,\ j \in \Omega_s$ must not be close. It is reasonable to put

$$c_s = \max \{c_j \,/\, j \in \Omega_s\}$$

and, after that, to assume

$$B_{js} = A_j - A_s\ ,\ \epsilon = 1$$

A structure of (2.8) type is generated formally by this scheme. It is clear that one can use an analogous formal approach to group the set of rows. However it is important to understand that the basic approaches to generation of the perturbed structure are the "heuristic" ones using intuitive presentations.

It is clear from (2.16) and (2.17) that it is not important how small the differences between the aggregated columns or rows are, but the smallness of the perturbations calculated on the primal and dual solutions is essential. Therefore, the efficiency of the aggregation procedures may be estimated only *a posteriori*.

2.3. Weakly controllable input–output characteristics

A mathematical model adequate for the problem of technological complex scheduling, taking into account the freedom to change some characteristics of processing units, was given by G. Dantzig and P. Wolfe [23]. In the same paper it was shown that such models are important to optimize the control of flow shop where chemical reactors are used as the basic processing units. The general model structure is given in the following form

$$\max \{c^T(u)y \,/\, A(u)y \leq b \,,\, y \geq 0 \,,\, u \in U\} \tag{3.1}$$

where the vector u is interpreted as a control of input–output coefficients and y is the activity level of all units.

It was shown in [27] that if

(a) $A(u)$, $c(u)$ are linear functions, $A(u)$ is $(n \times n)$-matrix, $c(u)$ is n-column;
(b) Each activity is controlled independently, i.e.,

$$u = (u_j) \;,\; u_j \in U_j \;,\; \prod_j U_j = U \;,\; c(u) = (c_j(u_j)) \;,\; A(u) = (A_j(u_j)) \tag{3.2}$$

(c) The sets U_j are polyhedrons,

then the problem (3.1) can be transformed into a linear program.† To do this, one can write all U_j in the form

$$U_j = \{u_j \,/\, u_j = \sum_k \lambda_{jk} u_{jk} \,,\, \sum_k \lambda_{jk} = 1 \,,\, \lambda_{jk} \geq 0\} \tag{3.3}$$

where u_{jk} are the extreme points, and use new variables $x_{jk} = \lambda_{jk} y_j$. In this instance, the program size generally increases, and the decomposition algorithm, including the generating columns procedure, is the most effective way to solve it. Otherwise, the standard simplex method is suitable if the number of extreme points is small (many examples are in [73]).

However, the Dantzig–Wolfe conditions mentioned above are restrictive in some situations. On the other hand, one can successfully replace these conditions by the weak controllability hypothesis, i.e, assuming that input–output coefficients can be changed by control to within a small value.

† If the Dantzig–Wolfe conditions are fulfilled, then the problem (3.1) and (3.4) is equivalent to a linear program of type (2.8). Each extreme point of any U_j generates a new activity, and the characteristics corresponding to all activities are similar.

To demonstrate this fact formally, one can introduce a small parameter ϵ denoting

$$c(u) = c + \epsilon\beta(u) \quad , \quad A(u) = A + \epsilon B(u) \tag{3.4}$$

Further, we can use the perturbation method and construct the following procedure based on Theorem 5.1 (Chapter 1) with estimates that will be justified at the end of this section.

(1) Solve the program

$$\max \{c^T y \,/\, Ay \leq b \,,\, y \geq 0\} \tag{3.5}$$

to find the optimal set Θ^* and the estimates λ^*, supposed to be unique.

(2) Solve the auxiliary problem

$$\max \{y^T[\beta(u) - B^T(u)\lambda^*] \,/\, y \in \Theta^* \,,\, u \in U\} \tag{3.6}$$

to find a suboptimal distribution y^* and a suboptimal control u^*. The corresponding value of the original program differs from the optimal one to within $O(\epsilon^2)$, and y^* is admissible for (3.1) to within $O(\epsilon)$.

(3) If y^* is the unique solution to (3.5) then

$$y^*(\epsilon) = y^* + \epsilon y_1^* + o(\epsilon) \tag{3.7}$$

To compute y_1^* one must find the index sets $\tilde{I}$, $\tilde{J}$,

$$\tilde{I} \simeq \{i \,/\, \sum_j a_{ij} y_j^* = b_i\} \quad , \quad \tilde{J} \triangleq \{j \,/\, y_j^* > 0\}$$

Then all components of y_1^* having indices $j \bar{\in} \tilde{J}$ are equal to zero and other ones form the vector $\tilde{y}_1^*$,

$$\tilde{y}_1^* = -\tilde{A}^{-1}\tilde{B}y^*$$

where

$$\tilde{A} = \{a_{ij} \,,\, i \in \tilde{I} \,,\, j \in \tilde{J}\} \quad , \quad \tilde{B} = \{b_{ij}(u^*) \,,\, i \in \tilde{I} \,,\, j = 1,\ldots,n\}$$

One can give two specific cases where that algorithm is simplified. In both cases it is supposed that (3.2) is fulfilled. Then, for any fixed $y \in \Theta^*$, problem (3.6) is separable into a number of mutually independent problems of how to choose the control of characteristics of each activity.

A. Let $\Theta^* = y^* = (y_j^*)$, i.e., let the solution to (3.5) be unique. Then (3.6) yields the following programs

$$\max \{(y^*)^T[\beta_j(u_j) - B_j^T(u_j)\lambda_*] \,/\, u_j \in U_j\} \tag{3.8}$$

A suboptimal decomposition of the original problem is then realized. As a heuristic approach, such a transformation was recommended in [64], and also the objectives of (3.8) were interpreted as the objective functions of the technological control.

It is a simple, but important practical result. Actually, it is a problem to state the objective function to choose the control adjustment point for a regulator of a processing unit if the latter is used as part of the technological complex. The objective in (3.8) is an objective function of the j-th processing unit, coordinated suboptimally with the objective function of the complex as a whole.

In this connection it was mentioned that one need not know the functions $\beta_j(u_j)$, $B_j(u_j)$ in an explicit form since their values can be obtained "on-line" and used as inputs to self-adjusting regulators of any type.

B. Let the polyhedron Θ^* be separable into k simplexes

$$\sum_{j\in\Omega_s} y_j = z_s^* \ , \ y_j \geq 0 \ , \ s = 1,\ldots,k \tag{3.9}$$

Such a situation arises where an industrial system includes k groups of similar processing units, and its operating conditions can be adjusted independently.

If the technological process of each processing unit, j, is considered as an independent activity (compare with Section 2.2), then the optimization problem for a complex as a whole can be written in the form

$$\max \{\sum_s \sum_{j\in\Omega_s} [c_s + \epsilon\beta_j(u_j)]y_j \,/\, \sum_s \sum_{j\in\Omega_s} [A_s + \epsilon B_j(u_j)]y_j \leq b \ , \ y_j \geq 0\}$$

and the auxiliary program is separable into k independent ones

$$\begin{aligned} \max \ \{ &\sum_{j\in\Omega_s} y_j[\beta_j(u_j) - B_j^T(u_j)\lambda^*] \,/\, \sum_{j\in\Omega_s} y_j = z_s^* \ , \ y_j \geq 0 \ , \\ &u_j \in U_j \ , \ j \in \Omega_s\} \end{aligned} \tag{3.10}$$

where the (z_s^*), λ^* can be determined from the solution to the aggregated program (2.11) and the dual program.

Having the solutions to (3.10), one can state the suboptimal load distribution inside each group and find suboptimal processing conditions for each processing unit. Note also that if suboptimal processing conditions are found, i.e., the program

$$\tilde{c}_j \triangleq \max \{\beta_j(u_j) - B_j^T(u_j)\lambda^* \,/\, u_j \in U_j\}$$

is solved, then the distribution of the activities can be found in explicit form

$$y_{j_s}^* = z_s^* \ , \ j_s \triangleq \arg\max \tilde{c}_j \ , \ y_j^* = 0 \ , \ j \neq j_s \ , \ j \in \Omega_s$$

i.e., the whole output is realized by the most effectively controlled technology (this is true, of course, if there are no additional constraints on the feasible capacities of the processing units).

In addition, consider one typical technological scheme in more detail. Let one processing unit, $j = 0$, produce a half-finished product, and let a number of other units, $j = 1,\ldots,n$, transform this product into different final goods. All the units are of the same model but their properties differ slightly and, moreover, depend on the technological control, which can be independently varied over a sufficiently wide range. It is necessary to distribute the active time T_j of each unit, i.e., to find the durations y_{js} of time intervals when unit j produces a product s. Moreover, one needs to find the optimal processing conditions. The switch-over losses are assumed to be negligible.

The model of optimal scheduling and control may be presented in the form

$$\max \{\sum_{s=1}^{k} \sum_{j=1}^{n} (c_s + \epsilon\beta_{js}(u_{js}))y_{js} - (c_0 + \epsilon\beta_0(u_0))\} \tag{3.11}$$

subject to

$$\sum_{s=1}^{k} \sum_{j=1}^{n} (a_s + \epsilon b_{js}(u_{js}))y_{js} \leq a_0 + \epsilon b_0(u_0)$$

$$\sum_{s=1}^{k} y_{js} \leq T_j \ , \ y_{js} \geq 0 \ , \ u_{js} \in U_{js} \ , \ u_0 \in U_0$$

The performance index characterizes the profit of the final process less the expenditure needed to produce the half-finished product. The constraint structure is determined by the following factors: (a) half-finished product input–output balance, (b) limited active time for each finite processing unit, and (c) specific processing conditions for the units. The description of input–output coefficients reflects the weakness of technological control influence and the smallness of differences between the finishing units. The peculiarity of the reduced problem, corresponding to (3.11), is that its solution does not yield a unique distribution or control levels. However, suboptimal values, z_s^*, of time intervals needed to produce each product, can easily be found as the solutions to the linear program

$$\max \{ \sum_{s=1}^{k} c_s z_s \,/\, \sum_{s=1}^{k} a_s z_s \leq a_0 \,,\; \sum_{s=1}^{k} z_s \leq T \,,\; z_s \geq 0 \} \tag{3.12}$$

where $T \triangleq \sum_j T_j$.

Let the solution to (3.12), $z^* = (z_s^*)$ be unique as well as the solution, $(\lambda_0^* \,,\, \lambda_1^*)$, to its dual. In accordance with the second stage of the perturbation method algorithm, one can construct the auxiliary program, having here the form

$$\max \{ \sum_{s=1}^{k} \sum_{j=1}^{n} [\beta_{js}(u_{js}) - \lambda_0^* b_{js}(u_{js})] y_{js} - [\beta_0(u_0) - \lambda_0^* b_0(u_0)] \} \tag{3.13}$$

subject to

$$\sum_{j=1}^{n} y_{js} = z_s^* \;,\; s = 1,\ldots,k \;,\; \sum_{s=1}^{k} y_{js} \leq T_j \;,\; j = 1,\ldots,n$$

$$y_{js} \geq 0 \;,\; u_{js} \in U_{js} \;,\; u_0 \in U_0$$

The method of solution is trivial. It is necessary to (a) find the optimal processing conditions for each unit independently by solving the problems

$$-\alpha_{js} \triangleq \max \{\beta_{js}(u_{js}) - \lambda_0^* b_{js}(u_{js}) \,/\, u_{js} \in U_{js}\} \tag{3.14}$$

$$\min \{\beta_0(u_0) - \lambda_0^* b_0(u_0) \,/\, u_0 \in U_0\} \tag{3.15}$$

and (b) solve the transportation program

$$\min \{\sum_s \sum_j \alpha_{js} y_{js} \,/\, \sum_{j=1}^{n} y_{js} = z_s^* \,,\; \sum_{s=1}^{n} y_{js} \leq T_j \,,\; y_{js} \geq 0\} \tag{3.16}$$

Note that the solvability condition

$$\sum_{s=1}^{k} z_s^* \leq \sum_{j=1}^{n} T_j$$

is fulfilled by (3.12). The solution obtained is suboptimal within to $O(\epsilon^2)$ (with respect to the objective function). Evidently, it can be improved by using an optimal basis.

In conclusion, we shall give the justification of the described scheme, bearing in mind the problem

$$\max \{f_0(y) + \epsilon f_1(y , u) / g_0(y) + \epsilon g_1(y , u) \leq 0 , u \in U\} \tag{3.17}$$

which is a generalization of (3.1) and (3.4). Here, $f_0(y)$, $g_1(y , u)$, $g_0(y)$, $g_1(y , u)$ are differentiable functions.

In accordance with the general scheme of Section 1.5, the calculation of the zero-order approximation includes two stages:

(1) Find y^* such that

$$f_0(y^*) = \max \{f_0(y) / g_0(y) \leq 0\} \tag{3.18}$$

Let y^* be unique. Let λ^* be a corresponding Lagrangian multipliers vector, determined by equation

$$\nabla f_0(y^*) = \sum_i \lambda_i^* g_{i0}(y^*) \quad , \quad \lambda^* = (\lambda_i^*) \tag{3.19}$$

(2) Solve the program

$$\max \{\Phi(u) / u \in U\} \tag{3.20}$$

where

$$\Phi(u) = f_1(y^* , u) - (\lambda^*)^T g_1(y^* , u)$$

As a result, we get a suboptimal solution y^*, u^* to the problem (3.17). The specific character of its structure permits us to give a more detailed estimate of the proximity of y^*, u^* and the solution $y^*(\epsilon)$, $u^*(\epsilon)$ to the perturbed problem than is possible within the general scheme of Section 1.5.

Theorem 3.1.

Let all the conditions of Lemma 5.3 (Chapter 1) be fulfilled. The complementarity slackness conditions are strictly fulfilled.

Moreover,†

$$\det \{\nabla\nabla L(y^*, \lambda^*)\} \neq 0 \tag{3.21}$$

where

$$L(y, \Lambda) \triangleq f_0(y) - \lambda^T g_0(y)$$

and where derivatives are calculated with respect to components y_i, λ_l, $l \in \omega$ and $\omega = \{l / \lambda_l^* > 0\}$. Let u^* be unique. Then the optimal value, $F^*(\epsilon)$, of the objective function in (3.17) is estimated on y^*, u^* to within $O(\epsilon^2)$

$$F^*(\epsilon) = f_0(y^*) + \epsilon\Phi(u^*) + O(\epsilon^2)$$

Moreover, the following representations are true

$$y^*(\epsilon) = y^* + \epsilon y_1^* + o(\epsilon) \tag{3.22}$$

$$\Phi(u^*) - \Phi(u^*(\epsilon)) = O(\epsilon) \tag{3.23}$$

where the method to find y_1^* is shown in the proof [see (3.30)].

Corollary. (a) If $\Phi(u)$ is strongly concave and U is convex, then

$$\|u^*(\epsilon) - u^*\| = O(\epsilon^{1/2}) \tag{3.24}$$

(b) If $\Phi(u)$ is linear and U is a polyhedral set, then

$$\|u^*(\epsilon) - u^*\| = O(\epsilon) \tag{3.25}$$

Proof. First of all, let us show that (3.22) is true. By Theorem 5.1 (Chapter 1), the following limits exist:

$$\lim_{\epsilon\to 0} y^*(\epsilon) = y^*, \quad \lim_{\epsilon\to 0} u^*(\epsilon) = u^* \tag{3.26}$$

† Not that condition (3.21) is fulfilled when

$$f_0(y) = c_0^T y, \quad g_0(y) = A_0 y$$

and the solution to the problem (3.5) is unique.

Since it is supposed that the complementarity conditions are strictly fulfilled, then (3.26) yields

$$\begin{aligned} &g_{l0}(y^*(\epsilon)) + \epsilon g_{l1}(y^*(\epsilon) , u^*(\epsilon)) = 0 \ , \ l \in \omega \\ &g_{l0}(y^*(\epsilon)) + \epsilon g_{l1}(y^*(\epsilon) , u^*(\epsilon)) < 0 \ , \ l \bar{\in} \omega \end{aligned} \tag{3.27}$$

for all sufficiently small $\epsilon \geq 0$.

On the other hand, (3.21) guarantees that the regularity conditions are fulfilled for small ϵ and, hence,

$$\begin{aligned} &\nabla f_0(y^*(\epsilon)) + \epsilon \nabla_y f_1(y^*(\epsilon) , u^*(\epsilon)) = \\ &= \sum_{l\in\omega} \lambda_l^*(\epsilon)\{\nabla g_{l0}(y^*(\epsilon)) + \epsilon \nabla_y g_{l1}(y^*(\epsilon) , u^*(\epsilon))\} \end{aligned} \tag{3.28}$$

Equations (3.27) and (3.28) can be presented in the form

$$\begin{aligned} &g_{l0}(y^*(\epsilon)) = -\epsilon g_{l1}(y^* , u^*) + o(\epsilon) \\ &\nabla f_0(y^*(\epsilon)) - \sum_{l\in\omega} \lambda_l^*(\epsilon) \nabla g_{l0}(y^*(\epsilon)) = \\ &= \epsilon\{-\nabla_y f_1(y^* , u^*) + \sum_{l\in\omega} \lambda_l^* \nabla_y g_{l1}(y^* , u^*)\} + o(\epsilon) \end{aligned} \tag{3.29}$$

The behavior of the functions $y^*(\epsilon)$, $\lambda^*(\epsilon)$ is given by the system (3.29) for small ϵ. Since (3.21) shows that the Jacobian of (3.21) is not equal to zero at the point $\epsilon = 0$, one can apply the implicit function theorem. The validity of (3.22) is verified now by standard means, y_1^* being found as a solution to the linear system

$$\begin{aligned} &[\nabla\nabla f_0(y^*) - \sum_{l\in\omega} \lambda_l^* \nabla\nabla g_{l0}(y^*)] y_1 - \sum_{l\in\omega} \nabla g_{l0}(y^*) \lambda_{l1} = \\ &= -\nabla_y f_1(y^* , u^*) + \sum_{l\in\omega} \lambda_l^* \nabla_y g_{l1}(y^* , u^*) \\ &[\nabla g_{l0}(y^*)]^T y_1 = -g_{l1}(y^* , u^*) \ , \ l \in \omega \end{aligned} \tag{3.30}$$

After that, the other assertions of the theorem appear as corollaries of Lemma 5.1 (Chapter 1), since $u^*(\epsilon) \in U$ and the value $\delta(\epsilon)$ is of the order of $O(\epsilon)$. Take the proof of the corollary (b). Put

$$U \triangleq \{u \mid Du \leq d\} \ , \ \Phi(u) \triangleq p^T u \tag{3.31}$$

Let $D = (\bar{D} , \bar{\bar{D}})$, $d = (\bar{d} , \bar{\bar{d}})$ and $\bar{D}u^* = \bar{d}$, $\bar{\bar{D}}u^* < \bar{d}$. Since $\bar{D}u^*(\epsilon) \leq \bar{d}$ we have

$$\bar{D}[u^*(\epsilon) - u^*] \leq 0 \tag{3.32}$$

and, by (3.23) and (3.31),

$$p^T[u^*(\epsilon) - u^*] = O(\epsilon) \tag{3.33}$$

Assume that (3.25) is false, i.e., that there exists a subsequence $\{\epsilon'\}$ such that

$$\lim_{\substack{\epsilon \to 0 \\ \epsilon \in \{\epsilon'\}}} \frac{\|\epsilon\|}{\|u^*(\epsilon) - u^*\|} = 0$$

Without loss of generality, one can suppose that, over $\{\epsilon'\}$

$$\lim_{\epsilon \to 0} \frac{u^*(\epsilon) - u^*}{\|u^*(\epsilon) - u^*\|} = \rho \ , \ \ \|\rho\| = 1$$

By dividing (3.32) and (3.33) into $\|u^*(\epsilon) - u^*\|$ and taking the limit, we get

$$\bar{D}\rho \leq 0 \ , \ \ p^T\rho = 0$$

However, there is a sufficiently small γ such that

$$u^* + \gamma\rho \in U \ , \ \ p^T(u^* + \gamma\rho) = p^T u^*$$

and this contradicts the uniqueness of u^*. The contradiction proves (3.25). The proof of (3.24) is trivial (see Section 1.3).

2.4. Input–output analysis

Input–output analysis is a classic approach to various problems of economics. The simplest and most basic model is well-known.

Let x_j be the total output of a product j realized by a mono-product industry j, $j = 1,\ldots,N$. Let y_j be the final consumption value of the product j. Having the input–output coefficients, a_{ij}, one can write the balance equations

$$x_i = \sum_{j=1}^{N} a_{ij}x_j + y_i \ , \ \ i = 1,\ldots,N \tag{4.1}$$

As usual, suppose that the matrix $A = \{a_{ij}\}$ is productive. A number of equivalent definitions of that property [57] is known. We shall indicate the following one.

Definition. Matrix A is productive if

$$(E - A)^{-1} y \geq 0 \quad \text{for all} \quad y \geq 0 \tag{4.2}$$

Under condition (4.2), the basic problem to find $x = (x_i) \geq 0$ such that (4.1) is true for any $y = y_i \geq 0$ is always solvable and

$$x = (E - A)^{-1} y \tag{4.3}$$

Hence, the computations of all the industries' output is realized by solving a system of linear equations. Practically, only the very large-size and (or) the ill-conditioned nature of the matrix $E - A$ may be a stumbling block to the application of standard numerical codes.

However, attempts to simplify computations using the specific structure of the input–output matrix are continued since a simplified computational procedure generally guarantees a simplification of the data set and, moreover, brings together formal schemes and traditional economic approaches.

The following basic hypotheses have been introduced by many authors in a more or less definite form.

A. Weakness of connections between macroindustries

Formally, this hypothesis is expressed by the assumption that matrix A is quasidiagonal, i.e.,

$$A = \bar{A} + \epsilon B \ , \ \bar{A} = \text{diag}\,(F_s \ , s = 1,\ldots,k) \tag{4.4}$$

where F_s is the productive matrix of input–output coefficients characterizing interactions between industries belonging to the same "macroindustry" and matrix ϵB corresponds to an exchange between macroindustries. The smallness of exchanges is effected by introducing the parameter ϵ. For $\epsilon = 0$, the problem separates into k independent ones. It is evident, too, that if ϵ is small enough then the solution to (4.3) and (4.4) can be presented in the form of a convergent series

$$x_s = x_s^{(0)} + \epsilon x_s^{(1)} + \epsilon^2 x_s^{(2)} + \cdots \tag{4.5}$$

where

$$x_s^{(0)} = (E_s - F_s)^{-1} y_s$$

$$x_s^{(t+1)} = (E_s - F_s)^{-1} B_s x^{(t)}$$

$$x^{(t)} = (x_s^{(t)}) \ , \ t = 1,2,\ldots$$

x_s, $s = 1,\ldots,k$ is the output vector of macroindustry s, i.e., $x_s = (x_i \ , \ i \in \Omega_s)$, E_s is the unit matrix of corresponding dimension, B_s is the submatrix of B consisting of the rows $i \in \Omega_s$.

Unfortunately, hypothesis A very seldom corresponds to the economic reality.

B. Similarity of characteristics of industries

Let us suppose that the vectors $A_j = (a_{ij})$ are similar for all industries included into the same group s. Formally, let

$$A_j = A_j(\epsilon) = A_s^0 + \epsilon B_j \ , \ j \in \Omega_s \ , \ s = 1,\ldots,k \tag{4.6}$$

This hypothesis results in the simplification of (4.1). Its solution can be realized using the aggregated values characterizing the group s ("macroindustry") as a whole.

Let us denote

$$z = (z_s) \ , \ z_s = \sum_{j \in \Omega_s} x_j \ , \ \bar{y} = (\bar{y}_s) \ , \ \bar{y}_s = \sum_{j \in \Omega_s} y_j$$

$$A^0 = \{A_s^0 \ , \ s = 1,\ldots,k\} \ , \ \bar{A} = \{\bar{a}_{rs} \ ; \ r,s = 1,\ldots,k\}$$

$$\bar{a}_{rs} = \sum_{i \in \Omega_r} a_{is}^0 \tag{4.7}$$

$$B = \{B_j \ , \ j = 1,\ldots,N\} \ , \ \bar{B} = \{\bar{b}_{sj} \ ; \ s = 1,\ldots,k \ , \ j = 1,\ldots,N\}$$

$$\bar{b}_{sj} = \sum_{i \in \Omega_s} b_{ij}$$

Theorem 4.1.

Let $\|\bar{A}\| < 1$, then the sequence $x^{(t)}$, $t = 0,1,\ldots$ determined by the equations

$$\begin{aligned} x^{(t+1)} &= A^0 z^{(t+1)} + \epsilon B x^{(t)} + y \\ z^{(t+1)} &= \bar{A} z^{(t+1)} + \epsilon \bar{B} x^{(t)} + \bar{y} \\ x^{(0)} &= A^0 z^{(0)} + y \\ z^{(0)} &= \bar{A} z^{(0)} + \bar{y} \end{aligned} \tag{4.8}$$

is convergent to (4.3) if the conditions (4.6) are satisfied, and $\epsilon \leq \bar{\epsilon}$ where

$$\bar{\epsilon} \frac{\|B\|}{1 - \|A^0 T\|} < 1 \tag{4.9}$$

$$T = (t_{sj}) \ , \ t_{sj} = \begin{cases} 1 \ , \ j \in \Omega_s \\ 0 \ , \ j \bar{\in} \Omega s \end{cases} \tag{4.10}$$

Proof. By virtue of (4.6), equation (4.1) has the form

$$x = \sum_s \sum_{j \in \Omega_s} (A_s^0 + \epsilon B_j) x_j + y$$

Summing the rows for $i \in \Omega_s$ we have

$$z_r = \sum_s \bar{a}_{rs} z_s + \epsilon \sum_s \sum_{j \in \Omega_s} \bar{b}_{rj} x_j + \bar{y}_r \ , \ r = 1,\ldots,k$$

Hence, (4.1) is equivalent to the system

$$\begin{aligned} z &= \bar{A} z + \epsilon \bar{B} x + \bar{y} \\ x &= A^0 z = \epsilon B x + y \end{aligned} \tag{4.11}$$

As usual, we shall look for the solution in the form of a power series

$$\begin{aligned} z &= z^{(0)} + \epsilon \varsigma^{(1)} + \epsilon^2 \varsigma^{(2)} + \cdots \\ x &= x^{(0)} + \epsilon \xi^{(1)} + \epsilon^2 \xi^{(2)} + \cdots \end{aligned} \tag{4.12}$$

Then, equating the coefficients with equal power of ϵ, we obtain

$$\begin{aligned} & z^{(0)} = \bar{A} z^{(0)} + \bar{y} \ , \ \varsigma^{(t+1)} = \bar{A} \varsigma^{(t+1)} + \bar{B} \xi^{(t)} \ , \ \varsigma^{(0)} \triangleq z^{(0)} \\ & x^{(0)} = A^0 z^{(0)} + y \ , \ \xi^{(t+1)} = A^0 \varsigma^{(t+1)} + B \xi^{(t)} \ , \ t \geq 0 \ , \ \xi^{(0)} \triangleq x^{(0)} \end{aligned}$$

and

$$\xi^{(t+1)} = \hat{B}\xi^{(t)} \quad , \quad \hat{B} = B + A^0(E_k - \bar{A})^{-1}\bar{B}$$

The condition $\|\hat{B}\| < 1/\epsilon$ guarantees the convergence of series (4.12) and thereby the convergence of the sequence $\{x^t\}$ since

$$x^{(t)} = x^{(0)} + \epsilon\xi^{(1)} + \cdots + \epsilon^{(t)}\xi^{(t)}$$

It remains to show that (4.9) yields the condition $\|\hat{B}\| < 1/\epsilon$. Let us first of all note that, by virtue of the definitions (4.7) and (4.11), we can write

$$\bar{A} = TA_0 \quad , \quad \bar{B} = TB$$

Hence, a representation

$$\hat{B} = B + A^0[E_k - TA^0]^{-1}TB$$

is valid. Let us state the validity of some simple estimates of $\|\hat{B}\|$. By the condition $\|TA_0\| < 1$ we have

$$\begin{aligned}
\|\hat{B}\| &\leq \|E_N + A^0[E_k - TA^0]^{-1}T\| \; \|B\| = \\
&= \|E_N + A^0[E_k + TA^0 + (TA^0)^2 + \cdots]T\| \; \|B\| = \\
&= \|E_N + A^0T + A^0TA^0T + A^0TA^0TA^0T + \cdots \| \; \|B\| \leq \\
&\leq (1 + \|A^0T\| + \|A^0T\|^2 + \cdots)\|B\| = \frac{\|B\|}{1 - \|A^0T\|} < \frac{1}{\epsilon}
\end{aligned}$$

This yields the desired result.

Note that one does not need to introduce the parameter ϵ explicitly if the scheme is applied to a concrete problem. One can put $\epsilon = 1$, and then (4.9) is directly reduced to the condition of small differences between vectors B_j belonging to the aggregatable group. Note also that the choice of basic vector A_s^0 for each group can be, to some extent, arbitrary. In particular, one can put

$$A_s^0 = \sum_{j\in\Omega_s} p_j A_j \quad , \quad \sum_{j\in\Omega_s} p_j = 1 \quad , \quad p_j > 0 \tag{4.13}$$

$$B_j = A_j - A_s^0 \tag{4.14}$$

Given this definition of A^0, B, one can write the condition (4.9) in the form

$$\frac{\|A - A^0 T\|}{1 - \|A^0 T\|} < 1$$

where the matrix T is determined by (4.10).

Presenting this condition more explicitly, we have

$$\{\max_s \max_{j\in\Omega_s} \sum_j |a_{ij} - \sum_{l\in\Omega_s} a_{il} p_l|\}(1 - \|A^0 T\|)^{-1} < 1 \tag{4.15}$$

which clearly shows the importance of small deviations of any vector within the same group as the basic one, this importance having been postulated when describing the original problem in the perturbed form.

Naturally, the proximity of vectors A_j may be violated by failing to choose the proper units of measurement. Therefore, one can improve the computational scheme by a preliminary normalization.

Let α_j be any norm of A_j, e.g., $\sum_j a_{ij}$, then pretransformation of (4.1) to the form

$$x_i' = \sum_{j=1}^{N} a_{ij}' x_j' + y_j' \tag{4.16}$$

where

$$x_i' = \alpha_i x_i \quad , \quad y_i' = \alpha_i y_i \quad , \quad \alpha_{ij}' = \frac{\alpha_i}{\alpha_j} a_{ij}$$

yields the coincidence of A_j' if A_j differs by a constant multiplier. However, as was outlined above, the success of the perturbation method is generally determined by a half-intuitive choice of reduced problem structure and, in this case, by the choice of corresponding vectors to aggregate them.

C. Structure within the basic and auxiliary industries

To formalize this popular hypothesis (see, e.g., [68]) one can represent the input–output matrix A in the form

$$A = A^0 + \epsilon B \tag{4.17}$$

where ϵ is a small parameter, B is arbitrary, but the reduced matrix A^0 has a specific structure:

$$\begin{array}{c} \overbrace{}^{z} \quad \overbrace{}^{\bar{x}_1} \quad \overbrace{}^{\bar{x}_k} \end{array}$$

$$\left|\begin{array}{ccccccc} & \bar{A}_0 & & & \bar{A}_1 & \cdots & \bar{A}_k \\ a_1 & 0 & 0 & 0 & A_1 & 0 & 0 \\ 0 & a_2 & 0 & 0 & 0 & & 0 \\ 0 & 0 & \vdots & 0 & 0 & & \\ 0 & 0 & 0 & a_k & 0 & 0 & A_k \end{array}\right| = A^0 \tag{4.18}$$

The variables $z \triangleq (z_s \ , s = 1,\ldots,k)$ can be interpreted as total outputs of basic industries, $s = 1,\ldots,k$, each vector $\bar{x}_s$, $s = 1,\ldots,k$, can be considered as the output vector of auxiliary industries connected with the s-th basic one. Hypothesis C means that products of auxiliary industries are used mainly by its basic one. It is also assumed that final goods are manufactured by the basic industries only. In accordance with (4.17) and (4.18), one can write (4.1) in the form

$$\begin{aligned} z &= \bar{A}_0 z + \sum_s \bar{A}_s \bar{x}_s + \epsilon B_0 x + y \\ \bar{x}_s &= a_s z_s + A_s \bar{x}_s + \epsilon B_s x \end{aligned} \tag{4.19}$$

The matrices A_s are assumed to be productive. The simplicity of the reduced system makes it possible to make effective use of the following iterative procedure

$$\begin{aligned} z^{(t+1)} &= \bar{A}_0 z^{(t+1)} + \sum_s \bar{A}_s \bar{x}_s^{(t+1)} + \epsilon B_0 x^{(t)} + y \\ \bar{x}_s^{(t+1)} &= a_s z_s^{(t+1)} + A_s \bar{x}_s^{(t+1)} + \epsilon B_s x_s^{(t)} \end{aligned} \tag{4.20}$$

from which

$$z^{(t+1)} = \bar{A} z^{(t+1)} + \epsilon \bar{B} x^{(t)} + y \tag{4.21}$$

where

$$\begin{aligned} \bar{A} &= \bar{A}_0 + (\bar{A}_s [E - A_s]^{-1} a_s \ , s = 1,\ldots,k) \\ \bar{B} &= B_0 + (\bar{A}_s [E - A_s]^{-1} B_s \ , s = 1,\ldots,k) \end{aligned}$$

The entries of matrix $\bar{A}$ of dimension $k \times k$ give a zero-order approximation of input–output coefficients reduced to the basic industries.

Matrix $E-\bar{A}$ must be nonsingular. A starting point is given by the system

$$\begin{aligned} z^{(0)} &= \bar{A}z^{(0)} + y \\ x_s^{(0)} &= A_s x_s^{(0)} + a_s z^{(0)} \end{aligned} \tag{4.22}$$

One can write the convergence conditions explicitly, but their meaning is clear in the constraints on the usage level of auxiliary products by the other basic industries.

To sum up the examples of perturbation method applications to input–output analysis, it is necessary to stress that a general iterative scheme

$$x^{(t+1)} = A^0 x^{(t+1)} + \epsilon B x^{(t)} + y \tag{4.23}$$

was used everywhere, and that variations were determined by the structural peculiarities of A^0 yielding, in turn, various ways for transforming the reduced system. A similar approach is also applicable to nonlinear models. The iterative procedure

$$x^{(t+1)} = A^0 x^{(t+1)} + \epsilon B(x^{(t)}) x^{(t)} + y \tag{4.24}$$

is effective enough if nonlinear perturbations appear as a result of attempts to improve some traditional linear model, and if matrix A^0 has various specific structures. Further, we shall give general convergence conditions for interative procedures of that type.

Theorem 4.2.

Let x^* be the unique solution to the equation

$$x = A(x)x + y$$

and let the entries of matrix $A(x)$ satisfy Lipschitz conditions with constant L. Let $A^{(t)}$, $t = 0,1,\ldots$, be any matrices such that

$$\|A^{(t)}\| < 1$$

Then the iterative procedure

$$\begin{aligned} x^{(t+1)} &= A^{(t)} x^{(t+1)} + B^{(t)} x^{(t)} + y \ , \ x^{(0)} = 0 \\ B^{(t)} &= A(x^{(t)}) - A^{(t)} \end{aligned} \tag{4.25}$$

converges to the solution if

$$\frac{\|B^{(t)}\| + \alpha L}{1 - \|A^{(t)}\|} \leq q < 1 \ , \ \alpha > 0 \ , \ t = 0,1,\ldots, \tag{4.26}$$

and

$$\|\delta^{(t)}\| \triangleq \|x^{(t)} - x^*\| \leq q^t \|\delta^{(0)}\|$$

Proof. By definition, for any t, we have

$$x^* = A^{(t)}x^* + B^{(t)}x^* + y + [A(x^*) - A(x^{(t)})]x^*$$

Hence,

$$\delta^{(t+1)} = A^{(t)}\delta^{(t+1)} + B^{(t)}\delta^{(t)} + [A(x^{(t)}) - A(x^*)]x^* \tag{4.27}$$

or

$$\delta^{(t+1)} = [E - A^{(t)}]^{-1}\{B^{(t)}\delta^{(t)} + [A(x^{(t)}) - A(x^*)]x^*\}$$

From the Lipschitz condition, one can write

$$\|[A(x^{(t)}) - A(x^*)]x^*\| \leq \sum_i\sum_j |a_{ij}(x^{(t)}) - a_{ij}(x^*)| \cdot |x_j^*| \leq$$
$$\leq L\sum_i\sum_j |x_j^*| \cdot \|\delta^{(t)}\| = L\alpha\|\delta^{(t)}\|$$

Therefore,

$$\|\delta^{(t+1)}\| \leq \|(E - A^{(t)})^{-1}\| \ \{\|B^{(t)}\| + L\alpha\} \ \|\delta^{(t)}\|$$

On the other hand,

$$\|(E - A^{(t)})^{-1}\| \leq \frac{1}{1 - \|A^{(t)}\|}$$

Thus, we have

$$\|\delta^{(t+1)}\| \leq \frac{\|B^{(t)} + L\alpha\|}{1 - \|A^{(t)}\|} \cdot \|\delta^{(t)}\|$$

which is equivalent to the desired result.

It is evident, too, that for

$$A(x) = A = \text{constant}$$

the convergence condition has the form $q < 1$ where

$$q = \frac{\|B^{(t)}\|}{1 - \|A^{(t)}\|} \tag{4.28}$$

All iteration procedures mentioned above correspond to the scheme (4.25) under $A(x) = A^{(0)} = \text{constant}$. It is of interest to note that the two-level iterative aggregation scheme (4.27, 4.28) is a special case of this scheme.

In fact, the relationships determining two-level iterative aggregation can be written in the form

$$\begin{aligned} x^{(t+1)} &= AP^{(t)}z^{(t+1)} + y \\ z^{(t+1)} &= TAP^{(t)}z^{(t+1)} + Ty \end{aligned} \tag{4.29}$$

where $T = (t_{sj})$ is "an aggregating matrix" given by (4.10) and where $P^{(t)} = \{p_{is}^t\}$ is a "weighting matrix" having entries

$$P_{is}^{(t)} = \begin{cases} x_i^t / \sum\limits_{r \in \Omega_s} x_r^t \ , & i \in \Omega_s \\ 0 \ , & i \bar{\in} \Omega_s \end{cases} \tag{4.30}$$

Let us denote $A^{(t)} \triangleq AP^{(t)}T$. It is not hard to see that

$$\|A^t\| \leq \|A\| < 1$$

and so it is possible to invert the matrix E–$A^{(t)}$. Let $\delta^{(t)} \triangleq x^{(t)} - x^*$. Then (4.29) yields

$$\delta^{(t+1)} = [E + AP^{(t)}(E - TAP^{(t)})^{-1}T - (E - A)^{-1}](E - A)x^*$$

or, after an identical transformation,

$$\delta^{(t+1)} = [(E - AP^{(t)}T)^{-1} - (E - A)^{-1}](E - A)x^* \tag{4.31}$$

On the other hand, by the definition of the matrices $P^{(t)}$, the following identity is true

$$P^{(t)}Tx^{(t)} = x^{(t)}$$

but this is equivalent to

$$[(E - A)^{-1} - (E - AP^{(t)}T)^{-1}](E - A)x^{(t)} = 0$$

Summing these identities and (4.31), the following recurrence relationship is found

$$\delta^{(t+1)} = (E - A^{(t)})^{-1}(A - A^{(t)})\delta^{(t)}$$

However, this equation coincides with (4.27) for $A(x) = A$ = constant, which proves the equivalence of the procedures.

Of course, one must stress that the two-level iterative aggregation scheme provides a concrete and possibly very effective way to choose the matrix A by means of the weighting matrix P calculated in the procedure. But if we try to estimate the convergence index *a priori*, then we shall get

$$q = \max_t \frac{\|B^{(t)}\|}{1 - \|A^{(t)}\|} \le \frac{\max_t \|A - A^{(t)}\|}{1 - \max_t \|A^{(t)}\|} \le \frac{\max_t \|A - A^{(t)}\|}{1 - \|A\|} \le$$

$$\le \frac{1}{1 - \|A\|} \max_t \max_s \max_{j\in\Omega_s} \sum_i |a_{ij} - \sum_{l\in\Omega_s} a_{il}p_{ls}^{(t)}| \le$$

$$\le \frac{1}{1 - \|A\|} \max_s \sum_i (\max_{j\in\Omega_s} a_{ij} - \min_{j\in\Omega_s} a_{ij})$$

and this estimate is no better than the estimate (4.15) for the perturbation scheme. Naturally, both estimates are far from precise, and do not imply that the number of iterations for (4.29) cannot be less than those for (4.8). However, it should be noted that procedure (4.29) needs fewer computations for each iteration. Therefore, it may be competitive.

2.5. Aggregation in optimization models based on input–output analysis

Bearing in mind the general goal of this book, one can consider the contents of the previous section as auxiliary to it. The main probelm is to investigate the optimal solution of programs based on some variants of an input–output model. One can select from three schemes.

A. Maximization of effect obtained by usage of final goods under fixed technology

Here, as opposed to classical input–output analysis, the final demand, y_j, for the output j is assumed not to be given *a priori* but must be chosen from the solutions to the following program

$$\max \{f(y) \,/\, x = Ax + y \,,\, y \geq 0 \,,\, r \geq x \geq 0\} \tag{5.1}$$

Components r_j of the vector r are interpreted as limiting levels of outputs j determined by the production capacity of industry j, $j = 1,\ldots,N$. The function $f(y)$ is usually supposed to be concave and, in particular, linear; the $N \times N$ matrix A is assumed to be productive.

The applicability of the perturbation method depends on the structure of the objective function and the constraints. Assuming a similarity of characteristics in some industry group, one can, in particular, obtain the following result.

Theorem 5.1.

Let $A_j = A_s + \epsilon B_j$, $j \in \Omega_s$, $A_0 = (A_s \,,\, s = 1,\ldots,k)$, A_0 be productive. Let

$$f(y) = \sum_s \sum_{j \in \Omega_s} (c_s + \epsilon\beta_j) y_j$$

Moreover, let $z^* = (z_s^*)$, $\bar{y}^* = (\bar{y}_s^*)$ be the unique and positive solution to the program

$$\Phi_0 \triangleq \max \{\sum_s c_s \bar{y}_s \,/\, z = \bar{A}z + \bar{y} \,,\, \bar{r} \geq z \geq 0 \,,\, \bar{y} \geq 0\} \tag{5.2}$$

where $\bar{A} = TA_0$, $\bar{r} = Tr$ and T is the aggregation matrix. Then the auxiliary problem separates out into a number of independent programs

$$\Phi_i \triangleq \max\{\sum_{j\in\Omega_s}\bar{\beta}_j x_j \,/\, \sum_{j\in\Omega_s} x_j = z_s^*\,,\ r_j \geq x_j \geq \Delta_j\,,\ j\in\Omega_s\}$$
$$s = 1,\ldots,k \tag{5.3}$$

where λ^* is the unique dual solution to (5.2),

$$\bar{\beta}_j = \beta_j - \lambda^{*T}B_j\ ,\ \Delta_j = \sum_s A_{js}z_s^*\ ,\ c = (c_j)\ ,\ c_j = c_s\ ,\ j\in\Omega_s$$

The optimal value of the objective (5.1) differs from

$$\Phi_0 + \epsilon(\sum_s \Phi_s - \sum_j \beta_j\Delta_j)$$

to within $O(\epsilon^2)$. If the solutions (x_j^0) to programs (5.3) are unique, then, for sufficiently small ϵ, the solution to (5.1) is given by the system

$$x = [A_0 + \epsilon B]x + y \tag{5.4}$$

where

$$x_j = r_j \quad \text{if} \quad x_j^0 = r_j$$
$$y_j = 0 \quad \text{if} \quad x_j^0 = \Delta_j$$

Proof follows immediately from Theorem 2.4 (Chapter 1) if one takes into consideration the specific features of matrix A_0. In fact, the reduced program [with respect to (5.1)] has the form

$$\max\{\sum_s c_s \sum_j y_j \,/\, x = \sum_s A_s \sum_{j\in\Omega_s} x_j + y\,,\ r\geq x\geq 0\,,\ y\geq 0\} \tag{5.5}$$

or

$$\max\{\sum_s c_s\bar{y}_s \left|\begin{array}{l} z_k = \sum_s a_{ks}z_s + \bar{y}_s\ ,\ \bar{r}\geq z\geq 0\ ,\ \bar{y}\geq 0 \\ x = \sum_s A_s z_s + y\quad ,\ r\geq x\geq 0\ ,\ y\geq 0 \end{array}\right.\}$$

Evidently, its solution is nonunique, and the optimal set can be determined as

$$\Theta_0^* = \{(x \,,\, y) \,/\, x - y = \sum_s A_s z_s^* \,,\, z_s^* = \sum_{j \in \Omega_s} x_j \,,\, \overline{y}_s^* = \sum_{j \in \Omega_s} y_j \,,$$
$$s = 1,\ldots,k \,,\, r \geq x \geq 0 \,,\, y \geq 0\}$$

The auxiliary program is written in the form

$$\max \{\sum_j (\beta_j y_j - \lambda^T B_j x_j) \,/\, (x \,,\, y) \in \Theta_0^*\}$$

which is equivalent to (5.3). The Lagrange multipliers $\lambda = (\lambda_i)$ satisfy the inequalities $\lambda_i \geq c_i$ and $\lambda_i = c_i$ if $y_i > 0$. However, for $\overline{y}^* > 0$ there exists a solution to reduced program (5.5) such tht $y_i > 0$ for all i, e.g.,

$$y_i = \frac{1}{|\Omega_s|} y_s^* \;,\; i \in \Omega_s$$

Since the multipliers are common to all solutions, we have $\lambda_i = c_i$ for all i. It should also be mentioned that the programs (5.3) can be solved explicitly, and it is suitable to solve the system (5.4) by means of the iterative procedure

$$x^{(t+1)} = A_0 x^{(t+1)} + y^{(t+1)} + \epsilon B x^{(t)}$$

by permitting the aggregation.

B. Model with controlled technology

Let us generalize the model (1.1), assuming that input–output coefficients of industry j can be independent of the choice of the control u_j from the feasible set U. Formally, we consider a program

$$F^*(\epsilon) \triangleq \max \{f(y\,,x\,,u) \,/\, x = A(u)x + y\,, r \geq x \geq 0\,, y \geq 0\,, u \in U\} \tag{5.6}$$

and we suppose additionally that controls are weak

$$A(u) = A_0 + \epsilon B(u) \tag{5.7}$$

Moreover, let us assume that the objective function is quasiseparable

$$f(y\,,x\,,u) = f_y(y) + f_x(x) + \epsilon f_1(x\,,y\,,u) \tag{5.8}$$

Applying the perturbation method to (5.6)–(5.8) one can prove the following result.

Theorem 5.2.

Let the functions f_y, f_x be twice differentiable, and let the function $f_1(\cdot)$ be differentiable in x and y. Let x^*, y^* be the unique solution to the reduced problem

$$\Phi_0^* = \max\{f_y(y) + f_x(x) \,/\, A^0x - x + y = 0\,, -x \leq 0\,, y \geq 0\,, x - r \leq 0\} \tag{5.9}$$

Let y^* be positive. Let $(\lambda^{(1)}, \lambda^{(2)}, \lambda^{(3)})$ be the unique Lagrange multipliers vectors associated with the three groups of constraints. If the complementarity slackness conditions are strictly fulfilled, then the Lagrange multipliers are determined by the following equations

$$\begin{aligned} &\lambda^{(1)} = \nabla f_y(y^*) \\ &-\sum_{j\in J_2} \lambda_i^{(2)} e_i + \sum_{i\in J_3} \lambda_i^{(3)} e_i = -[A_0 - E]^T \lambda^{(1)} + \nabla f_x(x^*) \end{aligned} \tag{5.10}$$

where

$$J_s \triangleq \{i \,/\, x_i^* = 0\}\ ,\ J_3 \triangleq \{i \,/\, x_i^* = r_i\}$$

If u^* is the unique solution to the program

$$\Phi_1^* \triangleq \max\{f_1(x^*\,,y^*\,,u) - (\nabla f_y(y^*))^T B(u) x^* \,/\, u \in U\} \tag{5.11}$$

then

$$\begin{aligned} F^*(\epsilon) &= \Phi_0^* + \epsilon \Phi_1^* + O(\epsilon^2) \\ x^*(\epsilon) &= x^* + \epsilon x_1^* + o(\epsilon) \\ y^*(\epsilon) &= y^* + \epsilon y_1^* + o(\epsilon) \end{aligned} \tag{5.12}$$

where the correction terms x_1^*, y_1^* can be found from the linear system

$$\begin{aligned} &\nabla\nabla f_y(y^*)y_1 - \varsigma^{(1)} = -\nabla_y f_1(x^* , y^* , u^*) \\ &\nabla\nabla f_x(x^*)x_1 - [A_0 - E]^T\varsigma^{(1)} + \sum_{i\in J_2} e_i\varsigma_i^{(2)} - \sum_{i\in J_3} e_i\varsigma_i^{(3)} = \\ &= B^T(u)\nabla f_y(y^*) - \nabla_x f_1(x^* , y^* , y^*) \end{aligned} \tag{5.13}$$

To prove this theorem one needs to make concrete the results of Section 2.3 (Theorem 3.1). It is not hard to write in detail the procedure of solution to (5.9) if the matrix A_0 and functions f_x, f_y have a special structure.

Let us consider a special case of the model B. Let the Danzig–Wolfe hypothesis be true, i.e., let U_j be polyhedrons, and

$$B(u) = Bu \quad , \quad B = \text{constant}$$

Then one can transform the relationships

$$x = (A_0 + \epsilon Bu)x + y \quad , \quad r \geq x \geq 0 \quad , \quad y \geq 0 \quad , \quad u_j \in U_j \tag{5.14}$$

into linear ones.

In fact, let $\{u_{jn}^* , n \in N_j\}$ be the collection of U_j extreme points. Then

$$\begin{aligned} &u_j \in U_j \quad \Leftrightarrow \quad u_j = \sum_{n\in N_j} p_{jn}u_{jn}^* \\ &\sum_{n\in N_j} p_{jn} = 1 \quad , \quad p_{jn} \geq 0 \end{aligned} \tag{5.15}$$

Substituting (5.15) into (5.14) and denoting

$$\xi_n = p_{jn}x_j \quad , \quad B_n = B_j u_{jn}^* \quad , \quad n \in N_j$$

yields

$$\begin{aligned} &\sum_{n\in N_j} \xi_n = \sum_j \sum_{n\in N_j} (a_{ij}^0 + \epsilon b_{in})\xi_n + y_i \\ &r_j \geq x_j = \sum_{n\in N_j} \xi_n \quad , \quad \xi_n \geq 0 \end{aligned} \tag{5.16}$$

At the same time, one can obtain an identical result assuming *a priori*, as was shown in [68], that there exists only a finite number, N_j, of technologies producing a product j. Then ξ_n can be interpreted as an output, produced by the technology n, and x_j is the total output. This type is apparently the most propagated variant of the optimizing input–output analysis, and one can see that Theorem 5.2 is applicable here, too.

Let us consider, in particular, the linear case

$$\min\{\sum_j \sum_{n\in N_j} (c_j+\epsilon\beta_n)\xi_n \,/\, \sum_{n\in N_i} \xi_n = \sum_j \sum_{n\in N_j} (a^0_{ij}+\epsilon b_{in})\xi_n + y_i\,,\, \xi_n \geq 0\} \quad (5.17)$$

The objective function can be interpreted as the total expenditure necessary to meet fixed final demand $y > 0$.

We shall present a procedure and its results in the form of a theorem that is a direct consequence of Section 1.2 (Theorem 2.2).

Theorem 5.3.

Let A^0 be productive, and let (x_i^*) be the solution to the aggregated system

$$x_i = \sum_j a^0_{ij} x_j + y_i \quad (5.18)$$

Let λ^* be the solution to the conjugate system

$$\lambda = A_0^T \lambda + c \quad (5.19)$$

and let n_j be unique for all j,

$$n_j \triangleq \arg \min_{n\in N_j} (\beta_n - B_n^T \lambda^*)$$

Then, for small ϵ, the solution to (5.17) coincides with the solution of the balance equation

$$\xi_{n_i} = \sum_j (a^0_{ij} + \epsilon b_{in_j})\xi_{n_j} + y_i \quad (5.20)$$

where one needs to put

$$\xi_n = 0 \;,\; n \neq n_j \;,\; n \in N_j$$

Note that sometimes one takes into account the capacity limitations for all kinds of technologies. A corresponding modification of Theorem 5.3 can be obtained as an exercise.

C. Dynamical Leontieff model

Some dynamic generalization of the model A can now be considered.

Let $u(t)$ be a new capacity constructed in period t; let $d(t)$ be the fixed part of final demand in the period t; and let $k(t)$ be a matrix of capital coefficients, $k_{ij}(t)$, denoting the quantity of i's output required to build one unit of capacity for the industry j. Then problem (5.1) can be generalized to describe the optimal development of the economic system

$$
\begin{aligned}
&\sum_{t=1}^{T} f_t(y_t) - \max \\
&x(t) = A(t)x(t) + K(t)u(t) + y(t) + d(t) \ , \ t = 1,\dots,T \\
&r(t) \geq x(t) \geq 0 \ , \ y(t) \geq 0 \ , \ u(t) \geq 0 \\
&r(t+1) = r(t) + u(t) \ , \ t = 1,\dots,T-1
\end{aligned}
\tag{5.21}
$$

A suboptimal aggregation scheme for (5.21) can be provided by the "similarity" hypothesis.

Theorem 5.4.

Let

$$A_j(t) = A_s^0(t) + \epsilon B_j(t) \ , \ j \in \Omega_s$$

$$A^0(t) = (A_s^0(t) \ , \ s = 1,\dots,k)$$

$$K_j(t) = K_s^0(t) + \epsilon K_j^1(t) \ , \ j \in \Omega_s$$

$$K^0(t) = (K_s^0(t) \ , \ s = 1,\dots,k)$$

$$f_t(y) = \sum_s \sum_{j \in \Omega_s} [c_s(t) + \epsilon \beta_j(t)] y_j$$

Let

$$z^*(t) = (z_s^*(t)) \ , \ \bar{y}^*(t) = (\bar{y}_s^*(t)) > 0 \ , \ \bar{u}^*(t) = (\bar{u}_s^*(t))$$

be the unique solution to the aggregated problem

$$\Phi_0 \triangleq \max \sum_n c_s(t)\bar{y}_s(t)$$
$$z(t) = \bar{A}(t)z(t) + \bar{K}(t)\bar{u}(t) + \bar{y}(t)\bar{d}(t)$$
$$\bar{r}(t) \geq z(t) \geq 0 \ , \ \bar{y}(t) \geq 0 \ , \ \bar{u}(t) \geq 0 \ , \ t = 1,\ldots,T$$
$$\bar{r}(t{+}1) = \bar{r}(t) + \bar{u}(t) \ , \ t = 1,\ldots,T{-}1 \tag{5.22}$$
$$\bar{A}(t) = \bar{T}A^0(t) \ , \ \bar{K}(t) = \bar{T}K^0(t) \ , \ \bar{d}(t) = \bar{T}d(t)$$
$$\bar{T} = \{t_{sj}\} \ ; \ t_{sj} = \begin{cases} 1 \ , \ j \in \Omega_s \\ 0 \ , \ j \bar{\in} \Omega_s \end{cases}$$

Let $u^*(t) = (u_j^*(t))$ where $u_j^*(t)$ are the unique solutions to the auxiliary problems

$$\Phi_s \triangleq \max \{\sum_t \sum_{j\in\Omega_s} [\bar{\beta}_j(t)x_j(t) - \gamma_j(t)u_j(t)]\} \tag{5.23}$$

subject to

$$\sum_{j\in\Omega_s} x_j(t) = z_s^*(t) \ , \ \sum_{j\in\Omega_s} u_j(t) = \bar{u}_s^*(t) \ , \ u_j(t) \geq 0$$
$$r_j(t) \geq x_j(t) \geq \Delta_j(t) \ , \ j \in \Omega_s \ , \ t = 1,\ldots,T$$
$$r_j(t{+}1) = r_j(t) + u_j(t) \ , \ j \in \Omega_s \ , \ t = 1,\ldots,T{-}1$$
$$\bar{\beta}_j(t) = \beta_j(t) - c^T(t)B_j(t) \ , \ \gamma_j = c^t(t)K_j^1(t)$$
$$\Delta(t) = (\Delta_j(t)) = d(t) + \sum_s A_s^0(t)z_s^* + \sum_s K_s^0(t)\bar{u}_s^*(t)$$

Then, for sufficiently small ϵ, the optimal value of the objective in (5.21) differs from

$$\Phi_0 + \epsilon(\sum_s \Phi_s - \sum_s \beta_j\Delta_j)$$

to within $O(\epsilon^2)$. If the solutions $x_j^0(t)$, $u_j^0(t)$, $r_j^0(t)$ to the problems (5.23) are unique, then the solution to (5.21) can be found from the system

$$\begin{aligned} x(t) &= A(t)x(t) + K(t)u(t) + y(t) + d(t) \\ r(t{+}1) &= r(t) + u(t) \end{aligned} \tag{5.24}$$

where it is necessary to put

$$x_j(t) = r_j(t) \quad \text{if} \quad x_j^0(t) = r_j^0(t)$$
$$y_j(t) = 0 \quad \text{if} \quad x_j^0(t) = \Delta_j(t)$$
$$u_j(t) = 0 \quad \text{if} \quad x_j^0(t) = 0$$

The proof is just as in Theorem 5.1. Note that the basic results relating to the proximity of the objective values are valid even if perturbations are nonlinear, although the particular solutins of the auxiliary problems are complicated.

If one uses another hypothesis about the structure of reduced matrices, e.g., the hypothesis of main industries' separation, one can obtain simplified suboptimal procedures as well (see, in particular, [68], where such a possibility was shown as an effective heuristic tool).

2.6. Aggregation in the interregional transportation problem with regard to price scales

Let us consider an interregional transportation system, taking into account the fact that shipping prices within each region are, as a rule, higher than those of interregional shipping. We shall express this peculiarity in the classic single-commodity transportation problem by introducing a small parameter.

Let Ω_{wi}, Ω_{di} be the sets of warehouses and destinations belonging to the region i, $i = 1,\ldots,k$. In accordance with the main idea, let us suppose that the shipping prices are given as follows

$$c_{lr}(\epsilon) = \epsilon c_{lr}^1 \ , \ c_{lr}^1 \geq 0 \ , \ l \in \Omega_{wi} \ , \ r \in \Omega_{di} \ , \ i = 1,2,\ldots,k$$
$$c_{lr}(\epsilon) = c_{ij} + \epsilon c_{lr}^1 \ , \ c_{ij} > 0 \ , \ l \in \Omega_{wi} \ , \ r \in \Omega_{dj} \ , \ i \neq j \ , \ i,j = 1,\ldots,k$$

We suppose, moreover, that the stock at the warhouse $l \in \Omega_{wi}$ is

$$a_l(\epsilon) = a_l + \epsilon a_l^1 \ , \ l = 1,\ldots,m_i \ , \ |\Omega_{wi}| \triangleq m_i$$

and that the total demand of the destination $r \in \Omega_{di}$ is

$$b_r(\epsilon) = b_r + \epsilon b_r^1 \ , \ r = 1,\ldots,n_i \ , \ |\Omega_{di}| \triangleq n_i$$

a_l, b_r, a_l^1, b_r^1 being positive.

Let the global equilibrium equation be fulfilled

$$\sum_l a_l(\epsilon) = \sum_r b_r(\epsilon) \tag{6.1}$$

and, moreover, for all regions, $i = 1,\ldots,k$,

$$\sum_{l\in\Omega_{wi}} a_l = \sum_{r\in\Omega_{di}} b_r \tag{6.2}$$

i.e., supply and demand equilibrium exists in each region if $\epsilon = 0$.

Let us write the general optimizationproblem as

$$\min \{\epsilon \sum_{i=1}^{k} \sum_{(l,r)\in\Omega_i} c^1_{lr} x_{lr} + \sum_{\substack{i,j=1 \\ i\neq j}}^{k} \sum_{\substack{l\in\Omega_{iw} \\ r\in\Omega_{jd}}} (c_{ij} + \epsilon c^1_{lr}) x_{lr} \,/\, x \triangleq \triangleq \{x_{lr}\} \,,\, x \in \Theta(\epsilon)\} \tag{6.3}$$

where $\Theta(\epsilon)$ is given by the following conditions

$$\sum_{r=1}^{n} x_{lr} = a_l + \epsilon a^1_l \;,\; l = 1,2,\ldots,m \tag{6.4}$$

$$\sum_{l=1}^{m} x_{lr} = b_r + \epsilon b^1_r \;,\; r = 1,2,\ldots,n \tag{6.5}$$

$$x_{lr} \geq 0 \;,\; l = 1,\ldots,m \;,\; r = 1,2,\ldots,n \tag{6.6}$$

The feasible set $\Theta(\epsilon)$ is nonempty and bounded by virtue of (6.1) for $\epsilon \in [0 \,,\, \epsilon_0]$, $\epsilon_0 > 0$. Any basic solution, i.e., a vertex of the polyhedron $\Theta(\epsilon)$, is the solution to the nonsingular system

$$\begin{aligned} &\sum_{(l,r)\in J} x_{lr} = a_l + \epsilon a^1_l \;,\; l = 1,\ldots,m \\ &\sum_{(l,r)\in J} x_{lr} = b_r + \epsilon b^1_r \;,\; r = 1,\ldots,n \end{aligned} \tag{6.7}$$

where J is the basic set of indices, $|J| = m + n - 1$. One of the equations (6.7) is a linear combination of others and may be omitted. The iverse matrix corresponding to the remaining system exists. Hence, its solution can be written in the form

$$\begin{aligned} &x_{lr}(\epsilon) = x^0_{lr} + \epsilon x^1_{lr} \;,\; l = 1,2,\ldots,m \;;\; r = 1,2,\ldots,n \\ &x^0_{lr} = x^1_{lr} = 0 \;,\; (l,r) \,\overline{\in}\, J \end{aligned} \tag{6.8}$$

where x^0_{lr}, x^1_{lr} are the solutions to the systems

$$\sum_{\substack{r \\ (l,r)\in J}} x^0_{lr} = a_l \ , \ l = 1,\ldots,m \ ; \quad \sum_{\substack{l \\ (l,r)\in J}} x^0_{lr} = b_r \ , \ r = 1,\ldots,n$$
$$\sum_{\substack{r \\ (l,r)\in J}} x^1_{lr} = a^1_l \ , \ l = 1,\ldots,m \ ; \quad \sum_{\substack{l \\ (l,r)\in J}} x^1_{lr} = b^1_r \ , \ r = 1,\ldots,n \tag{6.9}$$

Let us show that the optimal interregional shipping must be small.

Lemma 6.1. The necessary conditions for optimal solutions to any system of (6.7) type are that

$$x^0_{lr} = 0 \ , \ l \in \Omega_{wi} \ , \ r \in \Omega_{dj} \ , \ i \neq j \ , \ i,j = 1,\ldots,k \tag{6.10}$$

Proof. Let us suppose the contrary. Then, by the positivity of the c_{ij}, the optimal objective value F^*_ϵ satisfies the inequality

$$F^*_\epsilon > \alpha > 0 \tag{6.11}$$

where α is a constant independent of ϵ.

Further, let

$$x_{lr}(\epsilon) = \frac{a_l b_r}{d_i} + \epsilon \frac{a^1_l b^1_r}{D} \ , \ (l,r) \in \Omega_i \ , \ i = 1,\ldots,k$$
$$x_{lr}(\epsilon) = \epsilon \frac{a^1_l b^1_r}{D} \ , \ l \in \Omega_{wi} \ , \ r \in \Omega_{dj} \tag{6.12}$$

where

$$d_i \triangleq \sum_{l\in\Omega_i} a_l = \sum_{n\in\Omega_i} b_r \ , \ i = 1,2,\ldots,k \ , \ D \triangleq \sum_l a^1_l = \sum_r b^1_r$$

Solution (6.12) is feasible with respect to (6.4)–(6.6), and the objective value attainable with (6.12) is zero to within $O(\epsilon)$, which contradicts (6.11). QED.

The reduced program corresponding to (6.3)–(6.6) has the form

$$\max \{ \sum_{\substack{i,j=1 \\ i\neq j}} c_{ij} \sum_{\substack{l\in\Omega_{wi} \\ r\in\Omega_{dj}}} x_{lr} \}$$

$$\sum_{r=1}^{n} x_{lr} = a_l \ , \ l = 1,\ldots,m$$

$$\sum_{l=1}^{m} x_{lr} = b_r \ , \ r = 1,\ldots,n$$

Its solution set Θ^* is determined by the constraints

$$x_{lr} = 0 \ , \ l \in \Omega_{wi} \ , \ r \in \Omega_{dj} \ , \ i \neq j \ ; \ x_{lr} \geq 0 \ , \ \forall l,r$$
$$\sum_{r \in \Omega_{di}} x_{lr} = a_l \ , \ l \in \Omega_{wi} \ ; \ \sum_{l \in \Omega_{wi}} x_{lr} = b_r \ , \ r \in \Omega_{di}$$

By virtue of the linear dependence mentioned above, we do not have the unique dual solution here. At the same time, Lemma 6.1 permits us to obtain all results of interest.

Let us consider the following multiphase procedure.

(1) Solve the transportation problems for each region independently:

$$F_i \triangleq \min \sum_{(l,r) \in \Omega_i} c^1_{lr} x_{lr} \ , \ x_{lr} \geq 0 \ , \ (l,r) \in \Omega_i$$
$$\sum_{r \in \Omega_{di}} x_{lr} = a_l \ , \ l \in \Omega_{wi} \tag{6.13}$$
$$\sum_{l \in \Omega_{wi}} x_{lr} = b_r \ , \ r \in \Omega_{di}$$

The solution obtained determines the main part of the intraregional shipping. Let the optimal basis index set to the program i be $\sigma_i \in \Omega_i \triangleq \Omega_{wi} \times \Omega_{di}$ so that

$$x^0_{lr} > 0 \ , \ (l,r) \in \sigma_i \ , \ x^0_{lr} = 0 \ , \ (l,r) \bar{\in} \sigma_i \tag{6.14}$$

Non-degeneracy is assumed.

(2) Find the solution to the aggregated program

$$\tilde{F} \triangleq \min \{ \sum_{i,j=1}^{k} c_{ij} \tilde{x}_{ij} \} \ ; \ \tilde{x}_{ij} \geq 0 \ , \ i \neq j \ , \ i,j = 1,\ldots,k \tag{6.15}$$

$$\sum_{j=1}^{k} \tilde{x}_{ij} = \tilde{a}^1_i \ , \ i = 1,2,\ldots,k$$
$$\sum_{i=1}^{k} \tilde{x}_{ij} = \tilde{b}^1_j \ , \ j = 1,2,\ldots,k \tag{6.16}$$

where

$$c_{ii} = 0 \ , \ \tilde{a}^1_i = \sum_{l \in \Omega_{wi}} a^1_l \ , \ \tilde{b}^1_j = \sum_{r \in \Omega_{dj}} b^1_r \ , \ i,j = 1,2,\ldots,k \tag{6.17}$$

This solution gives the aggregated values of the interregional shipping

$$\tilde{x}'_{ij} \triangleq \sum_{\substack{l\in\Omega_{wi} \\ r\in\Omega_{dj}}} x'_{lr}$$

Let the set of basic variable indices be $\tilde{\sigma}$, i.e.,

$$i \neq j \ , \ \tilde{x}'_{ij} > 0 \ \Leftrightarrow \ (i,j) \in \tilde{\sigma} \tag{6.18}$$

(3) For each pair $(i,j) \in \tilde{\sigma}$ solve the elementary problem

$$\min \{\sum c^1_{lr} x_{lr} \,/ \sum_{\substack{l\in\Omega_{wi} \\ r\in\Omega_{dj}}} x_{lr} = \tilde{x}^1_{ij} \ , \ x_{lr} \geq 0\} \tag{6.19}$$

to determine the warehouse index in the region i and the destination index in the region j that realize the interregional connection for these regions. We can write the solutions in the explicit form

$$\begin{aligned} x^1_{lr} &= 0 \ , \ l \neq l^*_i \ , \ r \neq r^*_j \ , \ l \in \Omega_i \ , \ r \in \Omega_j \\ x^1_{lr} &= \tilde{x}'_{ij} \ , \ l = l^*_i \ , \ r = r^*_j \end{aligned} \tag{6.20}$$

where

$$(l^*_i \ , \ r^*_j) = \arg \min_{\substack{l\in\Omega_{wi} \\ r\in\Omega_{di}}} \{c^1_{lr}\} \ , \ i \neq j \tag{6.20$'$}$$

(4) Recompute the intraregional shipping balance for each region taking into account the export–import amounts found above. To that end, solve the systems

$$\begin{aligned} &\sum_{\substack{r \\ (l,r)\in\sigma_i}} x^1_{lr} = a^1_l - \sum_{r\bar{\in}\Omega_{di}} x^1_{lr} \ , \ l \in \Omega_{wi} \\ &\sum_{\substack{l \\ (l,r)\in\sigma_i}} x^1_{lr} = b^1_r - \sum_{l\bar{\in}\Omega_{wi}} x^1_{lr} \ , \ r \in \Omega_{di} \\ &x^1_{lr} = 0 \ , \ (l,r) \in \Omega_i \backslash \sigma_i \end{aligned} \tag{6.21}$$

By virtue of the assumed nondegeneracy, each system (6.21) can be replaced by the nonsingular one having $m_i + n_i - 1$ unknown variables.

(5) Now put

$$x_{lr}^1 = 0 \ , \ l \in \Omega_{wi} \ , \ r \in \Omega_{dj} \ , \ i \neq j \ , \ (i,j) \bar{\in} \tilde{\sigma}$$

and construct the final solution in the form

$$\begin{aligned} x_{lr}(\epsilon) &= x_{lr}^0 + \epsilon x_{lr}^1 \ , \ (l,r) \in \Omega_i \ , \ i = 1,2,\dots,k \\ x_{lr}(\epsilon) &= \epsilon x_{lr}^1 \ , \ l \in \Omega_{wi} \ , \ r \in \Omega_{wj} \ , \ i \neq j \ , \ i,j = 1,\dots,k \end{aligned} \tag{6.22}$$

The efficacy of the procedure is shown by the following result.

Theorem 6.1.

Let the solutions to (6.13) be nondegenerate, i.e., let

$$|\sigma_i| = n_i + m_i - 1 \ , \ i = 1,2,\dots,k \tag{6.23}$$

Then the solution $x(\epsilon)$ given by (6.22) is feasible with respect to (6.3)–(6.6), and the objective value achieved by it differs from the optimal one to within $O(\epsilon^2)$. The optimal solution, $x^*(\epsilon)$, may be written in the form

$$\begin{aligned} x_{lr}^*(\epsilon) &= x_{lr}^{0*} + \epsilon x_{lr}^{1*} \ , \ (l,r) \in \Omega_i \ , \ i = 1,\dots,k \\ x_{lr}^*(\epsilon) &= \epsilon x_{lr}^{1*} \ , \ l \in \Omega_{wi} \ , \ r \in \Omega_{dj} \ , \ i \neq j \ , \ i,j = 1,\dots,k \end{aligned} \tag{6.24}$$

where $\{x_{lr}^{0*}\}$, $(l,r) \in \Omega_i$ is the solution to the i-th program (6.13) and the aggregated variables

$$\tilde{x}_{ij}^{1*} \triangleq \sum_{\substack{l \in \Omega_{wi} \\ r \in \Omega_{di}}} x_{lr}^{1*} \ , \ i,j = 1,\dots,k$$

form the solution to (6.15) and (6.16).

Let us suppose additionally that the solution to program (6.13) are unique and, moreover, that the solution to (6.15) and (6.16) is unique and nondegenerate. Then, for $\epsilon \in [0 \ , \epsilon_0]$, (6.22) gives the optimal solution.

Proof. By (6.14) and (6.22), $x_{lr}(\epsilon) \geq 0$, $(l,r) \in \Omega_{wi} \times \Omega_{di}$ for small ϵ. Nonnegativeness of $x_{lr}(\epsilon)$ for $l \in \Omega_{wi}$, $r \in \Omega_{dj}$, $i \neq j$, follows from the nonnegativeness of x^1_{lr}. Taking into account (6.20) and (6.21) we have that, for $l \in \Omega_{wi}$, $i = 1,\ldots,k$

$$\sum_{r=1}^{n} x_{lr}(\epsilon) = \sum_{j=1}^{k} \sum_{r\in\Omega_{di}} x_{lr}(\epsilon) = \sum_{r\in\Omega_{di}} x^0_{lr} + \epsilon \sum_{j=1}^{k} \sum_{r\in\Omega_{di}} x^1_{lr} =$$
$$= a_l + \epsilon \sum_{r\in\Omega_{di}} x^1_{lr} + \epsilon \sum_{r\in\bar{\Omega}_{di}} x^1_{lr} = a_l + \epsilon a^1_l$$

and, for $r \in \Omega_{dj}$, $j = 1,\ldots,k$, analogously

$$\sum_{l=1}^{n} x_{lr}(\epsilon) = \sum_{i=1}^{k} \sum_{l\in\Omega_{wi}} x_{lr}(\epsilon) = \sum_{l\in\Omega_{wj}} x^0_{lr} + \epsilon \sum_{l\in\bar{\Omega}_{wj}} x^1_{lr} = b_r + \epsilon b^1_r$$

Thus (6.22) gives the feasible solution. By the direct substitution, obtain

$$F(x(\epsilon)) \triangleq \sum_{l,r} c_{lr}(\epsilon) x_{lr}(\epsilon) = \epsilon(\sum_{i=1}^{k} F_i + \tilde{F}) + O(\epsilon^2) \tag{6.25}$$

On the other hand, Lemma 6.1 implies the structure (6.24) of the optimal solution

$$x^*(\epsilon) = x^{0*} + \epsilon x^{1*}$$

Hence, the optimal objective value is

$$F^*_\epsilon = \epsilon(\sum_{i=1}^{k} \sum_{(l,r)\in\Omega_i} c^1_{lr} x^{0*}_{lr} + \sum_{\substack{i,j=1\\ i\neq j}}^{k} c_{ij} \sum_{\substack{l\in\Omega_{wi}\\ r\in\Omega_{dj}}} x^{1*}_{lr}) + \epsilon^2 \sum_{i,j=1}^{k} \sum_{\substack{l\in\Omega_{wi}\\ r\in\Omega_{dj}}} c^1_{lr} x^{1*}_{lr}$$

At the same time, the components $\{x^{0*}_{lr}\}$, $(l,r) \in \Omega_{wi} \times \Omega_{di}$ are feasible for the i-th program (6.13), and the aggregates

$$\tilde{x}^{1*}_{ij} \triangleq \sum_{l\in\Omega_{wi}} \sum_{r\in\Omega_{dj}} x^{1*}_{lr} \ , \ i,j = 1,2,\ldots,k$$

are feasible for (6.15) and (6.16). Hence

$$\sum_{(l,r)\in\Omega_i} c^1_{lr} x^{0*}_{lr} \geq F_i$$

$$\sum_{i,j=1}^{k} c_{ij} \sum_{\substack{l\in\Omega_{wi}\\ r\in\Omega_{dj}}} x^{1*}_{lr} \geq \tilde{F}$$

By definition,

$$F(x(\epsilon)) \geq F_\epsilon^*$$

Thus, $\{x_{lr}^{0*}\}$, $(l,r) \in \Omega_i$ is the optimal solution to the i-th program (6.13) and

$$\tilde{x}_{ij}^{1*} \triangleq \sum_{l \in \Omega_{wi}} \sum_{r \in \Omega_{dj}} x_{lr}^{1*} \ , \ i,j = 1,\ldots,k$$

is the optimal one to (6.15) and (6.16).

The first result of the theorem is proved. Further, under the condition that the solutions to (6.13), (6.15), and (6.16) are unique, we have

$$\begin{aligned} & x_{lr}^{0*} = x_{lr}^0 \ , \ (l,r) \in \Omega_i \ , \ i = 1,2,\ldots,k \\ & \sum_{\substack{l \in \Omega_{wi} \\ r \in \Omega_{dj}}} x_{lr}^{1*} = \tilde{x}_{ij}^1 \ , \ i,j = 1,\ldots,k \end{aligned} \tag{6.26}$$

where x_{lr}^0, x_{ij}^1 are determined by the investigated procedure. Let σ be the set of positive variables' indices in (6.22).

By definition,

$$|\sigma| = \sum_{i=1}^{k} |\sigma_i| + |\tilde{\sigma}|$$

The nondegeneracy of (6.15) and (6.16) means that $|\tilde{\sigma}| = k - 1$. This condition and (6.23) imply that

$$|\sigma| = \sum_{i=1}^{k} (m_i + n_i - 1) + (k - 1) = m + n - 1$$

i.e., $x(\epsilon)$ is a basic solution to (6.3)–(6.6). Let σ^* be the optimal basic index set. By (6.26),

$$\bigcup_{i=1}^{k} \sigma_i \bigcup \tilde{\tilde{\sigma}} \subset \sigma^*$$

where $\tilde{\tilde{\sigma}}$ is a subset of the pairs (l,r) such that $l \in \Omega_{wi}$, $r \in \Omega_{dj}$, $(i.j) \in \tilde{\sigma}$, with at least one element $(l,r) \in \tilde{\tilde{\sigma}}$, $l \in \Omega_{wi}$, $r \in \Omega_{dj}$ corresponding to each pair (i,j).

By definition,

$$|\tilde{\tilde{\sigma}}| \geq |\tilde{\sigma}| = k - 1 \Rightarrow \sum_{i=1}^{k} |\sigma_i| + |\tilde{\tilde{\sigma}}| \geq m + n - 1$$

However, σ^* has no more than $m + n - 1$ elements as the set of basic positive variables. Therefore,

$$(\bigcup_{i=1}^{k} \sigma_i) \bigcup \tilde{\tilde{\sigma}} = \sigma^* \quad , \quad |\tilde{\tilde{\sigma}}| = k - 1$$

The latter means that one and only one pair $(l,r) \in \tilde{\tilde{\sigma}}$ corresponds to the pair $(i,j) \in \tilde{\sigma}$ in the optimal basis. Moreover, this one does not include the pairs (l,r), $l \in \Omega_{wi}$, $r \in \Omega_{dj}$, $(i,j) \bar{\in} \tilde{\sigma}$ and the pairs $(l,r) \in \Omega_i/\sigma_i$ either. This implies that $\{x_{lr}^{1*}\}$, $(l,r) \in \Omega_i$ are the solutions to the systems (6.21), and a separation of $\tilde{\tilde{\sigma}}$ from $\tilde{\sigma}$ is realized in correspondence with (6.20) and (6.20′).

Thus, the procedure under consideration permits us to obtain the optimal basic index set. Since $x(\epsilon)$ is the basis solution corresponding to this set, then $x(\epsilon)$ is the optimal one. QED.

Formally, the problem described above is a successful example of perturbation method application in the case of dual nonuniqueness (and even nonboundedness). It is possible that the multiphase procedure obtained here may have a practical interpretation.

2.7. Optimization of discrete dynamic systems

Problems of the optimal control of discrete dynamic systems form an important class of mathematical programs. Their specific formal feature is a separation of variables into two groups. The first group includes the *state variables* $x^{(t)}$, $t = 0,1,\ldots,N$, which are connected by the recurrence relation

$$x^{(t+1)} = f(x^{(t)}, u^{(t)}, t) \quad , \quad t = 0,1,\ldots,N-1 \tag{7.1}$$

$x^{(0)}$ being a given value.

The second group consists of the *control variables* $u^{(t)}$, $t = 0,1,\ldots,N-1$, which are not connected directly apart from (7.1) but, for each t, are bounded by the condition

$$u^{(t)} \in U_t \quad , \quad t = 0,1,\ldots,N-1 \tag{7.2}$$

where U_t are compact sets.

As an objective one can take

$$\max G(x^N) \triangleq G^*$$

Equation (7.1) is treated as the discrete dynamic system description. The sequence $\{x^{(t)}\}$ is called the trajectory, which can be varied by means of the control laws $u^{(t)}$, applied at the successive instants t. The objective is a function of the "right end" of the trajectory. The sequences $x^{(t)}$, $u^{(t)}$ are usually called functions of the discrete argument t.

Let us consider the form of the perturbation method for such programs.

Let (7.1)–(7.3) be the reduced program with respect to the program

$$\begin{aligned} &x^{(t+1)} = f(x^{(t)}, u^{(t)}, x) + \epsilon\gamma(x^{(t)}, u^{(t)}, t) \ , \ t = 0,1,\dots,N-1 \\ &u^{(t)} \in U_t \ , \ t = 0,1,\dots,N-1 \\ &\max G(x^N) \triangleq G_\epsilon^* \end{aligned} \tag{7.4}$$

Let the solution to (7.1)–(7.3) be nonunique, i.e., there is a family of control functions $\{u^\beta\}$ such that any function of the discrete argument $u_\beta^{(t)}$ belonging to this family is an optimal control and, conversely, any optimal control is in $\{u^\beta\}$. Let $x_\beta^{(t)}$, $t = 1,\dots,N$, be the solution to (7.1) if $u^{(t)} = u_\beta^{(t)}$. Then the solutions set to the program (7.1)–(7.3) can be written in the form

$$\begin{aligned} \Theta_0^* = \ &\{x_\beta^{(t)} (t = 1,2,\dots,N) \ , \ u_\beta^{(t)} (t = 0,1,\dots,N-1) \, / \\ &u_\beta^{(t)} (t = 0,1,\dots,N-1) \in \{u^\beta\} \ , \ x_\beta^{(t+1)} = f(x_\beta^{(t)}, u_\beta^{(t)}, t) \ , \\ &t = 0,1,\dots,N-1 \ , \ x^0 \text{ is given}\} \end{aligned} \tag{7.5}$$

The conditions [(4.2), Chapter 1] that determine the Lagrange multipliers corresponding to each $\{x_\beta^{(t)}, u_\beta^{(t)}\}$ have the form of a so-called *conjugate* [with respect to (7.1)] *system*

$$\lambda_\beta^{(t)} = f_x^T(x_\beta^{(t)}, u_\beta^{(t)}, t)\lambda_\beta^{(t+1)} \ , \ t = N-1,N-2,\dots,1 \ , \ \lambda_\beta^{(N)} = \nabla G(x_\beta^{(N)}) \tag{7.6}$$

The auxiliary problem here takes the form [compare with (4.56), Chapter 1]

$$\max_\beta \sum_{\tau=0}^{N-1} (\lambda_\beta^{(\tau+1)})^T \gamma(x_\beta^{(\tau)}, u_\beta^{(\tau)}, \tau) \triangleq \Phi^* \tag{7.7}$$

The solution $\{u_\epsilon^{*(t)}, x_\epsilon^{*(t)}\}$ to the perturbed program is asymptotically close to solution (7.7)

$$\begin{aligned} &\lim_{\epsilon \to 0} u_\epsilon^{*(t)} = u^{*(t)} \ , \ t = 0,1,\dots,N-1 \\ &\lim_{\epsilon \to 0} x_\epsilon^{*(t)} = x^{*(t)} \ , \ t = 0,1,\dots,N-1 \end{aligned} \tag{7.8}$$

if the latter is unique, and the optimal objective value G_ϵ^* may be written in the form†

$$G_\epsilon^* = G^* + \epsilon\Phi^* + o(\epsilon) \tag{7.9}$$

† These results are analogous to those obtained in the Theorems 4,8 and 5.1 (Chapter 1). However, the equality constraints were not considered there. The peculiarities of the case under consideration are discussed later for a more general case in Theorem 2.4 (Chapter 4).

Let us consider some particular cases where the auxiliary program can be written comparatively simply and where some interesting results can be obtained.

1. Aggregatable control

Let the reduced and the perturbed programs be as follows

$$\begin{aligned}
&x^{(t+1)} = f(x^{(t)}, \psi^t(u^{(t)}), t) \;,\; t = 0,1,\dots,N-1 \;,\; x^{(0)} \text{ is given,}\\
&u^{(t)} \in U_t \;,\; \psi^t(u^{(t)}) \in \tilde{U}_t \,,\, t = 0,1,\dots,N-1 \;,\\
&\max G(x^N) \triangleq G^*
\end{aligned} \tag{7.10}$$

$$\begin{aligned}
&x^{(t+1)} = f(x^{(t)}, \psi^t(u^{(t)}), t) + \epsilon\gamma(x^{(t)}, u^{(t)}, t) \;,\\
&t = 0,1,\dots,N-1 \;,\; x^{(0)} \text{ is given,}\\
&u^{(t)} \in U_t \;,\; \psi^t(u^{(t)}) \in \tilde{U}_t \;,\; t = 0,1,\dots,N-1\\
&\max G(x^N) \triangleq G^*_\epsilon
\end{aligned} \tag{7.11}$$

where $\psi^t(\cdot)$ is a continuous mapping $R^m \to R^k$ $(m > k)$; and U_t, $\tilde{U}_t$, $t = 0,1,\dots,N-1$ are closed subsets of R^m, R^k.

Let the solution $\{\psi^{*(t)}, \tilde{x}^{*(t)}\}$ to the aggregated program

$$\begin{aligned}
&\tilde{x}^{(t+1)} = f(\tilde{x}^{(t)}, \psi^{(t)}, t) \;,\; t = 0,1,\dots,N-1 \;,\; x^0 \text{ is given,}\\
&\psi^{(t)} \in \tilde{U}_t \;,\; t = 0,1,\dots,N-1\\
&\max G(x^N)
\end{aligned} \tag{7.12}$$

be unique and suppose that $U^*_t \neq \emptyset$ for all t. Then any reduced optimal trajectory $\{\tilde{x}^{(t)}\}$ and the corresponding conjugate system's trajectory $\{\tilde{\lambda}^{(t)}\}$ are one and the same for all controls satisfying

$$u^t \in U^*_t \triangleq \{u \,/\, u \in U_t \,,\, \psi^t(u^{(t)}) = \psi^{*(t)}\} \;,\; t = 0,1,\dots,N-1 \tag{7.13}$$

The trajectories are determined by the equations

$$\begin{aligned}
&\tilde{x}^{(t+1)} = f(\tilde{x}^{(t)}, \psi^{*(t)}, t) \;,\; t = 0,1,\dots,N-1 \;,\; x^0 \text{ is given}\\
&\tilde{\lambda}^{(t)} = f^T_x(\tilde{x}^{(t)}, \psi^{*(t)}, t)\tilde{\lambda}^{(t+1)} \;,\; t = N-1,\dots,1 \;,\; \tilde{\lambda}^N = \nabla G(\tilde{x}^N)
\end{aligned} \tag{7.14}$$

Hence, the auxiliary program (7.7) is decomposed into N independent subprograms

$$\max \{(\tilde{\lambda}^{t+1})^T \gamma(\tilde{x}^t\,,\, u^{(t)}\,,\, t) \,/\, u^{(t)} \in U_t^* \} \tag{7.15}$$

In particular, let

$$\begin{aligned} &U_t = \{u \,/\, \check{u}^{(t)} \le u \le \hat{u}^{(t)}\} \quad,\quad t = 0,1,\ldots,N-1 \\ &\psi^t(u^{(t)}) \triangleq (\psi_i^t(u_i^{(t)})) \quad,\quad u^{(t)} \triangleq (u_i^{(t)}) \quad,\quad i = 1,\ldots,k \\ &\gamma(x\,,u^{(t)}\,,t) \triangleq \sum_{i=1}^{k} \gamma_i(x\,,u_i^{(t)}\,,t) \end{aligned} \tag{7.16}$$

Then each program (7.15) is separated into k subprograms

$$\max \{(\tilde{\lambda}^{(t+1)})^T \gamma_i(x^{(t)}\,,\, u_i^{(t)}\,,\, t) \,/\, \check{u}_i^{(t)} \le u_i^{(t)} \le \hat{u}_i^t\,,\, \psi_i^t(u_i^{(t)}) = \psi_i^{*(t)}\}$$

2. Weakly controllable systems

One must look for a solution to the optimal control problem

$$x^{(t+1)} = f(x^{(t)}\,,t) + \epsilon\gamma(x^{(t)}\,,u^{(t)}\,,t) \quad,\quad t=0,1,\ldots,N-1\,,x^{(0)} \text{ is given} \tag{7.17}$$

$$u^t \in U_t \quad,\quad t = 0,1,\ldots,N-1 \tag{7.18}$$

$$\max\ G(x^N) \tag{7.19}$$

Note that the reduced system (for $\epsilon = 0$) is uncontrollable

$$\bar{x}^{(t+1)} = f(\bar{x}^{(t)}\,,t) \quad,\quad t = 0,1,\ldots,N-1 \quad,\quad x^{(0)} \text{ is given} \tag{7.20}$$

The vector of Lagrange multipliers introduced formally is given by the conjugate system

$$\bar{\lambda}^{(t)} = f_x^T(\bar{x}^{(t)}\,,t)\bar{\lambda}^{(t+1)} \quad,\quad \bar{\lambda}^N = \nabla G(\bar{x}^N) \tag{7.21}$$

Here the perturbation method gives a very simple and effective result, which may be called *the asymptotic maximum principle.*

Theorem 7.1.

Let the functions $f(x^{(t)}, u^{(t)}, t)$ be differentiable in $x^{(t)}$, $t = 0,\dots,N-1$. Then a control $u^{*(t)}$ found as a solution to the N independent programs

$$\max\{\Phi^t(u^{(t)}) \,/\, u^{(t)} \in U_t\} \quad , \quad t = 0,\dots,N-1 \tag{7.22}$$

where

$$\Phi^t(u^{(t)}) \triangleq (\bar{\lambda}^{t+1})^T \gamma(\bar{x}^{(t)}, u^{(t)}, t)$$

and $\{\bar{x}^{(t)}\}$, $\{\bar{\lambda}^{(t)}\}$ are given by the equations (7.20) and (7.21), is a suboptimal one, i.e., the objective value achieved if one uses $u^{*(t)}$ in (7.11) differs from the optimal one by less than $o(\epsilon)$. The optimal objective value may be written in the form

$$G_\epsilon^* = G^* + \epsilon \sum_{\tau=1}^{N-1} \Phi^t(u^{*(t)}) + o(\epsilon)$$

Proof yields directly from Theorems 4.8 and 5.1 (Chapter 1).

Remark. Theorem 7.1 does not coincide with the well-known discrete maximum principle (see, e.g., [62, 33]). First of all, we do not mean to obtain here the optimal control but only a suboptimal one. On the other hand, the result is valid without any essential restrictions on the properties of functions $f(\cdot)$, $\gamma(\cdot)$, $G(\cdot)$, e.g., without the convexity conditions that are necessary in the standard case. Moreover, the theorem gives not only an optimality criterion but also a simple way to obtain the solution. Its approximate character can be feasible under the corresponding conditions. Lastly, if the usual maximum principle is applicable to the problem (7.17)–(7.18) for all ϵ, then the perturbation method also has some additional possibilities.

Theorem 7.2.

Let U_t be polyhedrons, and let

$$\gamma(x, u, t) = \gamma(x, t)u \tag{7.23}$$

Let $f(x, t)$, $\gamma(x, t)$ be differentiable in t, $t = 0,\dots,N-1$. If the solutions u^{*t} to the programs (7.22) are unique, then there exists $\hat{\epsilon} > 0$ such that, for $\epsilon \in [0, \hat{\epsilon}]$, the control u^{*t} given by the asymptotic maximum principle is the optimal solution to the problem (7.17)–(7.19).

Proof. Under the theorem conditions, the usual maximum principle implies that a control $\{u_\epsilon^{*(t)}\}$ is optimal only if, for each t, u_ϵ^{*t} is the solution to the program

$$\max \{(\lambda_\epsilon^{(t+1)})^T \gamma(x_\epsilon^{(t)}, u^{(t)}, t) \,/\, u^{(t)} \in U_t\} \tag{7.24}$$

where $x_\epsilon^{(t)}$, $\lambda_\epsilon^{(t)}$ satisfy the equations

$$\begin{aligned} &\lambda_\epsilon^{(t)} = [f_x^T(x_\epsilon^{(t)}, t) + \epsilon\gamma_x^T(x_\epsilon^{(t)}, u_\epsilon^{*(t)}, t)]\lambda_\epsilon^{(t+1)} \\ &\lambda_\epsilon^{(N)} \triangleq \nabla G(x_\epsilon^{(N)}) \end{aligned} \tag{7.25}$$

$$x_\epsilon^{(t+1)} = f(x_\epsilon^{(t)}, t) + \epsilon\gamma(x_\epsilon^{(t)}, u_\epsilon^{*(t)}, t) \;\;,\;\; t = 0,1,\ldots,N-1 \;\;,\;\; x_\epsilon^0 \triangleq x^0 \tag{7.26}$$

It is not hard to prove that

$$x_\epsilon^t = \bar{x}^t + O(\epsilon) \;\;,\;\; \lambda_\epsilon^t = \bar{\lambda}^t + O(\epsilon) \;\;,\;\; t = 1,2,\ldots,N \tag{7.27}$$

Hence, by the differentiability of $\gamma(\cdot)$, the programs (7.24) may be written in the form

$$\max \{(\bar{\lambda}^{(t+1)})^T \gamma(\bar{x}^{(t)}, t)u + O(\epsilon) \,/\, u^t \in U_t\} \tag{7.28}$$

Taking into account the uniqueness of $u^{*(t)}$, which is the solution to the linear program (7.22) differing from (7.28) in the $O(\epsilon)$ term only, we obtain

$$u_\epsilon^{*(t)} = u^{*(t)} \;\;,\;\; t = 0,1,\ldots,N \qquad \text{QED.}$$

In another particular case where the functions $\Phi^t(u)$ are strongly concave in u, a successive approximation scheme can be constructed. Let $u_0^{(t)} \triangleq u^{*(t)}$, $t = 0,\ldots,N-1$, and $x_1^{(t)}$, $\lambda_1^{(t)}$, $t = 1,\ldots,N$ be the solutions to the equations

$$\begin{aligned} &x_1^{(t+1)} = f(x_1^{(t)}, t) + \epsilon\gamma(x_1^{(t)}, u_0^{(t)}, t) \;\;,\;\; t = 0,1,\ldots,N \;\;,\;\; x_1^0 \triangleq x^0 \\ &\lambda_1^{(t)} = [f_x^T(x_1^{(t)}, t) + \epsilon\gamma_x^T(x_1^{(t)}, u_0^{(t)}, t)]\lambda_1^{(t+1)} \;\;,\;\; t = N-1,\ldots,1 \\ &\lambda_1^{(N)} \triangleq \nabla G(x_1^{(N)}) \end{aligned}$$

Let u_1^t be the solutions to the programs

$$\max \{(\lambda_1^{(t+1)})^T \gamma(x_1^{(t)}, u^{(t)}, t) \,/\, u^{(t)} \in U_t\}$$

Analogously, one can recursively find the solutions $\{x_i^{(t)}, \lambda_i^{(t)}\}$ $i = 2,\ldots$ to the equations

$$\begin{aligned}
&x_i^{(t+1)} = f(x_i^{(t)}, t) + \epsilon\gamma(x_i^{(t)}, t, u_{i-1}^{(t)}) \quad , \quad t = 0,1,\ldots,N-1 \quad , \quad x_i^0 \triangleq x^0 \\
&\lambda_i^{(t)} = [f_x^T(x_i^{(t)}, t) + \epsilon\gamma_x^T(x_i^{(t)}, u_{i-1}^{(t)}, t)]\lambda_i^{(t+1)} \quad , \quad t = N-1,\ldots,1 \\
&\lambda_i^N = \nabla G(x_i^N)
\end{aligned} \tag{7.29}$$

and the solutions $u_i^{(t)}$, $i = 2,\ldots$ to the programs

$$\max \{(\lambda_i^{(t+1)})^T \gamma(x_i^{(t)}, u^{(t)}, t) \,/\, u^{(t)} \in U_t\} \tag{7.30}$$

Theorem 7.3.

Let all functions in (7.17)–(7.19) be twice continuously differentiable. Let the sets U_t, $t = 0,\ldots,N-1$, be closed, bounded, and convex. Lastly, let the functions $\Phi_t(u)$ be strongly concave over U_t, $t = 0,\ldots,N-1$.

Then there exist positive constants k_1, k_2, k and $0 < \epsilon_0 < \dfrac{1}{k}$ such that, for $\epsilon \in [0, \epsilon_0]$,

$$\begin{aligned}
&\|u_i^t - u_\epsilon^{*t}\| \le k_1(k\epsilon)^{i+1} \quad , \quad t = 0,1,\ldots,N-1 \\
&\|x_i^t - x_\epsilon^{*t}\| \le k_2(k\epsilon)^{i+1} \quad , \quad t = 0,1,\ldots,N-1
\end{aligned} \tag{7.31}$$

where$\{u_\epsilon^{*t}, x_\epsilon^{*t}\}$ is the optimal solution to (7.17)–(7.19).

Proof is completely analogous to the results, which will be obtained later (see Chapter 6, Theorem 1.3 and 1.4), within a systematic investigation of the perturbation method approach to continuous, weakly controllable systems. Its presentation is not appropriate here.

As an example of its application, consider the problem of *decomposition in a dynamic Leontieff model.*

Let the mathematical model of economic development be written as in Section 2.5:

$$\begin{aligned}
&x(t) = A(t)x(t) + K(t)u(t) + d(t) \\
&r(t+1) = Dr(t) + u(t) \quad , \quad x(t) \le r(t)
\end{aligned} \tag{7.32}$$

$x(t) \ge 0$, $u(t) \ge 0$, $t = 0,1,\ldots,N$, $r(0)$ being given, but with another objective function

$$G[r(N)] - \max \tag{7.33}$$

This problem can be interpreted in the following way: one needs to maximize the final capacity levels $r(N)$ under the condition that the demands, $d(t)$ are met.

We shall suppose that the capital coefficients, $K(t)$, are large, formally, by setting

$$K(t) = \frac{1}{\epsilon} K_0(t) \tag{7.34}$$

If we introduce new control variables

$$v(t) \triangleq \frac{1}{\epsilon} u(t)$$

then (7.32) may be written as a weakly controllable system

$$\begin{aligned} r(t+1) &= Dr(t) + \epsilon v(t) \\ x(t) &= A(t)x(t) + K_0 v(t) + d(t) \\ x(t) &\leq r(t) \quad , \quad x(t) \, , \, v(t) \geq 0 \end{aligned}$$

In accordance with the algorithm described above, one can compute the uncontrolled change of the capacities, $\bar{r}(t)$, by the reduced equation

$$\bar{r}(t+1) = D\bar{r}(t) \quad , \quad \bar{r}(0) = (0)$$

and after that find the Lagrange multipliers $\{\bar{\lambda}(t)\}$ as the solution to the conjugate equation

$$\bar{\lambda}(t) = D^T \bar{\lambda}(t+1) \quad , \quad \bar{\lambda}(N) = \nabla G(\bar{r}(N))$$

A suboptimal control is now determined by the programs of (7.22) type, which have the following form

$$\max \bar{\lambda}^T(t) v(t)$$

such that

$$\begin{aligned} x(t) &= A(t)x(t) + K_0(t)v(t) + d(t) \\ 0 &\leq x(t) \leq \bar{r}(t) \quad , \quad v(t) \geq 0 \quad , \quad T = 1,\ldots,N-1 \end{aligned}$$

It is of interest to compare these programs with the static problem (5.1). One can see that the perturbation method allows us to transform the dynamic problem (7.32) and (7.33) into a sequence of independent ones where, at each step t, the capacity-building $v(t)$ is considered as a sort of scheduling demand, denoted $y(t)$ in (5.1). The Lagrange multipliers $\bar{\lambda}(t)$ play the role of estimates allowing the commensuration of activities of the various industries.

Note also that the optimal control problem (7.32) and (7.33) contains constraints on the state variables, i.e., it is in a class to which the standard maximum principle is not applicable.

The following section is devoted to other aspects of this difficult problem.

2.8. Control of weakly dynamic systems under state variable constraints

The constraints on variables that characterize the internal state of a system are decisive if one wants to design a control for some technology. It is often desirable to adjust operating conditions near critical limits (in temperature, pressure, concentration, etc.) since optimal performance indices can be achieved by such adjustments. The use of standard linear control laws is not very acceptable here. On the other hand, optimal design is difficult because of the inapplicability of the discrete maximum principle mentioned above. Computational schemes of mathematical programming can be used, but difficulties arise connected with the large size and, as a rule the nonconvexity of programs. Next we show a class of problems where the perturbation method permits us to bypass these obstacles.

Let the controlled system be weakly dynamic, i.e., the state at a step $t+1$ is only weakly dependent on the state at the previous step t. Such a situation takes place in stable technological units, e.g., in some chemical reactors.

We call that static models where this dependence is not taken into account are those most often used in economic investigations. Note also that if the system is stable one can choose the operating interval such that the influence of the initial state vanishes at the end of the interval and the new state is determined mainly by the control and external disturbances.

Let us describe the specific features of weakly dynamic systems by the following formal model

$$\begin{gathered} \max \, \{ \sum_{t=0}^{N-1} g(x^{t+1}, u^t) \,/\, x^{t+1} = \epsilon f(x^t, u^t) + \varphi(u^t, \omega^t) \;, \\ x^0 \le \bar{x} \;,\; x^{t+1} \le \bar{x} \;,\; u^t \in U \} \end{gathered} \tag{8.1}$$

where the objective can be interpreted as the income obtained from the controlled process.

The sequence $\{\omega^t\}$ represents the disturbance, and $\bar{x}$ is the vector of limiting values of the variables x^t. A heuristic approach suggests that, if ϵ is small, it is reasonable to choose the control bearing in mind the one-step income, i.e., as the solution to the programs

$$\max_u \{g[\varphi(u\,,\omega^t)\,,u]\,/\,\varphi(u\,,\omega^t) \le \bar{x}\,,\,u \in U\}\;,\;t = 0,1,\ldots,N-1 \tag{8.2}$$

This is the same result that the zero-order approximation of the perturbation method gives if, of course, the solution is unique for all t.

However, this suboptimal control is not strictly feasible, and its application is only possible if the constraints can be violated a little. If this is prohibited, then there is a slightly different heuristic approach. One can look for a suboptimal feasible control as the solution u^{*t} to the programs

$$\max_u \{g[\varphi(u,\omega^t),u]\,/\,\varphi(u,\omega^t) + \epsilon f(x^t,u) \le \bar{x}\,,\,u \in U\}\;,\quad t = 0,1,\ldots,N-1 \tag{8.3}$$

The obtained control u^{*t} depends on the state x^t and the disturbance ω^t at the same instant.

Let us suppose that these values can be measured by some devices. Then we obtain the control law as a combination of feedback and the feedforward. The simplest case of practical interest is where the control u^t is the scalar $(m = 1)$ and

$$\varphi(u\,,\omega^t) = \varphi_u(u) + \varphi_\omega(\omega^t)\;,\;\varphi_u(u) = \{\varphi_i(u)\}\;,\;f(x\,,u) = f(x) \tag{8.4}$$

where $\varphi_i(u)$, $i = 1,\ldots,N$, are monotonic functions (to be precise, assume that they are increasing ones).

The constraints in (8.3) have the simple form

$$u \le \min_i \varphi_i^{-1}[\bar{x}_i - \epsilon f_i(x^t) - \varphi_{\omega i}(\omega^t)]\;,\;\underline{u} \le u \le \bar{u} \tag{8.5}$$

If, moreover, the function $y(\cdot)$ is increasing in u, there is the explicit solution

$$u_0^t = \min\,\{\bar{u}\,,\,\min_i \varphi_i^{-1}[\bar{x}_i - \epsilon f_i(x^t) - \varphi_{\omega i}(\omega^t)]\} \tag{8.6}$$

This nonlinear control law can be called *the nearest bound control.* It can be realized by means of a simple device that is a very attractive feature from the practical point of view. Therefore, it is important to estimate the accuracy of such an approximation.

The estimates for the zero-order approximation error follow directly from the general results of the perturbation method (see, Section 1.5) and it is of no interest to state them here.

A dynamic programming approach is convenient for the estimation of (8.3) and (8.6) type control. As usual we can introduce the functions $B^\tau(x)$, $\tau = 1,\ldots,N$ by the definition

$$B^{\tau}(x) \triangleq \max \sum_{t=N-\tau}^{N-1} g(x^{t+1}\,,\,u^t) \tag{8.7}$$

subject to

$$\begin{aligned} &x^{N-\tau} = x \le \bar{x} \\ &x^{t+1} = \epsilon f(x^t\,,\,u^t) + \varphi(u^t\,,\omega^t)\ \ ,\ \ N-\tau \le t \le N-1 \\ &u^t \in U\ \ ,\ \ x^{t+1} \le \bar{x} \end{aligned} \tag{8.8}$$

Then the standard recursive relations are

$$\begin{aligned} B^{\tau}(x) &= \max\,\{g[\epsilon f(x\,,\,u) + \varphi(u\,,\omega^{N-\tau})\,,\,u] + \\ &\qquad + B^{\tau-1}[\epsilon f(x\,,\,u) + \varphi(u\,,\omega^{N-\tau})]\,/\,u \in U_d(x\,,\omega^{N-\tau})\} \\ B^0(x) &\triangleq 0 \end{aligned} \tag{8.9}$$

The feasible sets U_d can be described as follows

$$U_d[x\,,\omega^{N-\tau}] \triangleq \{u\,/\,u \in U\,,\epsilon f(x\,,\,u) + \varphi(u\,,\omega^{N-\tau}) \le \bar{x}\} \tag{8.10}$$

if one assumes that such sets are nonempty for all $x \le \bar{x}$, where the $B^{\tau}(x)$ are as defined. To that end, it is sufficient that there exists $u^* \in U$ such that $\varphi[u^*\,,\omega^{N-\tau}] < \bar{x}$ if, of course, U is bounded and ϵ is small.

The optimal objective value for the original program (8.1) coincides with $B^N(x^0)$, and the optimal control u^t is given as a function of x^t, ω^t by the solution to the program (8.9) for $\tau = N-t$. It is clear that the heuristic scheme described above follows from the rigorous scheme (8.9) if the "delay effect" expressed by the function $B^{\tau-1}(x)$ is omitted in each equation (8.9). Let us analyze in detail only the linear case where

$$g(x\,,\,u) = c^T x + d^t u\ \ ,\ \ f(x\,,\,u) = Ax\ \ ,\ \ \varphi(u\,,\omega) = bu + \omega \tag{8.11}$$

and A, b, c, d are constant matrices of corresponding dimension. We rewrite (8.9) and (8.10), taking into account (8.11)

$$\begin{aligned} B^{\tau}(x) = \ &\max\,\{\epsilon c^T A x + (c^T b + d^T)u + c^T \omega^{N-\tau} + \\ &+ B^{\tau-1}[\epsilon A x + bu + \omega^{N-\tau}]\,/\,u \in U_d(x\,,\omega^{N-\tau})\} \end{aligned} \tag{8.12}$$

$$U_d(x\,,\omega) = \{u\,/\,u \in U\,,\,bu \le \bar{x} - \epsilon A x - \omega\} \tag{8.13}$$

and formalize the basic result of this section.

Theorem 8.1.

Let U be a polyhedron. Let the programs

$$\max \{(c^T b + d^T)u \,/\, u \in U \,,\, bu \leq \bar{x} - \omega^t\} \;,\; t = 0,1,\ldots,N-1 \tag{8.14}$$

have a unique and nondegenerate solution. Then there exists an ϵ^0 such that, for $\epsilon \in [0\,,\epsilon_0]$, the optimal control in (8.1) and (8.11) is given by the solution to the following programs:

$$\max \{(c^T b + d^T)u \,/\, u \in U_d(x\,,\omega^t)\} \;,\; t = 0,1,\ldots,N-1 \tag{8.15}$$

Proof. By the boundedness of U, trajectories $\{x^t\}$ are bounded for any feasible control. Hence, there exists an ϵ_1 such that $U_d(x^t\,,\omega^t)$ are nonempty for $\epsilon \in [0\,,\epsilon_1]$, $t = 1,\ldots,N$. It is not hard to prove [84] that $B^\tau(x)$ are piecewise linear functions having the uniformly bounded subgradient

$$|B_x^\tau(x)| \leq \epsilon k \;,\; k = \text{constant} \tag{8.16}$$

The functions

$$L^\tau(u) \triangleq B^\tau[\epsilon Ax + bu + \omega^{N-\tau}] \tag{8.17}$$

are also piecewise linear and

$$|L_u^\tau(u)| \leq \epsilon \|b\|\, k$$

Thus, the optimal control

$$u^t(\omega^t\,,x^t) \;,\; t = 0,1,\ldots,N-1$$

is found as the solution to the programs

$$\max \{\bar{c}(u\,,\epsilon)u \,/\, u \in U_d(x^t\,,\omega^t)\} \tag{8.18}$$

and

$$|\bar{c}(u\,,\epsilon) - (b^t c + d)| \leq \epsilon \|b\|\, k \tag{8.19}$$

Note that the programs (8.14) are the reduced programs with respect to (8.18). By the assumptions of uniqueness and nondegeneracy of the solution to the reduced programs, there exists a $\epsilon_0 \leq \epsilon_1$ such that, for $\epsilon \in [o, \epsilon_0]$, the solutions to (8.18) are unique and coincide with the unique solution to (8.15). QED.

In the scalar case, it is possible to show a more complete result.

Theorem 8.2.

Let $b^T c + d > 0$, $B > 0$. If

$$\epsilon_0 \triangleq (\alpha_1 + \alpha_0 \beta^{-1})^{-1} \tag{8.20}$$

where

$$\begin{aligned} &\alpha_0 \triangleq \|c^T A\| + (b^T c + d) \max_i \|a_i\| \, b_i^{-1} \\ &\alpha_1 \triangleq \|A\| + \max_i \|a_i\| \, b_i^{-1} \ , \ \beta \triangleq \frac{b^T c + d}{\|b\|} \end{aligned} \tag{8.21}$$

then, for $0 \leq \epsilon \leq \epsilon_0$, $t = 0,\ldots,N-1$, $m = 1$, the solution to the problem (8.1), (8.11) is given by by the following formula

$$u_{\text{opt}}^t(x^t, \omega^t) = \min\, [\bar{u} \ , \min_i b_i^{-1}(\bar{x}_i^t - \epsilon a_i x^t - \omega_i^t)] \tag{8.22}$$

Proof. Let $\varsigma^{\tau-1}$ be the maximum value of $|B_x^{\tau-1}(x)|$. Assuming that

$$\|b\| \, \varsigma^{\tau-1} < b^T c + d \tag{8.23}$$

one can show that here the solution to the program of (8.12)-type coincides with (8.22) since the objective on the right-hand side of (8.12) is increasing. Hence, hypothesis (8.22) implies that

$$\begin{aligned} B^\tau(x) = \ & \epsilon c^T A x + c^T \omega^{N-\tau} + (b^T c + d) u_{\text{opt}}(x, \omega^{N-\tau}) + \\ & + B^{\tau-1}[\epsilon A x + \omega^{N-\tau} + b u_{\text{opt}}(x, \omega^{N-\tau})] \end{aligned}$$

One can now obtain the estimate that

$$\sup |B_x^\tau(x)| \triangleq \varsigma^\tau \leq \epsilon \alpha_0 + \varsigma^{\tau-1} \epsilon [\|A\| + \|b\| \max_i \|a_i\| \, b_i^{-1}]$$

or, using the notation (8.21),

$$\varsigma^{\tau} \leq \epsilon[\alpha_0 + \alpha_1 \varsigma^{\tau-1}] \tag{8.24}$$

so that

$$\varsigma^{\tau} \leq \beta \tag{8.25}$$

for all $\epsilon \in [0, \epsilon_0]$.

By induction, (8.23) is valid for all τ. Hence, (8.22) is valid, too. Note that the number of steps is not relevant.

In conclusion, note that the original statement is a conditional one from the practical point of view. Its main defect is the implicit introduction of the hypothesis that all disturbances can be determined at the very instant when the control is applied. But the disturbances can influence the process after the application of the control and are, in fact, indeterminate at this instant. Therefore, one must review this problem taking into account the stochastic nature of the disturbances. This will be done in Chapter 4.

CHAPTER 3

Singular Programs

This chapter is devoted to the investigation of the singular situation where a gap between the perturbed and reduced solutions may arise.

First (Section 3.1), it is supposed that the reduced programs are convex and have an unbounded set of Lagrange multipliers. (As mentioned in Section 1.2, this unboundedness is the origin of singular effects.) Conditions of regularizability are found here. Their fulfillment permits one to construct a modified reduced program, independent of ϵ, but having a solution such that the perturbed solution can be found in its vicinity for small ϵ.

A more refined analysis is carried out in Section 3.2. Some estimates are obtained by using the solutions to the auxiliary program similar to the one used in Chapter 1. For linear programs, the results are given in a simpler form.

It is well-known that Markov programming, i.e., problems of constructing a stationary optimal control for a finite-state stochastic dynamic system, is equivalent to linear programming with special structure. If the state set is divided into subsets in such a way that transition probabilities inside the subsets are essentially larger than probabilities of transition between states belonging to different subsets, then the corresponding linear program is singular. It is shown that the transitional process consists of two phases. First, an arbitrary initial probability distribution inside each subset is quickly transformed to a quasistationary distribution. In the second phase, the probabilities are redistributed slowly between the subsets "flowing" through weak connections. One can see that this two-stage process is the origin of the singularity manifest in the form of a gap between the perturbed and reduced solutions.

A specific hierarchical variant of the Howard procedure is suggested in Section 3.4 for solving the singularly perturbed Markov program. At each step, state weights are determined by means of an aggregated Markov chain with the original subsets as aggregated states. A finite-step convergence of the procedure to an optimal (or ϵ-optimal) strategy is proved.

Section 3.5 is devoted to an extension of the procedure described in Section 3.4 for general singularly perturbed programs. Specific properties of these programs are stated here, the first of which is the special structure of ϵ-expansions of simplex multipliers for any basis. Moreover, it is shown that one can find the principal terms of these expansions by solving an aggregated system whose size depends on a "singularity depth", characterized by a number of eigenvectors associated with the zero eigenvalue of a constraint matrix. This allows us to construct an original modification of the simplex method in which one needs to invert the aggregated basic matrix instead of the original ill-conditioned one. It is proved that, under some conditions, this modification gives an optimal basis in a finite number of steps. Unfortunately, the procedure may only be useful in the case where the reduced matrix has a special structure, e.g., where it is block-diagonal. However, it is of theoretical interest that well-known compact inverse methods used for solving large linear programs coincide with the modified simplex method mentioned above if a small parameter is artificially introduced in a sensible way (see Section 3.6).

3.1. Singularity and regularization in quasiconvex problems

As above, let us write the perturbed and reduced programs in the form

$$F^*(\epsilon) = \max \{f_0(x) + \epsilon f_1(x) \,/\, g_0(x) + \epsilon g_1(x) \le 0\} \tag{1.1}$$

$$F^*(0) = \max \{f_0(x) \,/\, g_0(x) \le 0\} \tag{1.2}$$

All our previous consideration was devoted to the investigation of situations where the optimal objective value in the perturbed program was a continuous function of ϵ on $[0, \epsilon']$, $\epsilon' > 0$ and, in particular, where

$$\lim_{\epsilon \to 0} F^*(\epsilon) = F^*(0)$$

At the same time another situation is also of interest.

Definition 1.1. The program (1.1) is singularly perturbed if

$$\lim_{\epsilon \to 0} F^*(\epsilon) \ne F^*(0) \tag{1.3}$$

Naturally, (1.3) implies that the limit points of the perturbed solutions do not belong to the reduced program solutions set.

As was established earlier (see Chapter 1), the gap in the objective value does not arise if the boundedness and the nonemptiness of the perturbed program solution set $\Theta^*(\epsilon)$ ensures that the set of reduced program Lagrange vectors $\Lambda(0)$ is also bounded,[†] i.e.,

---

[†] Here and later in this chapter, the reduced program is assumed to be convex and, hence, $\Lambda(0)$ does not depend on the choice of $x \in \Theta^*(0)$.

$$\Theta^*(\epsilon) \neq \phi \;\; ; \;\; \Theta^*(\epsilon) \subset D \Rightarrow \Lambda(0) \subset D_\Lambda \tag{1.4}$$

where D, D_Λ are compact sets,

$$\Lambda(0) = \{\lambda \,/\, \nabla f(\bar{x}) = \sum_{l \in \Omega(\bar{x})} \lambda_l \nabla g_l(\bar{x}) \;,\; \lambda_l \geq 0 \;,\; \bar{x} \in \Theta^*(0)\} \tag{1.5}$$

$\Omega(\bar{x})$ is the index set of the binding constraints at the point $\bar{x}$. Therefore we shall assume further that (1.4) is not met. This situation can be dealt with by the following lemma.

Lemma 1.1. Let $-f_0(x)$, $g_{0l}(x)$, $l = 1,\ldots,m$ be convex. Suppose that the set $\Lambda(0)$ is unbounded and that $v = \{v_l\}$, $l = 1,\ldots,m$ is an unlimited ray of this set

$$v \geq 0 \tag{1.6}$$

Then

$$\sum v_l g_{0l}(x) \geq 0 \;\;,\;\; \forall\, x \tag{1.7}$$

$$g_{0l}(x) = 0 \quad \text{if} \quad v_l > 0 \;\;,\;\; x \in \Theta(0) \tag{1.8}$$

where $\Theta(0)$ is the feasible set of (1.2).

Conversely, if $\Lambda(0)$ is not empty and a vector v satisfies the conditions (1.6)–(1.8), then it is an unlimited ray of $\Lambda(0)$.

Proof. For convex programs, the existence of the Lagrange vector is equivalent to the existence of the saddle point

$$f_0(x) - \sum_1^m (\bar{\lambda}_0 + \rho v_l) g_{0l}(x) \leq f_0(\bar{x}) \leq f_0(\bar{x}) - \sum_1^m \lambda_l g_{0l}(\bar{x}) \tag{1.9}$$

where

$$\{\bar{\lambda}_l\} \in \Lambda(0) \;\;,\;\; \bar{x} \in \Theta^*(0) \;\; ; \;\; \rho \geq 0 \;\;,\;\; \lambda_l \geq 0$$

Dividing by ρ and letting it tend to ∞, we obtain (1.7). The relation (1.8) follows from (1.7) by definition:

$$x \in \Theta(0) \Leftrightarrow g_{0l}(x) \leq 0 \;\;,\;\; l = 1,\ldots,m \Rightarrow \sum v_l g_{0l}(x) = 0 \;\;,\;\; x \in \Theta(0)$$

The converse statement is obvious. The unlimited rays constitute a polyhedral cone. Denote the extreme vectors of this cone by v_i, $i = 1,\ldots,k$ (any unlimited ray can be presented as a convex combination of v_i). Let us sum the constraints of (1.1) together with the weights v_{il}

$$\sum (g_{0l}(x) + \epsilon g_{1l}(x)) v_{il} \leq 0 \Rightarrow \epsilon \sum g_{1l}(x) v_{il} \leq -\sum g_{0l}(x) v_{il}$$

By Lemma 1.1, each vector v_i satisfies (1.6)–(1.8). Hence, we can obtain

$$\sum_{l=1}^{m} v_{il} g_{1l}(x) \leq 0 \tag{1.9'}$$

Thus, program (1.1) is equivalent to the program

$$\begin{aligned} F^*(\epsilon) = \ & \max \{ f_0(x) + \epsilon f_1(x) \, / \, g_0(x) + \\ & + \epsilon g_1(x) \leq 0 \, , \, \sum_{l=1}^{m} v_{il} g_{1l}(x) \leq 0 \, , \, i = 1,2,\ldots,k \} \end{aligned} \tag{1.10}$$

with the corresponding reduced program

$$\tilde{F}^*(0) \triangleq \max \{ f_0(x) \, / \, g_0(x) \leq 0 \, , \, \sum_{l=1}^{m} v_{il} g_{1l}(x) \leq 0 \, , \, i = 1,\ldots,k \} \tag{1.11}$$

Corollary (a criterion of the gap). Program (1.1) is singularly perturbed, i.e., the gap (1.3) occurs if for each

$$x \in \Theta^*(0)$$

there exists v_i, $i \in \overline{1,k}$ such that

$$\sum v_{il} g_{1l}(x) > 0$$

This is implied by

$$\overline{\lim_{\epsilon \to 0}} \, F^*(\epsilon) \leq \tilde{F}^*(0) < F^*(0) \tag{1.12}$$

Example. Consider the problem

$$F^*(\epsilon) = \max \{x_1 / x_1 + x_2 + \epsilon g_1(x) \leq 1 , -x_1 - x_2 + \epsilon g_2(x) \leq -1 , \\ x_1 \geq 0 , x_2 \geq 0\}$$

The reduced solution is

$$x_1^*(0) = 1 \ , \ x_2^*(0) = 0$$

The cone of unlimited rays consists of the single vector

$$v^T \triangleq (1,1)$$

and problem (1.11) is of the form

$$\tilde{F}^*(0) = \max \{x_1 / x_1 + x_2 = 1 , g_1(x) + g_2(x) \leq 0 , x_1 \geq 0 , x_2 \geq 0\}$$

Obviously, the gap arises if $g_1(1,0)+g_2(1,0) > 0$. In particular, this is so if

$$g_1(x) = x_1^2 - \frac{1}{4} \ , \ g_2(x) \equiv 0$$

with

$$\overline{\lim_{\epsilon \to 0}} \ F^*(\epsilon) \leq \tilde{F}^*(0) = \frac{1}{2} < 1 = F^*(0)$$

Let us formulate some regularity condition for program (1.11) allowing us to establish the connection between solutions to (1.10) and (1.11).

In accord with Lemma 1.1, all feasible solutions to problem (1.12) satisfy the conditions

$$g_{0l}(x) = 0 \ , \ l \in \omega$$

where ω is the index set of constraints† such that

$$\exists \ i \in \overline{1,k} \ , \ l \in \overline{1,m} \ , \ v_{il} > 0$$

† It is easy to see that ω is the index set of positive components of the vector $v = \frac{1}{k} \sum_1^m v_i$.

Consequently the admissible set of (1.11) can be described as follows:

$$\tilde{\Theta}(0) = \{x \,/\, g_{0l}(x) = 0 \,,\, l \in \omega \,;\, g_{0l}(x) \leq 0 \,,\, l \,\bar{\in}\, \omega \,, \quad \sum_{l=1}^{m} v_{il} g_{1l}(x) \leq 0 \,,\, i = 1,\ldots,k\} \subset \Theta(0) \tag{1.13}$$

Definition 1.2. The singularly perturbed program (1.1) is regularizable if, at each point x of the solution set $\tilde{\Theta}^*(0)$ of the program (1.11)

$$x \in \tilde{\Theta}^*(0)$$

there exists a vector $y(x)$ satisfying the relations

$$(\nabla g_{0l}(x))^T y(x) \leq 0 \;,\; y^T(x) \nabla\nabla g_{0l}(x) y(x) = 0 \;,\; l \in \omega \tag{1.14}$$

$$(\nabla g_{1l}(x))^T y(x) < 0 \;,\; l \in \omega \tag{1.15}$$

$$(\nabla g_{0l}(x))^T y(x) < 0 \;,\; l \in \Omega(x) \backslash \omega \tag{1.16}$$

where $\Omega(x)$ is, as above, the index set of the initial reduced problem constraints that are binding at x. Note that (1.14) implies, in particular, $\nabla\nabla g_{0l}(x) y(x) = 0$.

Conditions (1.14) and (1.16) are fulfilled, for example, if

$$\exists\, \bar{x} \;,\; \bar{x} \in \tilde{\Theta}(0) \;,\; g_{0l}(\bar{x}) < 0 \;,\; l \,\bar{\in}\, \omega$$

In this case the vector $y(x)$ can be taken as

$$y(x) \triangleq \bar{x} - x$$

and the inequalities (1.15) can be treated as constraints on the perturbing functions

$$(\nabla g_{1l}(x))^T \bar{x} < \nabla g_{1l}(x) x \;,\; l \in \omega \;,\; x \in \tilde{\Theta}^*(0)$$

Theorem 1.1.

Suppose that

(1) The functions $f_0(x)$, $f_1(x)$, $g_{1l}(x)$, $l = 1,\ldots,m$ are twice continuously differentiable, and the functions $g_{0l}(x)$, $l = 1,\ldots,m$ are thrice continuously differentiable.

(2) The solutions set $\Theta^*(\epsilon)$ of the perturbed program is nonempty and bounded, i.e., $\exists\, \epsilon' > 0$ such that for $0 \leq \epsilon \leq \epsilon'$

$$\emptyset \neq \Theta^*(\epsilon) \subset D \quad , \quad D \text{ is a compact set} \tag{1.17}$$

(3) Problem (1.1) is regularizable.

Then all the limit points of solutions to (1.1) belong to the solution set of problem (1.11)

$$\lim_{\epsilon \to 0} \Theta^*(\epsilon) \triangleq \Theta^{**} \subset \tilde{\Theta}^*(0) \tag{1.18}$$

and there exists the limit

$$\lim_{\epsilon \to 0} F^*(\epsilon) = \tilde{F}^*(0) \tag{1.19}$$

In particular, if x^* is the unique solution to (1.11), then, for any solution $x^*(\epsilon)$ to the perturbed program (1.1), there exists the limit

$$\lim_{\epsilon \to 0} x^*(\epsilon) = x^*$$

Proof. First, let us verify that, for any $x \in \tilde{\Theta}^*(0)$, the domain $\tilde{D}(x)$ is nonempty:

$$\begin{aligned} \emptyset \neq \tilde{D}(x) \triangleq \{x_1 / (\nabla g_{0l}(x))^T x_1 \leq g_{1l}(x) \,,\, l \in \Omega(x) \,;\, (\nabla g_{1l}(x))^T x_1 \leq 0 \,,\, l \in \omega \,; \\ x_1^T \nabla\nabla g_{0l}(x) x_1 + 2(\nabla g_{1l}(x))^T x_1 \leq 0 \,,\, l \in \omega\} \end{aligned} \tag{1.20}$$

To this end, we shall show that, for any $x \in \tilde{\Theta}^*(0)$, the relations

$$d_l \geq 0 \quad , \quad l \in \Omega(x) \quad , \quad d_l = 0 \quad , \quad l \bar{\in} \Omega(x) \quad , \quad \sum_{l \in \Omega(x)} d_l \nabla g_{0l}(x) = 0 \tag{1.21}$$

imply

$$-\sum_{l \in \Omega(x)} d_l g_{1l}(x) \geq 0 \tag{1.22}$$

If $d \triangleq \{d_l\}$ belongs to the polyhedral cone of the unlimited rays of Λ, then this follows from Lemma 1.1. On the other hand, by (1.21) we have

$$\sum_{l=1}^{m} d_l g_{0l}(x) \geq 0 \Rightarrow g_{0l}(x) = 0 \quad , \quad \text{if } d_l > 0 \quad , \quad x \in \Theta(0)$$

that is, by virtue of the second part of the same Lemma 1.1, an arbitrary d satisfying (1.21) belongs to that cone. Thus, one can state that there does not exist any vector d that simultaneously satisfies (1.21) and the inequality

$$-\sum_{l\in\Omega(x)} d_l g_{1l}(x) < 0$$

It follows (see Theorem 22.1 [114]) that[†]

$$\exists\, \bar{x}_1 \in R^n \ , \ (\nabla g_{0l}(x))^T \bar{x}_1 \le -g_{1l}(x) \ , \ l \in \Omega(x) \tag{1.23}$$

from which one can conclude that $\tilde{D}(x)$ is nonempty since, by (1.14)–(1.16), for a sufficiently large ρ

$$\bar{x}_1 + \rho y(x) \in \tilde{D}(x)$$

Let us now show that, for all small ϵ,

$$x + \epsilon x_1 + \epsilon^2 \rho y(x) \in \Theta(\epsilon) \tag{1.23'}$$

where $x \in \tilde{\Theta}^*(0)$, $x_1 \in \tilde{D}(x)$ and ρ is sufficiently large.

We have

$$\begin{aligned} &g_{0l}(x + \epsilon x_1 + \epsilon^2 \rho y(x)) + \epsilon g_{1l}(x + \epsilon x_1 + \epsilon^2 \rho y(x)) = \\ &= g_{0l}(x) + \epsilon[(\nabla g_{0l}(x))^T x_1 + g_{1l}(x)] + \\ &+ \frac{1}{2}\epsilon^2[x_1^T \nabla\nabla g_{0l}(x) x_1 + \rho 2(\nabla g_{0l}(x))^T y(x) + 2(\nabla g_{1l}(x))^T x_1 + \\ &+ \frac{1}{6}\epsilon^3[\varphi_l(x, x_1) + \rho(\nabla g_{1l}(x))^T y(x)] + o(\epsilon^3) \ , \ l \in \omega \end{aligned} \tag{1.24}$$

where $\varphi_l(x, x_1)$ is a clearly defined function of x and x_1. Taking into account (1.14)–(1.16) and the definition of $\tilde{D}$, we can write

$$g_{0l}(x + \epsilon x_1 + \epsilon^2 \rho y(x)) + \epsilon g_{1l}(x + \epsilon x_1 + \epsilon^2 \rho y(x)) \le 0 \ , \ l \in \omega$$

with

$$\rho > \max_{l\in\omega} |(\nabla g_{1l}(x))^T y(x)|^{-1} \cdot |\varphi_l(x, x_1)|$$

In exactly the same way, we obtain

$$g_{0l}(x + \epsilon x_1 + \epsilon^2 \rho y(x)) + \epsilon g_{1l}(x + \epsilon x_1 + \epsilon^2 \rho y(x)) \le 0 \ , \ l \in \Omega(x)\backslash\omega$$

[†] The fact that $\bar{x}_1$ exists may be established immediately using the separability theorem for convex sets.

Since, obviously,

$$g_{0l}(x + O(\epsilon)) + O(\epsilon) < 0 \ , \ l \bar{\in} \Omega(x)$$

(1.23′) is established. According to (1.23′) for $x \in \tilde{\Theta}^*(0)$

$$\begin{aligned} & f_0(x + \epsilon x_1 + \epsilon^2 \rho y(x)) + \epsilon f_1(x + \epsilon x_1 + \epsilon^2 \rho y(x)) \leq \\ & \leq F^*(\epsilon) \Rightarrow \tilde{F}^*(0) \triangleq f_0(x) \leq \underline{\lim}_{\epsilon \to 0} F^*(\epsilon) \end{aligned} \tag{1.25}$$

On the other hand, since $g_{0l}(x)$, $g_{1l}(x)$ are continuous, any partial limit $\bar{x}$ of a perturbed solution

$$x^*(\epsilon) \in \Theta^*(\epsilon) \ , \ \lim_{\epsilon' \to 0} x^*(\epsilon') = \bar{x} \in \Theta^{**}$$

satisfies the inclusion

$$\bar{x} \in \tilde{\Theta}(0) \Rightarrow \tag{1.26}$$

$$\overline{\lim_{\epsilon \to 0}} \ F^*(\epsilon) = f_0(\bar{x}) \leq \tilde{F}^*(0) \tag{1.27}$$

The conjunction of (1.25), (1.26), and (1.27) leads to the statements of the theorem.

3.2. The auxiliary problem in the singular case

Let us consider the problem under more restrictive conditions that allow us to extend the results of the previous section.

Lemma 2.1. Let (a) the program (1.1) be regularizable; (b) the functions

$$\tilde{g}_{1i}(x) \triangleq \sum_{l=1}^{m} g_{1l}(x) v_{il} \tag{2.1}$$

be convex; and (c) any vector z satisfying the conditions

$$\begin{aligned} & (\nabla g_{0l}(x))^T z \leq 0 \ , \ l \in \Omega(x) \\ & (\nabla \tilde{g}_{1i}(x))^T z \leq 0 \ , \ i \in \tilde{\Omega}(x) \end{aligned} \tag{2.2}$$

also satisfy the inequality

$$(\nabla f_0(x))^T z \leq 0 \tag{2.3}$$

where $x \in \tilde{\Theta}^*(0)$, $\tilde{\Omega}(x)$ is the index set of those constraints $\tilde{g}_{1i}(x) \leq 0$ that are binding at the point x.

Then the set P of Lagrange vectors of the program (1.11) is nonempty:

$$P \triangleq \{(\lambda, \mu) / \nabla f_0(x) = \sum_{l=1}^{m} \lambda_l \nabla g_{0l}(x) + \sum_{i=1}^{k} \mu_i \nabla \tilde{g}_{1i}(x) \; , \; \lambda_l \geq 0 \; , \; l \in \Omega(x) \; ; \\ \lambda_l = 0 \; , \; l \bar{\in} \Omega(x) \; ; \; \mu_i \geq 0 \; , \; i \in \tilde{\Omega}(x) \; ; \; \mu_i = 0 \; , \; i \bar{\in} \tilde{\Omega}(x)\} \tag{2.4}$$

Moreover, the function

$$\Phi(x) \triangleq \{f_1(x) + \min_{\lambda} \{-\sum_{l=1}^{m} \lambda_l g_{1l}(x) / (\lambda, \mu) \in P\}\} \tag{2.5}$$

is continuous on $\tilde{\Theta}^*(0)$.

Proof. The fact that the set P is nonempty follows from [28, Section 2.4]. Since (1.11) is a convex program, this set does not depend on x and may be written in the form

$$P = P_b + K$$

where P_b is a bounded polyhedron and K is a cone of unlimited rays. It is easy to establish that

$$K = \{(v, \varsigma) : v = \{v_l\} \geq 0 \; , \; l = 1, \ldots, m \; ; \; \varsigma = \{\varsigma_i\} \geq 0 \; , \; i = 1, \ldots, k \; , \\ \sum_{l=1}^{m} v_l \nabla g_{0l}(x) + \sum_{i=1}^{k} \varsigma_i \nabla \tilde{g}_{1i}(x) = 0\} \tag{2.6}$$

from which, by (1.15) and (2.1), a vector $(v, \varsigma) \geq 0$ belongs to K if

$$\varsigma = 0 \Rightarrow \sum_{l \in \Omega(x)} v_l \nabla g_{0l}(x) = 0$$

The latter implies (1.22) with d_l replacing v_l and, consequently, we have

$$\min \{-\sum_{l=1}^{m} \lambda_l g_{1l}(x) / (\lambda, \mu) \in P\} = \min \{-\sum_{l=1}^{m} \lambda_l g_{1l}(x) / (\lambda, \mu) \in P_b\} \tag{2.7}$$

The boundedness of P_b and the continuity of $g_{1l}(x)$ leads to the continuity of the optimal value (2.7) as a function of x and thus to the continuity of $\Phi(x)$.

Under the conditions of Lemma 1.2, one can introduce the auxiliary program

$$\tilde{F}_1^* \triangleq \max \{\Phi(x) \,/\, x \in \tilde{\Theta}^*(0)\} \tag{2.8}$$

The set $\tilde{\Theta}^*(0)$ is closed so, if it is bounded, program (2.8) is solvable.

Theorem 2.1.

Let the conditions of the Theorem 1.1 and Lemma 2.1 be fulfilled and let

$$g_{0l}(x) = a_l^T x - b_l \ , \ l \in \omega \tag{2.9}$$

where $a_l \in R^n$, $b_l \in R^1$.

Let there exist a vector $\bar{x}_1$, belonging to $\tilde{D}(\bar{x})$, where $\bar{x}$ is simultaneously a solution to (2.8) and a dual solution to (2.7) under $x = \bar{x}$, i.e., $\bar{x}_1$ is a solution to the program

$$\begin{aligned} \max_{x_1} \ & \{(\nabla f_0(\bar{x}))^T x_1 \,/\, (\nabla g_{0l}(\bar{x}))^T x_1 \leq -g_{1l}(\bar{x}) \ , \ l \in \Omega(\bar{x}) \ ; \\ & (\nabla \tilde{g}_{1i}(\bar{x}))^T x_1 \leq 0 \ , \ i \in \tilde{\Omega}(\bar{x})\} \end{aligned} \tag{2.7'}$$

Then:

(1) The limit points of the solutions to the perturbed problem (1.1) are contained in the solution set of auxiliary problem (2.8)

$$\lim_{\epsilon \to 0} \Theta^*(\epsilon) \triangleq \Theta^{**} \subset \arg\max \{\Phi(x) \,/\, x \in \tilde{\Theta}^*(0)\} \tag{2.10}$$

In particular, if the solution x^* to the auxiliary problem is unique, then there exists the limit

$$\forall \, x^*(\epsilon) \in \Theta^*(\epsilon) \ , \ \lim_{\epsilon \to 0} x^*(\epsilon) = x^*$$

(2) The optimal objective value of the perturbed problem admits the expansion

$$F^*(\epsilon) = \tilde{F}^*(0) + \epsilon \tilde{F}_1^* + o(\epsilon) \tag{2.11}$$

(3) For all sufficiently small ϵ and some fixed positive ρ

$$\tilde{x}(\epsilon) \triangleq \bar{x} + \epsilon\bar{x}_1 + \epsilon^2\rho y(x) \in \Theta(\epsilon) \tag{2.12}$$

$$f_0(\tilde{x}(\epsilon)) + \epsilon f_1(\tilde{x}(\epsilon)) = F^*(\epsilon) + o(\epsilon) \tag{2.13}$$

Proof. This is similar to the proof of Theorem 4.8 (Section 1.4). Therefore, we shall only indicate its main ideas.

First, as in (1.24), the inclusion (2.12) is established, hence

$$f_0(\tilde{x}(\epsilon)) + \epsilon f_1(\tilde{x}(\epsilon)) \leq F^*(\epsilon) \ , \ f_0(\tilde{x}(\epsilon)) + \epsilon f_1(\tilde{x}(\epsilon)) = \tilde{F}^*(0) + \epsilon\tilde{F}_1^* + o(\epsilon) \Rightarrow$$
$$\Rightarrow \tilde{F}^*(0) + \epsilon\tilde{F}_1^* + o(\epsilon) \leq F^*(\epsilon)$$

Then [see (4.53) and (4.54), Section 1.4] the existence of the saddle point of the Lagrange function

$$\tilde{L}(x,\lambda) = f_0(x) - \sum_{l=1}^{m} \lambda_l g_{0l}(x) - \sum_{i=1}^{k} \mu_i \tilde{g}_{1i}(x)$$

is taken into account.

Remark. Note that under condition (2.9) the inequality (1.9′) is equivalent to the relations

$$\sum_{l=1}^{m} v_{li} a_l = 0 \ , \ \sum_{l=1}^{m} v_{li} b_l = 0 \ , \ i = 1,\dots,k$$

and the regularization conditions (1.14)–(1.16) are of the form

$$a_l^T y(x) \leq 0 \ , \ l \in \omega$$
$$(\nabla g_{1l}(x))^T y(x) < 0 \ , \ l \in \omega$$
$$(\nabla g_{0l}(x))^T y(x) < 0 \ , \ l \in \Omega(x) / \omega$$

In conclusion, consider the linear program presented in standard form

$$\max\ \{(c_0^T + \epsilon c_1^T)x \,/\, (A_0 + \epsilon A_1)x = b_0 + \epsilon b_1 \ , \ x \geq 0\} \tag{2.14}$$

where $A_0(m \times n)$, $A_1(m \times n)$, $c_0(n \times 1)$, $c_1(n \times 1)$, $b_0(m \times 1)$, $b_1(m \times 1)$ are matrices.

As above, the necessary condition for program (2.14) to be singularly perturbed is the unboundedness of the Lagrange vectors set.

Lemma 2.2. Assume that the program

$$\max \{c_0^T x \,/\, A_0 x = b_0 \,,\, x \geq 0\} \triangleq F^*(0) \tag{2.15}$$

has a positive feasible solution

$$\bar{x} > 0 \;,\; A_0 \bar{x} = b_0 \tag{2.16}$$

Then the solutions set to the problem

$$\min \{\lambda^T b_0 / \lambda^T A_0 \geq c_0^T \,,\, \lambda \geq 0\} \triangleq F^*(0) \tag{2.16$'$}$$

is unbounded if and only if there exists a vector v such that

$$v^T A_0 = 0 \;,\; v^T b_0 = 0 \tag{2.17}$$

An arbitrary vector v satisfying (2.17) is an unlimited ray of the solutions set of (2.16′).

Proof. From the fact that the solutions set of (2.16′) is unbounded, it follows that there exists a vector v such that[†]

$$v^T b_0 = 0 \;,\; v^T A_0 \geq 0 \tag{2.17$'$}$$

from which, taking into account (2.16),

$$v^T (A_0 \bar{x} - b_0) = 0 \Rightarrow v^T A_0 \bar{x} = 0 \Rightarrow v^T A_0 = 0$$

The converse is obvious.

Assume further that the dimension of the subspace spanned by the vectors v satisfying (2.17) is equal to k. Assume that the columns of the matrix $V(n \times k)$ constitute a basis for this subspace. Problem (2.14) is obviously equivalent to the problem

[†] Let $\{\lambda^{(t)}\}$ be a sequence such that $\|\lambda^{(t)}\|^{\to\infty}$ if $t \to \infty$, $\lambda^{(t)} \in \Lambda \Leftrightarrow b_0^T \lambda(t) = F^*(0)$, $A_0^T \lambda^{(t)} \geq c_0$, $\lambda^{(t)} > 0 \Rightarrow b_0^T \lambda^{(t)} \|\lambda^{(t)}\|^{-1} = F^*(0) \|\lambda^{(t)}\|^{-1}$; $A_0^T \lambda^{(t)} \|^{(t)}\|^{-1} \geq c_0 \|\lambda^{(t)}\|^{-1}$. Let v be a partial limit of $\lambda^{(t)} \|\lambda^{(t)}\|^{-1}$. Then (2.17′) is obtained if $t \to \infty$.

$$F^*(\epsilon) \triangleq \max \ \{(c_0^T + \epsilon c_1^T)x \,/\, (A_0+\epsilon A_1)x = b_0 + \epsilon b_1 \,, \quad V^T A_1 x = V^T b_1 \,, \, x \geq 0\} \tag{2.18}$$

with the reduced one

$$\tilde{F}^*(0) \triangleq \{c_0^T x \,/\, A_0 x = b_0 \,, \ V^T A_1 x = V^T b_1 \,, \, x \geq 0\} \tag{2.19}$$

Theorem 2.3.

Suppose that

(a) $\exists\ \bar{x} > 0 \ , \ A_0 \bar{x} = b_0 \ , \ V^T A_1 \bar{x} = V^T b_1$ (2.20)

(b) $\text{rank}\left\{\begin{matrix} A_0 & b_0 \\ V^T A_1 & V^T b_1 \end{matrix}\right\} = m$ (2.21)

(c) $\emptyset \neq \Theta^*(\epsilon) \subset D \ , \ 0 \leq \epsilon \leq \epsilon' \ , \ \epsilon' > 0$ (2.22)

where $\Theta^*(\epsilon)$ is the set of solutions to (2.14) and D is a compact set in R^n.

Then

(1) The auxiliary program

$$\max_{x \in \tilde{\Theta}^*(0)} \ \{c_1^T x + \min_{\lambda,\mu} \{-\lambda^T(A_1 x - b_1) \,/\, (\lambda\,,\mu) \in P\}\} \tag{2.23}$$

is solvable, in which $\tilde{\Theta}^*(0)$ is the solutions set of (2.19) and P is the solutions set of the program

$$\min \{\lambda^T b_0 + \mu^T V^T b_1 \,/\, \lambda^T A_0 + \mu^T V^T A_1 \geq c_0^T\} \tag{2.24}$$

this being the dual to (2.19).

(2) All the solutions to the perturbed problem (2.14) converge (with $\epsilon \to 0$) to the solutions set of the auxiliary problem (2.23), and the other statements of Theorem 2.4 (Section 1.2) referring to the connections between solutions to (2.19) and (2.18) are valid here.

Proof. By virtue of (2.21), there exists a submatrix $\{\bar{A}_0, \bar{b}_0\}(m-k \times n+1)$ of the matrix $\{A_0, b_0\}(n \times n+1)$ such that

$$\operatorname{rank}\begin{Bmatrix} \bar{A}_0 & \bar{b}_0 \\ V^T A_1 & V^T b_1 \end{Bmatrix} = m \tag{2.25}$$

Hence, for sufficiently small ϵ,

$$\operatorname{rank}\begin{Bmatrix} \bar{A}_0 + \epsilon \bar{A}_1 & b_0 + \epsilon b_1 \\ V^T A_1 & V^T b_1 \end{Bmatrix} = m \tag{2.26}$$

where $\bar{A}_1$, $\bar{b}_1$ are the submatrices of A_1, b_1 corresponding to $\bar{A}_0$, $\bar{b}_0$. The relation (2.26) implies that (2.18) is equivalent to the program

$$\begin{aligned} F^*(\epsilon) \triangleq \max \{(c_0^T + \epsilon c_1^T)x \,/\, (\bar{A}_0 + \epsilon \bar{A}_1)x = \bar{b}_0 + \epsilon \bar{b}_1 \,, \\ V^T A_1 x = V^T b_1 \,, \, x \geq 0\} \end{aligned} \tag{2.27}$$

The reduced program

$$\tilde{F}^*(0) \triangleq \max \{c_0^T x \,/\, \bar{A}_0 x = \bar{b}_0 \,, \, V^T A_1 x = V^T b_1 \,, \, x \geq 0\} \tag{2.28}$$

is equivalent to (2.19).

Lemma 2.2 and relation (2.26) allow us to conclude that the solution set of the program dual to (2.28) is bounded. Therefore, the connections between the solutions to (2.27) and (2.28) are given by Theorem 2.4 (Section 1.2). Let us show that the objective in (2.23) coincides with that in the auxiliary problem arising from Theorem 2.4 for programs (2.27) and (2.28).

It is not hard to prove that

$$\lambda = \begin{matrix} n-k \\ k \end{matrix} \begin{pmatrix} \bar{\lambda} \\ 0 \end{pmatrix} + V\alpha \;\;,\;\; \mu = \bar{\mu} \tag{2.29}$$

where (λ, μ) solves the dual program to (2.19), $(\bar{\lambda}, \bar{\mu})$ solves the dual program to (2.28), and α is a vector of arbitrary constants. It is assumed in (2.29) that the matrices A_0, A_1, b_0, b_1 are written in the form

$$A_0 = \begin{pmatrix} \bar{A}_0 \\ \bar{\bar{A}}_0 \end{pmatrix} \;,\; A_1 = \begin{pmatrix} \bar{A}_1 \\ \bar{\bar{A}}_1 \end{pmatrix} \;,\; b_0 = \begin{pmatrix} \bar{b}_0 \\ \bar{\bar{b}}_0 \end{pmatrix} \;,\; b_1 = \begin{pmatrix} \bar{b}_1 \\ \bar{\bar{b}}_1 \end{pmatrix}$$

Consequently, we have

$$\lambda^T(-A_1x + b_1) = \bar{\lambda}^T(-\bar{A}_1x + \bar{b}_1) + \alpha^T V^T(-A_1x + b_1) \tag{2.30}$$

However,

$$-V^T A_1 x + V^T b_1 = 0$$

This, together with (2.30), completes the proof.

3.3. An approximate aggregation of Markov chains with incomes

This section is devoted to an important class of singularly perturbed linear programs that illustrate connections between singular perturbations in mathematical programming and those in optimal control problems where (see Chapter 5) we focus on the natural division of process rates into "slow" and "fast". Here, we also deal with the preliminaries of the construction of a new type of aggregation procedure, which differs essentially from all the others developed so far.

First of all, let us consider the standard formulation of the problem.

Consider a controlled Markov chain with incomes. Suppose that a finite set I_r of possible controls is assigned to each state r, $(r = 1,2,\ldots,n)$, with the choice of the control $\beta_r \in I_r$ implying, by definition, (a) a choice of the r-th row of the transition matrix $\{p_{r1}^{\beta_r},\ldots,p_{rn}^{\beta_r}\}$ and (b) the choice of the incomes $\{g_{r1}^{\beta_r},\ldots,g_{rn}^{\beta_r}\}$ that are obtained in transitions from the state r to the other states of the system. In particular, this leads to the choice of a one-step expected income

$$g_r^{\beta_r} \triangleq \sum_{l=1}^{n} p_{rl}^{\beta_r} g_{rl}^{\beta_r}$$

(We shall say that a control strategy $\beta = \{\beta_1,\ldots,\beta_n\}$ is "chosen" if in each state r some control $\beta_r \in I_r$ is fixed. If this is so, the action of the system is described by a homogeneous Markov chain with the transition matrix

$$P^{\beta}(n \times n) \triangleq \{p_{rl}^{\beta_r}\} \quad , \quad r,l = 1,2,\ldots,n$$

and the vector of one-step expected incomes

$$g^{\beta} \triangleq \{g_r^{\beta_r}\} \quad , \quad r = 1,\ldots,n$$

If the process starts at state r, the mathematical expectation $M_r^\beta(t)$ of the income over t steps is defined by the equation

$$M_r^\beta(t) = \sum_{\tau=1}^{t} \sum_{l=1}^{n} p_{rl}^\beta(\tau) g_l^{\beta_l}$$

where $p_{rl}^\beta(\tau)$ is the conditional probability of being at the instant τ in state l provided that the process is begun from the state r. Assume that the matrix P^β is ergodic, i.e., there exist the limits

$$\lim_{\tau\to\infty} p_l^\beta(\tau) = p_l^\beta \ , \ p^\beta \triangleq \{p_l^\beta\} \ , \ l = 1,\ldots,n$$

which do not depend on initial conditions and form the unique invariant measure of the Markov chain

$$(P^\beta)^T p^\beta = p^\beta \ , \ \sum_{l=1}^{n} p_l^\beta = 1$$

It can be shown (see [83] and [84]) that $M_r^\beta(t)$ can be written as

$$M_r^\beta(t) = \sigma^\beta t + \mu_r^\beta + \gamma_r(t) \ , \ \gamma_r(t) \underset{t\to\infty}{\to 0} \tag{3.1}$$

Here, σ^β does not depend on the initial conditions and, further, will be referred to as the rate of income growth

$$\sigma^\beta = \sum_{r=1}^{n} p_r^\beta g_r^{\beta_r} \tag{3.2}$$

The vector of initial weights $\mu^\beta \triangleq \{\mu_r^\beta\}$, $r = 1,2,\ldots,n$ is determined to within a constant by the system

$$\sigma^\beta I - g^\beta = (P^\beta - E)\mu^\beta \tag{3.3}$$

with $I(n \times 1)$ being the unit vector.

The standard statement of the Markov programming problem is: find the control strategy β^* yielding the maximal value of the rate of income growth. The following result is known (see, for example, [49]).

Lemma 3.1. The choice of control strategy β^* that maximizes σ^β

$$\sigma^* \triangleq \sigma^{\beta^*} \triangleq \max_\beta \sigma^\beta \tag{3.4}$$

is equivalent to solution of the linear program

$$\sigma^* = \max \sum_{r=1}^{n} \sum_{\beta_r \in I_r} g_r^{\beta_r} \pi_r^{\beta_r} \tag{3.5}$$

where the feasible set is given by the equations

$$\sum_{\beta_r \in I_r} \pi_r^{\beta_r} - \sum_{l=1}^{n} \sum_{\beta_l \in I_l} p_{lr}^{\beta_l} \pi_l^{\beta_l} \tag{3.6}$$

$$\pi_r^{\beta_r} \geq 0 \ , \ \beta_r \in I_r \ , \ r = 1,2,\ldots,n \tag{3.7}$$

$$\sum_{r=1}^{n} \sum_{\beta_r \in I_r} \pi_r^{\beta_r} = 1 \tag{3.8}$$

This equivalence is interpreted in the following way. Each optimal control strategy $\beta^* \triangleq \{\beta_1^*,\ldots,\beta_n^*\}$ fixes some optimal basis of (3.5)–(3.8), with the basic variables being the solution to the system of rank n

$$\pi_r^{\beta_r^*} - \sum_{l=1}^{n} p_{lr}^{\beta_l^*} \pi_l^{\beta_l^*} = 0 \ , \ \sum_{r=1}^{n} \pi_r^{\beta_r^*} = 1 \tag{3.9}$$

Conversely, an arbitrary optimal basic solution to (3.5)–(3.8) has only one nonzero variable $\pi_r^{\beta_r}$, $\beta_r \in I_r$, and its index gives some optimal (at the state r) control.

Equations (3.5)–(3.8) can be regarded as the problem of constructing optimal probability distributions, with $\pi_r^{\beta_r}$ as the joint final probability of being in state r with the control β_r.

Hence, the lemma show, in particular, that "pure" strategies which do not use random choice are optimal.

Suppose now that the set of system states decomposes into k ($k < n$) subsets Ω_i, $i = 1,2,\ldots,k$ within which the states are strongly connected while the connections between the subsets are weak. Let us formalize this hypothesis in the language of perturbation theory.

Consider the class of controlled Markov systems, each characterized by the transition matrices

$$P^{\beta}(\epsilon) = \{p_{lr}^{\beta_l}(\epsilon)\}$$

with a fixed ϵ from the segment $0 \le \epsilon \le \epsilon_0$. Moreover, assume that for $\beta_l \in I_l$, $l \in \Omega_i$, $i = 1,\ldots,k$

$$p_{lr}^{\beta_l} = \begin{cases} a_{lr}^{\beta_l} + \epsilon b_{lr}^{\beta_l} , & r \in \Omega_i \\ \epsilon b_{lr} , & r \in \Omega_j , \ i \ne j \end{cases} \tag{3.10}$$

As the transition matrix $P^{\beta}(\epsilon)$ is assumed to be stochastic for any ϵ, $0 \le \epsilon \le \epsilon_0$, the following conditions must be fulfilled:

$$\begin{aligned} & a_{lr}^{\beta_l} \ge 0 , \quad l,r \in \Omega_i , \quad l \in I_l ; \\ & b_{lr}^{\beta_l} \ge 0 \text{ if } a_{lr}^{\beta_l} = 0 \\ & (\text{in particualr, } b_{lr}^{\beta_l} \ge 0 \text{ if } l \in \Omega_i , \ r \in \Omega_j , \ i \ne j) ; \\ & \sum_{r\in\Omega_i} a_{lr}^{\beta_l} = 1 , \quad \beta_l \in I_l , \quad l \in \Omega_i , \quad i = 1,\ldots,k \\ & \sum_{r=1}^{n} b_{lr}^{\beta_l} = 0 , \quad \beta_l \in I_l , \quad l = 1,\ldots,n \end{aligned} \tag{3.11}$$

By virtue of (3.10) and (3.11), program (3.5)–(3.8) may be a singularly perturbed one. A basis of the left (corresponding to zero) eigen subspace of the constraint matrix (3.6) with $\epsilon = 0$ is given by the vectors $\hat{v}_i(n+1,1)$, $i = 1,\ldots,k$

$$\hat{v}_{ir} = 1 , \quad r \in \Omega_i ; \quad \hat{v}_{ir} = 0 , \quad r \bar{\in} \Omega_i ; \quad \hat{v}_{i,n+1} = 0 \tag{3.10'}$$

The program generating (3.5)–(3.8) is, obviously, reduced to k independent subprograms

$$\begin{aligned} & \max \sum_{r\in\Omega_i} \sum_{\beta_r\in I_r} g_r^{\beta_r} \pi_r^{\beta_r} \\ & \sum_{\beta_r\in I_r} \pi_r^{\beta_r} - \sum_{l\in\Omega_i} \sum_{\beta_l\in I_l} a_{lr}^{\beta_l} \pi_l^{\beta_l} = 0 , \quad r \in \Omega_i \\ & \sum_{r\in\Omega_i} \sum_{\beta_r\in I_r} \pi_r^{\beta_r} = 1 , \quad \pi_r^{\beta_r} \ge 0 , \quad \beta_r \in I_r , \quad r \in \Omega_i \end{aligned} \tag{3.11'}$$

where $i = 1,2,\ldots,k$.

However, in accordance with the result of the previous section, to obtain an approximate solution one should introduce additional constraints of the form

$$\sum_{r\in\Omega_i}\sum_{l=1}^{n}\sum_{\beta_l\in I_l} b_{lr}^{\beta_l}\pi_l^{\beta_l} = 0 \quad , \quad i = 1,\ldots,k \tag{3.11''}$$

Evidently, a gap in the objective values arises at $\epsilon = 0$ if the solutions to (3.11′) do not satisfy (3.11″).

Now consider this phenomenon from an alternative point of view. In accordance with Lemma 3.1, each i-th subprogram is equivalent to the problem of Markov programming for the chain with transition matrix $\{a_{lr}^{\beta_l}\}$, $\beta_l \in I_l$, $l,r \in \Omega_i$.

Let us show that the control strategy obtained as the composition of the optimal control strategies for such independent subproblems is generally not even suboptimal for the entire system, i.e., for the controlled Markov chain with n states and transition matrix given by (3.10). Note, first of all, that, for any control strategy β, the transition matrix P^β of the perturbed chain is of the form

$$P^\beta = A^\beta + \epsilon B^\beta \tag{3.12}$$

where A^β is a stochastic block-diagonal matrix

$$A^\beta = \text{diag}\,\{A^{1\beta},\ldots,A^{k\beta}\} \tag{3.13}$$

and $A^{i\beta}(n_i \times n_i)$ $(n_i = |\Omega_i|)$ are stochastic matrices. We shall further assume that the matrices $A^{i\beta}$ are indecomposable and regular, i.e., for each of them the unit eigenvalue has a unit multiplicity and that there are no other eigenvalues differing from unit with unit modulus. So, for $i = 1,2,\ldots,k$, the Markov chain with the transition matrix $A^{i\beta}$ will have the final positive probability vector

$$(A^{i\beta})^T p^{i\beta} = p^{i\beta} \quad , \quad \sum_{r\in\Omega_i} p_r^{i\beta} = 1 \tag{3.14}$$

$$p^{i\beta} > 0 \tag{3.15}$$

which is independent of the initial conditions. As far as the matrix A^β is concerned we can conclude that it will have a k-multiple unit eigenvalue with corresponding left and right eigenvectors v_i^β, v_i, $i = 1,\ldots,k$

$$\begin{aligned} v_{ir}^\beta &= p_r^{i\beta} \quad , \quad r \in \Omega_i \;\; ; \;\; v_{ir}^\beta = 0 \quad , \quad r \bar{\in} \Omega_i \\ v_{ir} &= 1 \quad , \quad r \in \Omega_i \;\; ; \;\; v_{ir} = 0 \quad , \quad r \bar{\in} \Omega_i \end{aligned} \tag{3.16}$$

Representation (3.12) divides the corresponding Markov process into two phases: the phase of fast redistribution of probabilities inside the subsets Ω_i and the phase of slow transition of probabilities from one subset to another. During the first "short" phase, only strong connections (given by $A^{i\beta}$, $i = 1,\dots,k$) are essential. During the second "long" phase, small perturbations begin to influence the process, which is approximately described by a Markov chain with the transition matrix $Q^\beta(k \times k)$

$$Q_{ij}^\beta \triangleq \epsilon \sum_{r\in\Omega_i} p_r^{i\beta} \sum_{l\in\Omega_j} b_{rl}^{\beta_r} \;, \; i \neq j \;;$$
$$Q_{ii}^\beta \triangleq 1 - \sum_{j\neq i} Q_{ij}^\beta = 1 + \epsilon \sum_{r\in\Omega_i} p_r^{i\beta} \sum_{l\in\Omega_i} b_{rl}^{\beta_r} \tag{3.17}$$

each subset Ω_i being associated with a state i of this chain. Note, that, as defined, the matrix $Q^\beta(\epsilon)$ is stochastic and can be written as

$$Q^\beta = E + \epsilon(M^\beta)^T \;, \; M^\beta = V_L^T(B^\beta)^T V_\beta \tag{3.18}$$

where $V_l(n \times k)$, $V_\beta(n \times k)$ are matrices with columns v_i, v_i^β $i = 1,\dots,k$, satisfying the correlation

$$V_L^T V_\beta = E \tag{3.19}$$

The validity of such a description of the process is confirmed by the following lemma.

Lemma 3.2. Let the matrix P^β be indecomposable. Then the matrix Q^β is also indecomposable and the right eigenvector p^β of the matrix $(P^\beta)^T$

$$(P^\beta)^T p^\beta = p^\beta \;, \; \sum_{r=1}^{n} p_r^\beta = 1$$

is in the form

$$p^\beta = V_\beta \tilde{p}^\beta + O(\epsilon) \tag{3.20}$$

where $\tilde{p}^\beta(k \times 1)$ is the right eigenvector of $(Q^\beta)^T$:

$$(Q^\beta)^T \tilde{p}^\beta = \tilde{p}^\beta \;, \; \sum_{i=1}^{k} \tilde{p}_i^\beta = 1$$

Remark. If we premultiply (3.20) by V_L^T and take into account (3.12), we shall obtain

$$V_L^T p^\beta = \tilde{p}^\beta + O(\epsilon)$$

i.e., the final probabilities of the subsets Ω_i of the Markov chain with transition matrix $P^\beta(\epsilon)$ are to within $O(\epsilon)$ of the final probabilities of the Markov chain with transition matrix $Q^\beta(\epsilon)$. The final probabilities of the disaggregated states can be obtained by writing (3.20) in the scalar form

$$p_r^\beta = \tilde{p}_i^\beta \cdot p_r^{i\beta} \quad , \quad r \in \Omega_i$$

Proof. The fact that the matrix Q^β is indecomposable follows from (3.17) and (3.15). According to the condition of the lemma, the system

$$\begin{aligned} &(A^\beta + \epsilon B^\beta - E)^T u^\beta(\epsilon) = 0 \\ &u_1^\beta(\epsilon) = 1 \quad , \quad u^\beta(\epsilon) \triangleq (u_i^\beta(\epsilon)) \end{aligned} \tag{3.21}$$

has a unique solution. Using the Cramer rule, we can expand it into the ϵ power series

$$u^\beta(\epsilon) = \frac{u_{-l}^\beta}{\epsilon^l} + \frac{u_{-(l-1)}^\beta}{\epsilon^{l-1}} + \cdots + \frac{u_{-1}^\beta}{\epsilon} + u_0^\beta + \cdots \tag{3.22}$$

where l is a natural number.

Substituting (3.22) into (3.21), one has

$$(A^\beta - E)^T u_{-l}^\beta = 0 \tag{3.23}$$

$$(A^\beta - E)^T u_{-(l-1)}^\beta + (B^\beta)^T u_{-l}^\beta = 0 \tag{3.24}$$

$$\vdots \qquad\qquad \vdots$$

$$(A^\beta - E)^T u_{-1}^\beta + (B^\beta)^T u_{-2}^\beta = 0 \tag{3.25}$$

$$(A^\beta - E)^T u_0^\beta + (B^\beta)^T u_{-1}^\beta = 0 \tag{3.26}$$

From (3.23) it follows that

$$u_{-l}^\beta = V_\beta \gamma \tag{3.27}$$

where $\gamma(k \times 1)$ is a vector. The substitution of (3.27) into (3.24) and premultiplication by V_L^T leads to

$$M^\beta \gamma = 0 \;\Leftrightarrow\; (Q^\beta)^T \gamma = \gamma$$

However, the fact that Q^β is indecomposable means that either all components of γ are equal to zero or that they are all positive. Suppose that $\gamma > 0$, then

$$u_{-m}^\beta > 0$$

and all components of $u^\beta(\epsilon)$ tend to infinity as $1/\epsilon^l$, which contradicts (3.21). Consequently,

$$\gamma = 0 \;\Rightarrow\; u_{-l}^\beta = 0$$

Similarly, we obtain:

$$\begin{aligned} u_{-(l-1)}^\beta &= \cdots = u_{-1}^\beta = 0 \;; \\ u_0^\beta &= V_\beta \gamma \;,\; (Q^\beta)^T \gamma = \gamma \;,\; \gamma > 0 \end{aligned} \tag{3.28}$$

Thus, the solution to (3.21) takes the form

$$u^\beta(\epsilon) = u_0^\beta + O(\epsilon) \tag{3.29}$$

Since the matrix P^β is indecomposable and, hence, its eigenvector associated with a unit eigenvalue is unique, the relation (3.29) is transformed into (3.20) by means of renormalization.

Definition. The Markov chain with the transition matrix Q^β and the vector of expected one-step incomes $\tilde{g}^\beta \triangleq \{\tilde{g}_i^\beta\}$, $i = 1,2,\ldots,k$

$$\tilde{g}_i^\beta \triangleq \sum_{r \in \Omega_i} p_r^{i\beta} g_r^{\beta_r} \tag{3.30}$$

is called the aggregated chain with incomes corresponding to the control strategy β.

Lemma 3.3. Let the matrix P^β be indecomposable. Then the rate of income growth $\sigma^\beta(\epsilon)$ of the perturbed Markov chain corresponding to the control strategy β is to within $O(\epsilon)$ of that of the aggregated chain

$$\sigma^\beta(\epsilon) = \tilde{\sigma}^\beta + O(\epsilon) = \sum_{i=1}^{k} \tilde{p}_i^\beta \tilde{g}_i^\beta + O(\epsilon) \tag{3.31}$$

Proof. In accordance with (3.2) and Lemma 3.2,

$$\sigma^\beta(\epsilon) = \sum_{i=1}^{k} \sum_{r\in\Omega_i} p_r^\beta g_r^{\beta_r} = \sum_{i=1}^{k} \tilde{p}_i^\beta \sum_{r\in\Omega_i} p_r^{i\beta} g_r^{\beta_r} + O(\epsilon) \triangleq$$

$$\triangleq \sum_{i=1}^{k} \tilde{p}_i^\beta g_i^\beta + O(\epsilon) = \tilde{\sigma}^\beta + O(\epsilon)$$

Thus, the lemma is proved.

With the adopted notation, the solution of each i-th program (3.11′) is equivalent to finding the control strategy maximizing $\tilde{g}_i^\beta$. On the basis of Lemmas 3.3 and 3.2, one can conclude that $\sigma^\beta(\epsilon)$ is, to within $O(\epsilon)$, equal to the weighted sum of g_i^β, with the weights being the final probabilities of the subsets Ω_i depending principally on the connections between blocks. Consequently, the choice of the control strategy that yields the maximum for each $\tilde{g}_i$, $i = 1,\ldots,k$ does not, generally speaking, maximize the sum $\sum_{i=1}^{k} \tilde{p}_i^\beta \tilde{g}_i^\beta$, for the probability distribution between the subsets Ω_i may appear to be unprofitable, i.e., the probabilities of subsets with relatively small $\tilde{g}_i^\beta$ can be large. This leads to a gap between the solutions to (3.6)–(3.8) and (3.11′).

We have considered the Markov programming problem where the choice of control strategy should ideally maximize the expected rate of income growth. There are, at the same time, other criteria. Sometimes the expectation of the full discounted incomes

$$h_r^\beta = \sum_{t=1}^{\infty} \alpha^t \sum_{l=1}^{n} p_{rl}^\beta(\tau) g_l^{\beta_l} \tag{3.32}$$

is regarded as the objective function. Here, the previous notation is used, and α denotes the discount factor. Note that it may also be interpreted as the probability that the process will continue if it is initially supposed to cease with the probability α. It is easy to show (see, for example, [74]) that the vector $h^\beta \triangleq \{h_r^\beta\}$, $r = 1,2,\ldots,n$, is the solution to the linear system

$$h^\beta = g^\beta + \alpha P^\beta h^\beta \Rightarrow \tag{3.33}$$

$$h^\beta = (E - \alpha P^\beta)^{-1} g^\beta \tag{3.34}$$

Thus, in this case, the problem is reduced to that of the choice of control strategy maximizing each component of h^β.

Assume now that for arbitrary control strategies the transition matrices are of the form (3.12). Then, according to (3.34),

$$h^\beta(\epsilon) = (E - \alpha(A^\beta + \epsilon B^\beta))^{-1} g^\beta$$

Since $0 < \alpha < 1$ and A^β is stochastic, we obtain the expansion in ϵ

$$h^\beta(\epsilon) = h^\beta + \epsilon h^{\beta 1} + \cdots \tag{3.35}$$

where

$$h^\beta = (E - \alpha A^\beta) g^\beta$$

This means that the control strategy that solves the reduced problem will be approximately, to within $O(\epsilon)$, optimal for the perturbed chain. As in (3.11), the reduced problem is decomposed here into k independent subproblems, each consisting of finding the controls for the states from Ω_i that maximize the corresponding components of

$$h^{i\beta} = (E - A^{i\beta})^{-1} g^{i\beta}$$

where $g^{i\beta} \triangleq \{g_r^{\beta_r}\}$, $r \in \Omega_i$, $i = 1,2,\ldots,k$.

We see that the singular effect is absent here. This is explained by the fact that the interactions between blocks begin to appear only after a sufficiently long period of time while, because of discounting, the end of the trajectory does not influence the objective function. Hence, the correspondence between the reduced and perturbed problems is caused by the regularizing influence of the discount factor.

Note that, if α is close to unity and, consequently, the matrices $E - \alpha A^\beta$ are ill-conditioned, the maiximization of (3.32) should be replaced by the maximization of the rate of income growth (see, also, [120]). In this case the problem becomes a singularly perturbed one. A general aggregation procedure will be developed for such problems in the next section.

3.4. An approximation algorithm for Markov programming

Let us construct a suboptimal control strategy that approximately maximizes the rate of income growth in the controlled Markov chain characterized by (3.12)–(3.20), (3.30) and (3.31).

Without reference to perturbation theory, Howard [84] proposed a technique adaptable to such problems where the initial weights defined by (3.3) take a key role similar to that of the simplex multipliers in the standard simplex method. For the basic matrices (3.12), the system (3.3) is rewritten in the form

$$\sigma^\beta(\epsilon) I - g^\beta = (A^\beta + \epsilon B^\beta - E_n)\mu^\beta(\epsilon) \tag{4.1}$$

Lemma 4.1. Let the matrix P^β be indecomposable. Then

$$\mu^\beta(\epsilon) = \frac{\mu^\beta_{-1}}{\epsilon} + \mu^\beta_0 + O(\epsilon) \tag{4.2}$$

where

$$\mu^\beta_{-1} = V_L \alpha^\beta \tag{4.3}$$

and $\alpha^\beta(k \times 1)$ is a vector defined up to a constant by the system

$$\tilde{\sigma}^\beta I - \tilde{g}^\beta = (M^\beta)^T \alpha^\beta \tag{4.4}$$

In (4.1) and (4.4) I is the unit vector of dimension $(n \times 1)$ and $(k \times 1)$, respectively. Further, the following relations are valid:

$$\tilde{\sigma}^\beta I_{n_i} - \bar{g}^{i\beta} = (A^{i\beta} - E_{n_i})\mu_0^{i\beta} \tag{4.5}$$

$$\begin{aligned} &\bar{g}^{i\beta} \triangleq \{\bar{g}^\beta_r\} \ , \ r \in \Omega_i \ , \ |\Omega_i| \triangleq n_i \\ &\bar{g}^\beta_r \triangleq g_r^{\beta_r} + \sum_{j=1}^{k} \alpha^\beta_j \sum_{l \in \Omega_j} b^\beta_{rl} \end{aligned} \tag{4.6}$$

where $\mu_0^{i\beta}$ are the subvectors of μ_0^β $(\mu_0^\beta = (\mu_0^{i\beta}), i = 1,\ldots,k)$ corresponding to the subset Ω_i, and $\tilde{\sigma}^\beta$ is the rate of income growth for the aggregated chain with control strategy β.

Proof. Since the matrix $A^\beta + \epsilon B^\beta$ is indecomposable, the solution to (4.1), which was originally defined up to a constant, will be unique provided it satisfies the restriction

$$\mu_1^\beta(\epsilon) = 0 \tag{4.7}$$

Using the Cramer rule, one can expand the solution to (4,1) and (4.7) into the power series

$$\mu^\beta(\epsilon) = \frac{\mu^\beta_{-l}}{\epsilon^l} + \frac{\mu^\beta_{-(l-1)}}{\epsilon^{l-1}} + \cdots + \frac{\mu^\beta_{-1}}{\epsilon} + \mu^\beta_0 + \cdots \tag{4.8}$$

Substituting (4.8) into (4.1) and taking into account the expansion

$$\sigma^\beta(\epsilon) = \tilde{\sigma}^\beta + \epsilon\tilde{\sigma}_1^\beta + \cdots \tag{4.9}$$

arising from Lemmas 3.2 and 3.3, we have

$$(A^\beta - E_n)\mu_{-l}^\beta = 0 \tag{4.10}$$

$$(A^\beta - E_n)\mu_{-(l-1)}^\beta + B^\beta\mu_{-l}^\beta = 0 \tag{4.11}$$

$$\vdots$$

$$(A^\beta - E_n)\mu_{-1}^\beta + B^\beta\mu_{-2}^\beta = 0 \tag{4.12}$$

$$(A^\beta - E_n)\mu_0^\beta + B^\beta\mu_{-1}^\beta = 0 \tag{4.13}$$

From (4.10) it follows that

$$\mu_{-l}^\beta = V_L\gamma \tag{4.14}$$

where γ $(k \times 1)$ is a vector. The substitution of (4.14) into (4.11) and premultipication by V_β^T lead to

$$(M^\beta)^T\gamma = 0 \Rightarrow Q^\beta\gamma = \gamma$$

But, in accordance with Lemma 3.2, the matrix Q^β is indecomposable. Consequently, either all components of γ are equal to zero or none are. Suppose the latter. Then all the components of μ_{-l}^β are nonzero, and so all components of $\mu^\beta(\epsilon)$ tend to infinity as $1/\epsilon^l$. But the latter contradicts (4.7). Thus,

$$\gamma = 0 \Rightarrow \mu_{-l}^\beta = 0$$

In exactly the same way one can establish

$$\begin{aligned} &\mu_{-(l-1)}^\beta = \cdots = \mu_{-2}^\beta = 0 \\ &\mu_{-1}^\beta = V_L \cdot \alpha \end{aligned} \tag{4.15}$$

To obtain (4.4), it is now sufficient to substitute (4.15) into (4.13) and premultiply the result by V_β^T. The subsystems (4.5) are equivalent to (4.13) since, according to (4.15),

$$B^\beta\mu_{-1}^\beta = B^\beta V_L\alpha = \{\sum_{j=1}^{k}\alpha_j \sum_{l\in\Omega_j} b_{rl}\} \;,\; r = 1,2,\ldots,n$$

The lemma is proved.

For the problem under consideration, application of the well-known Howard procedure results in a hierarchical modification in which the iterations decompose into the action of a coordinating center dealing with the aggregated matrix M^β and into the actions of disjoint "blocks" ("subsystems"), operating with the matrices $A^{i\beta}$.

To describe this in more detail, assume that at some iteration a certain control strategy β is chosen so that the matrices $Q^\beta(M^\beta)$, $A^{i\beta}$, $\tilde{g}^\beta$, $g^{i\beta}$, $i = 1,\ldots,k$ [see (3.17) and (3.30)] are fixed. Assume further that only the matrices $Q^\beta(M^\beta)$, $\tilde{g}^\beta$ are available to the coordinator while each "block" i "knows" only the matrices $A^{i\beta}$, $g^{i\beta}$ and the corresponding sets of admissible controls $\beta_r \in I_r$, $r \in \Omega_i$. Then the next iteration proceeds as follows.

At the first stage, the center:

(1) Calculates the final probability vector $\tilde{p}^\beta = \{\tilde{p}_i^\beta\}$, $i = 1,\ldots,k$ of the aggregated process

$$(Q^\beta)^T\tilde{p}^\beta = \tilde{p}^\beta \ , \ \sum_{i=1}^{k} \tilde{p}_i^\beta = 1 \tag{4.16}$$

(2) Finds the vector $\alpha^\beta \triangleq \{\alpha_i^\beta\}$, $i = 1,\ldots,k$ of initial weights and the rate of the income growth $\tilde{\sigma}^\beta$ for the aggregated chain

$$\tilde{\sigma}^\beta I - \tilde{g}^\beta = (M^\beta)^T\alpha^\beta \ \Rightarrow \tag{4.17}$$

$$\tilde{\sigma}^\beta = \sum_{i=1}^{k} \tilde{g}_i^\beta \tilde{p}_i^\beta \tag{4.18}$$

(3) Transmits information about α^β, $\tilde{\sigma}^\beta$ to the "blocks".

At the second stage, each i-*th "block"*:

(1) Changes its one-step expected income by the formula

$$\bar{g}_r^{\gamma_r} \triangleq g_r^{\gamma_r} + \sum_{j=1}^{k} \alpha_j^\beta \sum_{l\in\Omega_j} b_{rl}^{\gamma_r} \ , \ r = 1,\ldots,n \tag{4.19}$$

adapting them for the objectives of the system on the whole. Note that such an adaptation makes it profitable for the block i to choose controls that raise the probability of transition from Ω_i into any of the subsets Ω_j characterized by greater weights α_j^β.

(2) Finds the initial values $\mu_0^{i\beta}$ from the system

$$\tilde{\sigma}^\beta I_{n_i} - \bar{g}^{i\beta} = (A^{i\beta} - E_{n_i})\mu_0^{i\beta} \tag{4.20}$$

(3) For each $r \in \Omega_i$ solves the problems

$$\max_{\gamma_r \in I_r} \{ \sum_{l \in \Omega_i} a_{rl}^{\gamma_r} \mu_{0l}^\beta + \bar{g}_r^{\gamma_r} \} \triangleq \max_{\gamma_r \in I_r} \Delta_r(\beta, \gamma_r) \tag{4.21}$$

and replaces β_r by γ_r maximizing the objective in (4.21). If the control β_r itself maximizes in (4.21), it is kept constant: $\beta_r \triangleq \gamma_r$.

(4) Constructs, using new controls γ_r^*, the new chain with the transition matrix $A^{i\gamma}$ and finds the final probabilities

$$A^{i\gamma} p^{i\gamma} = p^{i\gamma} \ , \quad \sum_{r \in \Omega_i} p_r^{i\gamma} = 1 \tag{4.22}$$

(5) Calculates, according to (3.17) and (3.30), the i-th row of the aggregated matrix

$$Q_{ij}^\gamma \triangleq \epsilon \sum_{r \in \Omega_i} p_r^{i\gamma} \sum_{l \in \Omega_j} b_{rl}^{\gamma_r} \ , \quad i \neq j$$

$$Q_{ii}^\gamma \triangleq 1 + \epsilon \sum_{r \in \Omega_i} p_r^{i\gamma} \sum_{l \in \Omega_i} b_{rl}^{\gamma_r}$$

and the i-th aggregated income

$$\tilde{g}_i^\gamma = \sum_{r \in \Omega_i} p_r^{i\gamma} g_r^{\gamma_r}$$

(6) Transmits the latter information to the center.

This finishes the iteration and the next one begins. The process is repeated until two values of $\tilde{\sigma}^\beta$ obtained at consecutive iterations are equal.

Theorem 4.1.

Let the matrix P^β be indecomposable. Then the procedure converges in a finite number of steps to the control strategy β^* that maximizes the aggregated rate of income growth

$$\tilde{\sigma}^{\beta^*} \triangleq \max_{\beta} \tilde{\sigma}^{\beta}$$

and so maximizes the rate of income growth for the perturbed problem to within $O(\epsilon)$.

If β^* is uniquely defined, it will be the optimal control strategy for the perturbed problem given that ϵ is sufficiently small.

Proof. Let $\tilde{\sigma}^{\beta}$ and $\sigma^{\beta+1}$ be the rates of income growth of the aggregated chains corresponding to the control strategies that were obtained at consecutive iterations. Using (4.5) one can write

$$\begin{aligned}\tilde{\sigma}^{\beta+1} - \tilde{\sigma}^{\beta} = & \{g_r^{\beta+1} + \sum_{j=1}^{k} \alpha_j^{\beta+1} \sum_{l\in\Omega_j} b_{rl}^{\beta+1} + \sum_{l\in\Omega_i} a_{rl}^{\beta+1}\mu_{0l}^{\beta+1} - \mu_{0r}^{\beta+1}\} - \\ & - \{g_r^{\beta} + \sum_{j=1}^{k} \alpha_j^{\beta} \sum_{l\in\Omega_j} b_{rl}^{\beta} + \sum_{l\in\Omega_i} a_{rl}^{\beta}\mu_{0l}^{\beta} - \mu_{0r}^{\beta}\}\end{aligned}$$

or, with new notation,

$$\begin{aligned}\tilde{\sigma}^{\beta+1} - \tilde{\sigma}^{\beta} = & \nu_r^{\beta+1} + \sum_{j=1}^{k} (\alpha_j^{\beta+1} - \alpha_j^{\beta}) \sum_{l\in\Omega_j} b_{rl}^{\beta+1} + \\ & + \sum_{l\in\Omega_i} (\mu_{0l}^{\beta+1} - \mu_{0l}^{\beta})\, a_{rl}^{\beta+1} - (\mu_{0r}^{\beta+1} - \mu_{0r}^{\beta})\end{aligned} \tag{4.23}$$

where

$$\begin{aligned}\nu_r \triangleq \; & g_r^{\beta+1} + \sum_{j=1}^{k} \alpha_j^{\beta} \sum_{l\in\Omega_i} b_{rl}^{\beta+1} + \sum_{l\in\Omega_i} a_{rl}^{\beta+1}\mu_{0l}^{\beta} - \\ & - g_r^{\beta} - \sum_{j=1}^{k} \alpha_j^{\beta} \sum_{l\in\Omega_j} b_{rl}^{\beta} - \sum_{l\in\Omega_i} a_{rl}^{\beta}\mu_{0l}^{\beta}\end{aligned} \tag{4.24}$$

Let us multiply (4.23) by $p_r^{i(\beta+1)}$ and sum them over all $r \in \Omega_i$. Then multiply the result by $\tilde{p}_i^{\beta+1}$ and sum over $i \in \overline{1,k}$. We obtain

$$\tilde{\sigma}^{\beta+1} - \tilde{\sigma}^{\beta} = \sum_{i=1}^{k} \tilde{p}_i^{\beta+1} \sum_{r\in\Omega_i} p_r^{i(\beta+1)} \nu_r^{\beta+1} \tag{4.25}$$

By (4.21),

$$\nu_r^{\beta+1} \geq 0 \Rightarrow \tilde{\sigma}^{\beta+1} \geq \tilde{\sigma}^{\beta} \tag{4.26}$$

Since the number of different control strategies is finite, it is sufficient to show that if

$$\tilde{\sigma}^{\beta} = \tilde{\sigma}^{\beta+1} \tag{4.27}$$

at two consecutive iterations, $\tilde{\sigma}^{\beta}$ is the maximal one among all other possible aggregated rates. Assume this is false, i.e., that in spite of (4.27), there exists a control strategy ς such that

$$\tilde{\sigma}^{\varsigma} > \tilde{\sigma}^{\beta} \tag{4.28}$$

In a way similar to (4.25), one can establish that

$$\tilde{\sigma}^{\varsigma} - \tilde{\sigma}^{\beta} = \sum_{i=1}^{k} \tilde{p}_i^{\varsigma} \sum_{r \in \Omega_i} p_r^{i\varsigma} \nu_r^{\varsigma}$$

However, it follows from (4.27) and (4.21) that

$$\nu^{\varsigma} \leq \nu^{\beta+1} = 0 \Rightarrow \tilde{\sigma}^{\varsigma} \leq \tilde{\sigma}^{\beta}$$

which contradicts the supposition. QED.

Remark 1. Note that all the results are applicable to controlled Markov processes of Poisson type, i.e., to the processes described by the differential equations

$$\begin{aligned} \dot{P}_{\beta} &= (\Lambda^{\beta}(\epsilon))^T P_{\beta} \\ \Lambda^{\beta}(\epsilon) &= \Lambda + B\epsilon \;\;, \;\; \Lambda = \text{diag}\,\{\Lambda_1,\ldots,\Lambda_k\} \end{aligned} \tag{4.29}$$

Remark 2. A similar procedure can be used for the optimization of closed Leontief models composed of weakly connected subsystems.

Remark 3. The procedure described above is similar to the Dantzig–Wolfe method if the latter is applied to the regularized reduced problem (3.11′) and (3.11″), with (3.11″) playing the role of connecting constraints.

3.5. An iterative algorithm for suboptimization

Consider the perturbed linear program

$$F^*(\epsilon) = \max\,\{(c_0^T + \epsilon c_1^T)x \,/\, (A + \epsilon B)x = b_0 + \epsilon b_1 \;,\; x \geq 0\} \tag{5.1}$$

Assume that there exists a vector v such that

$$v^T A = 0 \ , \ v^T b_0 = 0 \tag{5.2}$$

According to Theorem 2.3, to obtain an approximate solution an additional constraint

$$v^T B x = v^T b_1$$

must be added, which may lead to a gap of $F^*(\epsilon)$ at $\epsilon = 0$. This may also be seen from the behavior of the dual solution that define the derivative of $F^*(\epsilon)$. Consider a basis $A_b + \epsilon B_b$ of (5.1)

$$\det\,(A_b + \epsilon B_b) \neq 0 \tag{5.3}$$

The vector of simplex multipliers associated with $A_b + \epsilon B_b$ is defined by the system

$$(A_b + \epsilon B_b)^T \lambda^b(\epsilon) = c_0^b + \epsilon c_1^b \tag{5.4}$$

where c_0^b, c_1^b are subvectors of c_0, c_1, corresponding to the basic variables.

Since the rows of A are linearly dependent, those of A_b are linearly dependent also and

$$\det\, A_b = 0$$

Using the Cramer rule, one can write the solution of (5.4) in the form

$$\lambda^b(\epsilon) = \frac{\lambda_{-l}^b}{\epsilon^l} + \frac{\lambda_{-(l-1)}^b}{\epsilon^{l-1}} + \cdots + \frac{\lambda_{-1}^b}{\epsilon} + \lambda_0^b + \epsilon\lambda_1^b + \cdots \tag{5.5}$$

where l is some natural number. Thus, the expansion of simplex multipliers contains negative powers of ϵ. This means, in particular, that when $A^b + \epsilon B^b$ is an optimal basis, solutions of the problem dual to (5.1) tend to ∞ with $\epsilon \to 0$, and the aforementioned gaps may appear.

It is easy to formulate a condition allowing us to refine the expansion (5.5).

Lemma 5.1. Let $\{v_1^L, \ldots, v_k^L\}$, $\{v_1^b, \ldots, v_k^b\}$ be bases for the left and right eigen subspaces of the matrix A. Assume that

$$\det\, M_b \neq 0 \ , \ M_b \triangleq V_L^T B_b V_b \tag{5.6}$$

where V_L, V_b are $(m \times k)$-matrices constructed of v_i^L, $i = 1,\dots,k$, v_i^b, $i = 1,\dots,k$, respectively.

Then $\lambda^b(\epsilon)$ is presented in the form

$$\lambda^b(\epsilon) = \frac{\lambda^b_{-1}}{\epsilon} + \lambda^b_0 + \epsilon\lambda^b_1 + \cdots \tag{5.7}$$

and, moreover,

$$\lambda^b_{-1} = V_L\tilde{\lambda}^b \tag{5.8}$$

where λ_b $(k \times 1)$ is the solution to the system

$$M_b^T\tilde{\lambda}^b = V_b^T c_0^b \;\Leftrightarrow\; \tilde{\lambda}^b = M_b^{-T} V_b^T c_0^b \tag{5.9}$$

The vector λ_0^b is a particular solution to the system

$$(A_b)^T\lambda_0^b = c_0^b - B_b^T\lambda^b_{-1} \tag{5.10}$$

Proof. Substituting (5.5) into (5.4) and equating coefficients in equal powers of ϵ, we have

$$A_b^T\lambda^b_{-l} = 0 \tag{5.11}$$

$$A_b^T\lambda^b_{-(l-1)} + B_b^T\lambda^b_{-l} = 0 \tag{5.12}$$

$$\vdots \qquad\qquad \vdots$$

$$A_b^T\lambda^b_{-1} + B_b^T\lambda^b_{-2} = 0 \tag{5.13}$$

$$A_b^T\lambda_0^b + B_b^T\lambda^b_{-1} = 0 \tag{5.14}$$

Acording to (5.11), $\lambda^b_{-l} = V_L\alpha$ where $\alpha \triangleq \{\alpha_i\}$, $i = 1,\dots,k$. This, together with (5.11) premultiplied by V_b^T, leads to

$$M_b^T\alpha = 0 \Rightarrow \alpha = 0 \Rightarrow \lambda^b_{-l} = 0$$

Similarly, it is easy to verify that

$$\lambda_{-(l-1)} = \cdots = \lambda_{-2} = 0$$

Hence, the relation (5.13) takes the form

$$(A_b)^T \lambda_{-1}^b = 0 \Rightarrow \lambda_{-1}^b = V_L \alpha \ , \ \alpha \triangleq \{\alpha_i\} \ , \ i = 1,\ldots,k \tag{5.15}$$

Finally, the use of the latter in (5.14) gives, after premultiplication by V_b^T,

$$M_b^T \alpha = V_b^T c_0^b \Rightarrow \alpha = c_0^b$$

Thus, (5.7)–(5.9) are proved and (5.10) is equivalent to (5.14).

Suppose now that the columns of the matrix V_L form a basis of $\ker A^T$, i.e., the left eigen subspace of A corresponding to the zero eigenvalue

$$V_L^T A = 0 \tag{5.16}$$

Let us calculate estimates $\Delta_r(\epsilon)$ of the nonbasic columns

$$\Delta_r(\epsilon) = (\lambda^b(\epsilon))^T (A_r + \epsilon B_r) - c_{r0} - \epsilon c_{r1} = \Delta_r + \epsilon \Delta_{r1} + \cdots \tag{5.17}$$

where

$$\Delta_r \triangleq (\lambda_{-1}^b)^T b_r + (\lambda_0^b)^T A_r - c_{r0} \ , \ \Delta_{r1} = (\lambda_0^b)^T B_r - c_{r1}$$

There are no negative powers in (5.17) since, by Lemma 5.1,

$$(\lambda_{-1}^b)^T A_r = (\tilde{\lambda}^b)^T V_L^T A_r = 0$$

In Section 1.2 a computational procedure was proposed for regular linear programming problems. The main feature distinguishing this procedure from an ordinary simplex method is that, at every iteration, instead of column estimates, the dominant terms of power expansions (5.17) are used. As was pointed out, the procedure appears to converge in a finite number of iterations. The (5.17) expansions enable us to construct a similar procedure for the singularly perturbed linear programs. However, generally speaking, the strategy thereby obtained will not be the optimal one for the reduced program since the dominant terms of the column estimates essentially depend on the perturbing matrix B.

To construct such a procedure, let us show that the other quantities used in forming a new basis can be developed into the power series of ϵ:

$$X^r(\epsilon) = X_0^r + \epsilon X_1^r + \cdots \tag{5.18}$$

$$x^b(\epsilon) = x_0^b + \epsilon x_1^b + \cdots \tag{5.19}$$

where $X^r(\epsilon)$ is the solution of the system

$$(A_b + \epsilon B_b)X^r(\epsilon) = A_r + \epsilon B_r \tag{5.20}$$

representing the coefficients of the expansion of the column $A_r + \epsilon B_r$ over the columns of the basis, and where $x^b(\epsilon)$ is the basic solution

$$(A_b + \epsilon B_b)x^b(\epsilon) = b_0 + \epsilon b_1 \tag{5.21}$$

Lemma 5.2. Let the conditions of Lemma 5.1 be satisfied, and let the columns of the matrix V_L form a basis for the left eigen subspace of A corresponding to the zero eigenvalue.

Then (5.18) and (5.19) are true, with X_0^r in (5.18) taking the form

$$X_0^r = d^r + \sum_{i=1}^{k} y_i^r v_i^b \triangleq d^r + V_b y^r \tag{5.22}$$

where d^r is a particular solution to the system

$$A_b d^r = A_r \tag{5.23}$$

and $y^r \triangleq \{y_i^r\}$, $i = 1,\ldots,k$ is the solution to the system

$$M_b y^r = V_L^T B_r - V_L^T B_b d_r \tag{5.24}$$

Proof. First, we show the validity of (5.18). Similarly to (5.7), we can write down for some natural number l

$$X^r(\epsilon) = \frac{X_{-l}^r}{\epsilon^l} + \frac{X_{-(l-1)}^r}{\epsilon^{l-1}} + \cdots + \frac{X_{-1}^r}{\epsilon} + X_0^r + \epsilon X_1^r + \cdots \tag{5.25}$$

The following relations arise, as in Lemma 1.1, after the substitution of (5.25) into (5.20):

$$\begin{aligned} &X_{-l}^r = X_{-(l-1)}^r = \cdots = X_{-2}^r = 0 \\ &A_b X_{-1}^r = 0 \end{aligned} \tag{5.26}$$

$$A_b X_0^r + B_b X_{-1}^r = A_r \tag{5.27}$$

$$A_b X_1^r + B_b X_0^r = B_r \tag{5.28}$$

In accordance with (5.26), $X_{-1}^r = V_b \beta$, $\beta \triangleq \{\beta_i\}$, $i = 1,2,\ldots,k$. Substituting this into (5.27) and premultiplying by V_L^T, we have

$$M_b \beta = 0 \Rightarrow \beta = 0 \Rightarrow X_{-1}^r = 0$$

The equations (5.27) and (5.28) take the form

$$A_b X_0^r = A_r \tag{5.29}$$

$$A_b X_1^r + B_b X_0^r = B_r \tag{5.30}$$

Multiply the latter by V_L^T

$$V_L^T B_b X_0^r = V_L^T B_r \tag{5.31}$$

The solution of the system (5.29) and (5.31) is unique. Assume the contrary, i.e., that there exists a nontrivial vector u such that

$$A_b u = 0 \ , \ \|u\| = 1 \tag{5.32}$$

$$V_L^T B_b u = 0 \tag{5.33}$$

By (5.32), $u = V_b \beta$ and, consequently,

$$(V_L^T B_b V_b)\beta = 0 \Leftrightarrow M_b \beta = 0 \Rightarrow \beta = 0 \Rightarrow u = 0$$

which contradicts the definition of u.

Let us prove now that the solution to (5.29) and (5.30) can be written in the form (5.22). Any vector

$$X_0^r = d^r + V_b y \tag{5.34}$$

will satisfy (5.29) with an arbitrary vector $y \triangleq \{y_i\}$, $i = 1,2,\ldots,k$. Substitution into (5.34) leads to the system (5.24), which has a unique solution. Thereby (5.18), (5.22)–(5.24) are proved. The proof of (5.19) is identical.

Consider the procedure detailed above. As in the ordinary simplex method we shall, at each iteration, construct a new basis from that obtained in the proceeding iteration and the results thereof.

Assume that the following conditions are fulfilled:

(1) The rank of the $(m \times n)$-matrix $A + \epsilon B$ equals m, i.e., its rows are linearly independent if $0 < \epsilon \leq \hat{\epsilon}$.

(2) The feasible set $\Theta(\epsilon)$ in (5.1) is not empty, and there exists a bounded set $\bar{\Theta} \subset R^n$ such that, if $0 < \epsilon \leq \hat{\epsilon}$, then

$$\emptyset \neq \Theta(\epsilon) \subset \bar{\Theta}$$

(3) Each basis $A_b + \epsilon B_b$ of (5.1) satisfies the requirement:

$$\ker A_b^T = \ker A^T \Rightarrow \operatorname{rank} A_b = \operatorname{rank} A$$

(4) For each basis $A_b + \epsilon B_b$

$$\det M_b \neq 0 \ , \ M_b \triangleq V_L^T B_b V_b$$

where V_L $(m \times k)$, V_b $(m \times k)$ are respectively, the matrices, with columns forming bases for the left and right eigen subspaces of A_b corresponding to the zero eigenvalue.

(5) For each basis $A_b + \epsilon B_b$ the basic solution (5.21) is positive by zero-order approximation, i.e., $x_0^b > 0$ in the expansion (5.19).

At the start of iteration t, assume that we know the basis $A_t + \epsilon B_t$, the zero-order approximation x_0^t of the basic solution, and the matrices V_L, V_t formed by bases of $\ker A_t^T$ and $\ker A_t$, correspondingly. Then iteration t proceeds as follows:

A. Calculate an approximate vector of simplex multipliers associated with the basis.

(1) Construct the aggregated matrix

$$M_t(k \times k) \triangleq V_L^T B_b V_t$$

and weighted cost coefficients

$$\tilde{c}^t = V_t^T c_0^t$$

where c_0^t is the subvector of c_0 corresponding to the basis under consideration.

(2) Solve the system

$$M_t^T \tilde{\lambda}^t = \tilde{c}^t$$

and find the vector λ_{-1}^t by the formula

$$\lambda_{-1}^t = V_L \tilde{\lambda}^t$$

(3) Augment the vector c_0 by the vector ν^t

$$\nu^t = c_0 - B^T \lambda_{-1}^t$$

(4) Determine a particular solution to the system

$$A_t^T \lambda_0^t = \nu^{bt}$$

where ν^{bt} is the subvector of ν^t corresponding to the given basis.

B. Determine the column entering the basis. For each nonbasic variable, form the column estimates

$$A_t^T \lambda_0^t = \nu^{bt}$$

and find

$$\Delta_{r^*}^t \triangleq \min_{r \bar{\in} \sigma_t} \Delta_r^t \ , \ \Delta_r^t \triangleq (\lambda_0^t)^T A_r - \nu^t \triangleq (\lambda_0^t)^T A_r + (\lambda_{-1}^t)^T B_r - c_{0r}$$

If $\Delta_{r^*}^t \geq 0$, the calculations cease. Otherwise, r determines which column is introduced into the basis. σ_t is the index set of basic variables at iteration t.

C. Compute the transformed column $X_0^t \triangleq \{X_{0l}^t\}$, $l = 1,\ldots,N$.

(1) Find a particular solution d^t to the system

$$A_t d^t = A^{r^*}$$

(2) Find the solution y^t to the system

$$M_t y^t = V_L^T B_{r^*} - V_L^T B_t d^t$$

(3) Construct X_0^t

$$X_0^t = d^t + V_t y^t$$

D. Determine the vector to be taken out of the basis. Choose l^* according to

$$\min_{X_{0l}^t > 0} \frac{x_{0l}^t}{X_{0l}^t} \triangleq \frac{x_{0l^*}^t}{X_{0l^*}^t}$$

and remove the column with this number from the basis.

E. Prepare the initial data for a new cycle.

(1) Construct the new basis $A_{t+1} + \epsilon B_{t+1}$ replacing the column $A_{l^*} + \epsilon B_{l^*}$ with the column $A_{r^*} + \epsilon B_{r^*}$.
(2) Compute the zero-order approximation x_0^{t+1} of the new basic solution.
(3) Find the matrix V_{t+1} of right eigenvectors of A_{t+1} corresponding to the zero eigenvalues. The matrix of left eigenvectors of A_{t+1} coincides with V_L.

This finishes the iteration and the process is repeated from step A.

As far as the calculations are concerned, the procedure we suggest differs from the standard simplex method in that, first, the inverse of the ill-conditioned matrix $A_b + \epsilon B_b$ $(m \times m)$ is replaced at each iteration by the inverse of the aggregated nonsingular matrix M_b $(k \times k)$ of small dimension $(k < m)$, and, second, by solving systems of the type

$$A_b y = d \tag{5.35}$$

Such an approach appears to be efficient when the structure of the matrix A_b permits some simplification of the systems (5.35) — for example, where it is block-diagonal and where (5.35) decomposes into a series of subsystems.

Theorem 5.1.

Let the conditions 1–5 shown above be fulfilled. Then the procedure converges in a finite number of steps N to some point x_0^N in the reduced program's feasible set independent of which solutions are taken in A(4) and C(1). The procedure is completed if the column estimates are nonnegative

$$\Delta_r^N \geq 0 \quad , \quad r \in \sigma_N \tag{5.36}$$

There exists an $\hat{\epsilon}_1 > 0$ such that, if $0 < \epsilon \leq \hat{\epsilon}_1$, the matrices $A_t + \epsilon B_t$ $(t = 0,1,\ldots,N)$ generated by the procedure are bases of the program (5.1). In particular the matrix $A_N + \epsilon B_N$ obtained at the last iteration N is that basis. The associated basic solution $x^N(\epsilon)$

$$(A_N + \epsilon B_N)x^N(\epsilon) = b \tag{5.37}$$

satisfies the inequality

$$(c_0^N + \epsilon c_1^N)^T x^N(\epsilon) \geq F^*(\epsilon) - M \cdot \epsilon \tag{5.38}$$

where $c_0^N + \epsilon c_1^N$ is the subvector of $c_0 + \epsilon c_1$ corresponding to $x^N(\epsilon)$, i.e., the value of the objective function is equal to the optimal one to within $O(\epsilon)$. In the case where the inequalities (5.36) are strictly satisfied, there exists $\hat{\epsilon}_2 > 0$ such that, for $0 < \epsilon \leq \hat{\epsilon}_2$, the solution to (5.1) is unique, its nonzero components being given by $x^N(\epsilon)$.

Proof. Let the basis $A_t + \epsilon B_t$ of (5.1) be known. It is easy to show that the program (5.1) can be transformed into

$$\begin{aligned} F^*(\epsilon) \triangleq \max \{ f^t(\epsilon) - \sum_{r \bar{\in} \sigma_t} \Delta_r^t(\epsilon) x_r \,/\, \sum_{r \bar{\in} \sigma_t} X^r(\epsilon) x_r \leq \\ \leq x^t(\epsilon) \ , \ x_r \geq 0 \ , \ r \bar{\in} \sigma_t \} \end{aligned} \tag{5.39}$$

where $X^r(\epsilon)$, $x^t(\epsilon)$, $\Delta_r^t(\epsilon)$ are defined in (5.20), (5.21) and (5.11) with

$$A_b + \epsilon B_b \triangleq A_t + \epsilon B_t$$

and

$$f^t(\epsilon) \triangleq (c_0^t + \epsilon c_1^t)^T x^t(\epsilon)$$

In the case where

$$\min_{r \bar{\in} \sigma_t} \Delta_r^t \triangleq \Delta_{r^*}^t \geq 0$$

according to (5.17) and condition 2,

$$F^*(\epsilon) \leq f^t(\epsilon) + \epsilon(\max_{x \in \Theta} \|x\|) \cdot \sum_{r \bar{\in} \sigma_t} |\Delta_{r1}^t + \epsilon \Delta_{r2}^t + \cdots | \tag{5.40}$$

and this is equivalent to (5.38) for sufficiently small ϵ [less than the radius of convergence in (5.40)] if the corresponding notation is used. Similarly, one can obtain a result concerning the strict optimality of $x^t(\epsilon)$ under the condition that $\Delta_{r^*}^t$ is strictly positive.

Suppose now that

$$\Delta_{r^*}^t < 0 \tag{5.41}$$

Let us construct the solution to (5.39) in the form

$$x_{r^*} = \Theta \geq 0 \ , \ x_r = 0 \ , \ r \neq r^* \ , \ r \bar{\in} \sigma_t$$

The variable Θ must satisfy the constraints

$$X^{r^*}(\epsilon)\Theta \leq x^t(\epsilon)$$

Using the notation $X^t(\epsilon)$ instead of $X^{r^*}(\epsilon)$, as it usually does in the ordinary "simplex", put

$$\Theta(\epsilon) = \min_{X_l^t(\epsilon)>0} \frac{x_l^t(\epsilon)}{X_l^t(\epsilon)} \triangleq \frac{x_{l^*}^t(\epsilon)}{X_{l^*}^t(\epsilon)} \tag{5.42}$$

The vector with the coordinates

$$\begin{aligned} &x_r^{t+1}(\epsilon) = 0 \ , \ r \bar{\in} \sigma_t \ , \ r \neq r^* \ ; \ x_{r^*}^{t+1}(\epsilon) = \Theta(\epsilon) = \frac{x_{l^*}^t(\epsilon)}{X_{l^*}^t(\epsilon)} \\ &x_r^{t+1}(\epsilon) = x_r^t(\epsilon) - X_r^t(\epsilon)\frac{x_{l^*}^t(\epsilon)}{X_{l^*}^t(\epsilon)} \ , \ r \in \sigma_t \ , \ r \neq l^* \ ; \ x_{l^*}^{t+1}(\epsilon) = 0 \end{aligned} \tag{5.43}$$

will be the basic feasible solution to (5.1) and the basis $A_{t+1} + \epsilon B_{t+1}$ corresponding to it is obtained from $A_t + \epsilon B_t$ by replacing column $A_{l^*} + \epsilon B_{l^*}$ with column $A_{r^*} + \epsilon B_{r^*}$.

In accordance with Lemma 5.2, one can prove that there exists $\hat{\epsilon}_t > 0$ such that, for $0 < \epsilon \leq \hat{\epsilon}_t$, the index l^* defined in (5.42) coincides with the one defined by

$$\min_{X_{0l}^t \geq 0} \frac{x_{0l}^t}{X_{0l}^t} \triangleq \frac{x_{0l^*}^t}{X_{0l^*}^t} \tag{5.44}$$

where x_0^t, X_0^t are the main terms of the corresponding expansions (5.19) and (5.18). This latter is due to the fact that some of the components of $X_{0l^*}^t$ are positive.

Suppose the contrary. Then, taking into account that $x_0^t > 0$ and (5.42), it is not difficult to establish that the optimal value of the objective function (5.39) grows as $1/\epsilon$ when $\epsilon \to 0$. However, this contradicts condition 2. Moreover, the l^* determined by (5.44) is unique since, otherwise, several components among $x_r^{t+1}(\epsilon)$, $r \in \sigma_{t+1}$ would have zeros as the zero-order approximation, and this contradicts condition 5 of the theorem.

Thus, if (5.41) is true, the procedure under consideration forms the new basis and

$$x_{r^*0}^{t+1} = \Theta_0 = \frac{x_{0l^*}^t}{X_{0l^*}^t} > 0 \ , \ f_0^{t+1} = f_0^t - \Delta_{r^*}^t \Theta_0$$

where $f_0^t (f_0^{t+1})$ is the zero-order approximation to the value of the objective function at iteration $t(t+1)$. Hence,

$$f_0^{t+1} > f_0^t$$

which proves the convergence of the procedure, since the number of vertices of the polyhedron $\Theta(\epsilon)$ is finite.

The formulae of step E of the procedure follow from (5.43), after the substitution of the expansions (5.18) and (5.19). The theorem is proved. A way to form the values $\hat{\epsilon}_1$, $\hat{\epsilon}_2$ and M mentioned in the conditions of the theorem is contained in the proof.

However, in the practical application of the described procedure, there is no need to evaluate these constants *a priori*. Let it be required, for instance, to find the solution to the system

$$\max \{p^T x \, / \, Px = d \ , \ x \geq 0\} \tag{5.45}$$

First, write the matrices in the form

$$P = A + B \ , \ p = c + c_1 \ , \ d = b + b_1$$

where the rows of A are linearly dependent and its structure ensures that the construction of particular solutions to (5.35) is essentially simpler than that of the inverses of the bases in (5.45). Upon application to the singularly perturbed program

$$\max \{(c_0 + \epsilon c_1)^T x \,/\, (A + \epsilon B)x = b + \epsilon b_1 \,,\, x \geq 0\}$$

the procedure converges for any ϵ and, in particular, for $\epsilon = 1$. The square matrix P_b is thereby fixed for the program (5.45).

If the inequality

$$x^b \triangleq P_b^{-1} b \geq 0 \tag{5.46}$$

is fulfilled, P_b is a basis for (5.45). To determine whether it is the optimal one, it is sufficient to check that the column estimates are nonnegative

$$\Delta_r \triangleq c_b^T P_b^{-T} P_r - c_r \geq 0 \tag{5.47}$$

where c_b is the subvector of c corresponding to the basic variables and P_r, $r \bar{\in} \sigma$ are the nonbasic columns of P. As follows from the theorem, if B is sufficiently small, condition (5.46) will be fulfilled and the error in (5.47) will also be small. This latter result means that the indicated basis is localized in some small neighborhood of the optimum and, consequently, is efficient as an initial approximation for solving the problem by any precise method. Such an interpretation corresponds to the intuitively clear supposition that, when far from the optimum, there is no particular need to select carefully the columns improving the current basis: one can make do with column estimates approximated to within $O(\epsilon)$. For the given class of quasi-irregular linear programs, the realization of this idea takes the form of an iterative aggregation.

The connection of this form with well-known schemes of the compact inverse method will now be shown.

3.6. An artificial introduction of singular perturbations in compact inverse methods

The use of variants of the simplex method with the compact inverse is one of the foremost approaches to large-scale linear programming problems. Arranging the calculations to take advantage of structure leads to very efficient iterations and thereby diminishes the computational effort and memory requirements.

The presentation of inverses arising from the products of smaller-dimension matrices generates a number of such methods. The procedure of the previous section appears to be one among these methods if singular perturbations are introduced in a certain artificial way. To show this, let us first consider the application of singular perturbations to the solution of linear nonsingular systems of equations. Let G_b $(m \times m)$ be a basis for some linear program where it is necessary to find the solution of the system

$$G_b x = g \quad , \quad \det G_b \neq 0 \tag{6.1}$$

We shall divide the matrix G into the three submatrices

$$G_1^b(s \times m) \quad , \quad G_2^b(s \times m) \quad , \quad G_3^b(m{-}2s \times m) \quad , \quad m \geq 2s$$

With a corresponding partition of the vector g $(m \times 1)$, the system (6.1) can be presented as follows:

$$\begin{aligned} G_1^b x &= g_1 \\ G_2^b x &= g_2 \\ G_3^b x &= g_3 \end{aligned} \tag{6.2}$$

Multiplying the second group of equations in (6.2) by $1/\epsilon$ leads to the system

$$\begin{aligned} (G_2^b + \epsilon G_1^b)x &= g_2 + \epsilon g_1 \\ G_2^b x &= g_2 \\ G_3^b x &= g_3 \end{aligned} \tag{6.3}$$

which is equivalent to (6.2) for any $\epsilon \neq 0$. Using the notation

$$A_b \triangleq \begin{pmatrix} G_2^b \\ G_2^b \\ G_3^b \end{pmatrix} \; , \; B_b \triangleq \begin{pmatrix} G_1^b \\ 0 \\ 0 \end{pmatrix} \; , \; b_0 \triangleq \begin{pmatrix} g_2 \\ g_2 \\ g_3 \end{pmatrix} \; , \; b_1 = \begin{pmatrix} g_1 \\ 0 \\ 0 \end{pmatrix} \tag{6.4}$$

we can write this in the form

$$(A_b + \epsilon B_b)x = b_0 + \epsilon b_1 \tag{6.5}$$

The matrix A_b is singular. The basis of $\ker A_t^T$ is determined by the columns of the $(m \times n)$-matrix V_L

$$V_L = \begin{pmatrix} E_s \\ -E_s \\ 0 \end{pmatrix} \tag{6.6}$$

where E_s $(s \times s)$ is the identity matrix of dimension s. To determine the right eigenvectors we write G_b in the following way

$$G_b \triangleq \begin{pmatrix} G^b_{1d} & G^b_{1N} \\ G^b_{2d} & G^b_{2N} \\ G^b_{3d} & G^b_{3N} \end{pmatrix} \tag{6.7}$$

where the submatrixes G^b_{2N} $(s \times m-s)$, G^b_{3N} $(m-2s \times m-s)$ form the nonsingular matrix G^b_N $(m-s \times m-s)$

$$G^b_N \triangleq \begin{pmatrix} G^b_{2N} \\ G^b_{3N} \end{pmatrix}, \quad \det G^b_N \neq 0 \tag{6.7$'$}$$

It is always possible to obtain such a representation by rearranging the columns of the matrix G_b for

$$\operatorname{rank} \begin{pmatrix} G^b_2 \\ G^b_3 \end{pmatrix} = m - s$$

The representation (6.7) together with the notation

$$G^b_d \triangleq \begin{pmatrix} G^b_{2d} \\ G^b_{3d} \end{pmatrix}$$

mean that the columns of the $(m \times s)$ matrix

$$V_b \triangleq \begin{pmatrix} -E_s \\ (G^b_N)^{-1}\, G^b_d \end{pmatrix} \tag{6.8}$$

form the basis of $\ker A_b$. In accordance with the notation of the previous section, let us construct the aggregated matrix M_b $(s \times s)$

$$M_b \triangleq V_L^T B_b V_b = (G_{1d}^b \; G_{1N}^b) \begin{pmatrix} -E_s \\ (G_N^b)^{-1} G_d^b \end{pmatrix} = G_{1N}^b (G_N^b)^{-1} G_d^b - G_{1d}^b \tag{6.9}$$

Lemma 6.1. If the matrix G_b is nonsingular, the matrix M_b is nonsingular as well.

Proof. The solution x of system (6.1) exists, is unique, and does not depend on ϵ. Hence, system (6.5) is equivalent to two systems:

$$A_b x = b_0 \Leftrightarrow G_2^b x = g_2 \;,\quad G_3^b x = g_3 \tag{6.10}$$

$$B_b x = b_1 \Leftrightarrow G_1^b x = g_1 \Leftrightarrow V_L^T B_b x = V_L^T b_1 \tag{6.11}$$

The general solution to (6.10) is of the form

$$x_{\text{gen}} = x_{\text{part}} + V_b y \tag{6.12}$$

where x_{part} is a particular solution to (6.10); for example,

$$x_{\text{part}} = \begin{pmatrix} 0 \\ (G_N^b)^{-1} \bar{g} \end{pmatrix} \begin{matrix} s \\ m-2 \end{matrix} \;,\quad \bar{g} \triangleq \begin{pmatrix} g_2 \\ g_3 \end{pmatrix} \tag{6.13}$$

where y $(s \times 1)$ is an arbitrary vector. Substituting (6.12) in (6.11), we obtain

$$M_b y = V_L^T b_1 - V_L^T B_b x_{\text{part}} \tag{6.14}$$

The solution to (6.14) exists because the solution to (6.10) and (6.11) exists.

The result will be proved if we prove that the solution to (6.14) is unique. Suppose the contrary, i.e., assume there exist two solutions, y_1 and y_2,

$$y_1 \neq y_2 \Rightarrow V_b y_1 \neq V_b y_2 \Rightarrow x_1 \triangleq x_{\text{part}} + V_b y_1 \neq x_{\text{part}} + V_b y_2 \triangleq x_2$$

However, x_1, x_1, as defined, satisfy (6.10) and (6.11). Consequently, they are solutions to (6.1) that contradicts the uniqueness of its solution. The lemma is proved.

The proof contains the following algorithm to solve (6.1).

(1) Transform system (6.1) into the form (6.3).
(2) Write (6.7) and invert the matrix G_N^b $(m-s \times m-s)$, which is, by definition, square and nonsingular.
(3) Construct the aggregated matrix M_b $(s \times s)$ and a particular solution x_{part} of the system (6.10) by the formulae (6.9) and (6.13), respectively.

(4) Find the solution y of the nonsingular system (6.14).
(5) Obtain the solution according to formula (6.12).

All this can be summed up by writing

$$x \triangleq (G_b)^{-1} g = \begin{pmatrix} 0 \\ (G_N^b)^{-1} \bar{g} \end{pmatrix} + \begin{pmatrix} -E_s \\ (G_N^b)^{-1} G_d^b \end{pmatrix} M_b^{-1} (g_1 - G_{1N}^b (G_N^b)^{-1} \bar{g}) \tag{6.15}$$

where $\bar{g}$ is as defined in (6.13). In the same way the inverse G_b^{-1} can easily be expressed:

$$G_b^{-1} = \begin{pmatrix} 0 & 0 \\ 0 & (G_N^b)^{-1} \end{pmatrix} + \begin{pmatrix} -E_s \\ (G_N^b)^{-1} G_d^b \end{pmatrix} M_b^{-1} (E_s \, , \, -G_{1N}^b (G_N^b)^{-1}) \tag{6.16}$$

The application of the described technique to the solution of (6.1) is efficient if G is originally divided into two parts: one having a particular simple structure, and the other as a perturbation that can be denoted as $(G_2^T, G_3^T)^T$ and G_1, respectively. For instance, if the matrix $(G_2^T, G_3^T)^T$ is block-diagonal and, consequently, so is the matrix $G_N^b \triangleq \{G_N^{b1}, \ldots, G_N^{bk}\}$, then the inverse of G_N^b is decomposable into k independent inverses of the matrices G_N^{bi} of lower dimensions.

Consider now the linear program

$$\max \{c^T x \,/\, Gx = g \, , \, x \geq 0\} \tag{6.17}$$

where G $(m \times n)$ is a matrix $(n > m)$. Suppose, as above, that G is divided into the three submatrices G_1 $(s \times n)$, G_2 $(s \times n)$, G_3 $(2s \times n)$. The corresponding singularly perturbed problem takes the form

$$\max \{c^T x \,/\, (A + \epsilon B) x = b_0 + \epsilon b_1 \, , \, x \geq 0\} \tag{6.18}$$

where

$$A = \begin{pmatrix} G_2 \\ G_2 \\ G_3 \end{pmatrix} , \quad B = \begin{pmatrix} G_1 \\ 0 \\ 0 \end{pmatrix} , \quad b_0 = \begin{pmatrix} g_2 \\ g_2 \\ g_3 \end{pmatrix} , \quad b_1 = \begin{pmatrix} g_1 \\ 0 \\ 0 \end{pmatrix} \tag{6.19}$$

Programs (6.17) and (6.18) are equivalent. In particular, each basic solution to (6.17) is the basic solution to (6.18) and vice versa. Hence, by the lemma, supposing that (6.17) is nonsingular, the basic solutions to (6.18) satisfy the conditions of Theorem 5.1, ensuring the convergence of the procedure described in the previous section.

Let us show that the sequence of basic solutions to (6.18) generated by this procedure coincides with that of the program (6.17) generated by the standard simplex-method procedure provided that the process starts from the same initial basis. Let the basic solutions to (6.17) and (6.18) coincide up to iteration t, i.e., in particular let the bases G_t, $A_t + \epsilon B_t$ satisfy the relations

$$A_t = \begin{pmatrix} G_2^t \\ G_2^t \\ G_3^t \end{pmatrix}, \quad B_t = \begin{pmatrix} G_1^t \\ 0 \\ 0 \end{pmatrix}, \quad G_t = \begin{pmatrix} G_1^t \\ G_2^t \\ G_3^t \end{pmatrix}, \quad \det G_t \neq 0 \tag{6.20}$$

The result will be proved if the column leaving the basis and that taking its place are the same for both programs. According to (6.20) and (6.19) the basic solutions to the systems

$$(A_t + \epsilon B_t)x^t = b_0 + \epsilon b_1$$
$$G_t x^t = g$$

coincide. In the same way the solutions to the systems

$$(A_t + \epsilon B_t)X^r = A^r + \epsilon B^r$$
$$G_t X^r = G^r$$

coincide, where

$$A^r \triangleq \begin{pmatrix} G_2^r \\ G_2^r \\ G_3^r \end{pmatrix}, \quad B^r \triangleq \begin{pmatrix} G_1^r \\ 0 \\ 0 \end{pmatrix}$$

and G^r is a nonbasic column of the matrix G

$$G^r \triangleq \begin{pmatrix} G_1^r \\ G_2^r \\ G_3^r \end{pmatrix} \tag{6.21}$$

Let us prove the coincidence of the column estimates. In accordance with the standard simplex procedure, the column estimates are given in (6.17) by the relations

$$\Delta_r = (c^t)^T G_t^{-1} G^r - c_r \ , \ r \bar{\in} \sigma_t \tag{6.22}$$

where c^t is the subvector of c corresponding to the basic variables. Using the representation (6.16) for G_t^{-1} and taking into account (6.21), we can write

$$\begin{aligned}\Delta_r = \ & (c_N^t)^T (G_N^t)^{-1} \begin{pmatrix} G_2^r \\ G_3^r \end{pmatrix} + (c^t)^T \begin{pmatrix} -E_s \\ (G_N^t)^{-1} G_d^t \end{pmatrix} M_t^{-1} G_1^r - \\ & - (c^t)^T \begin{pmatrix} -E_s \\ (G_N^t)^{-1} G_d^t \end{pmatrix} M_t^{-1} G_{1N}^t (G_N^t)^{-1} \begin{pmatrix} G_2^r \\ G_3^r \end{pmatrix} - c_r\end{aligned} \tag{6.23}$$

(c_N^t is composed of the last $m - s$ components of c^t).

Let us now calculate the approximate column estimates in (6.18) following the procedure of Section 3.5. We have that

$$\Delta_r = \lambda_{-1}^T B^r + \lambda_0^T A^r - c_r \tag{6.24}$$

where

$$\lambda_{-1}^T = \tilde{\lambda}^T V_L^T \ , \ \tilde{\lambda}^T M_t = c_t^T V_t \tag{6.25}$$

and where λ_0 is a particular solution to the system

$$\lambda_0^T A_t = (c^t)^T - \lambda_{-1}^T B_t \tag{6.26}$$

Equations (6.25), (6.6) and (6.8) give

$$\lambda_{-1}^T = c_t^T V_t M_t^{-1} V_L^T = c_t^T \begin{pmatrix} -E_s \\ (G_N^t)^{-1} G_d^t \end{pmatrix} M_t^{-1} (E_s , -E_s , 0) \tag{6.27}$$

Hence, the system (6.26) takes the form

$$\lambda_0^T = \begin{pmatrix} G_2^t \\ G_2^t \\ G_3^t \end{pmatrix} = (c^t)^T - (c^t)^T \begin{pmatrix} -E_s \\ (G_N^t)^{-1} G_d^t \end{pmatrix} M_t^{-1} G_1^t \tag{6.28}$$

Its particular solution can clearly be written in the form

$$\lambda_0^T = (\overset{s}{x}\,,\ \overset{m-s}{\bar{\lambda}_0^T}) \tag{6.29}$$

where

$$\bar{\lambda}_0^T = \left[(c_N^t)^T - (c^t)^T \begin{pmatrix} -E_s \\ (G_N^t)^{-1} G_d^t \end{pmatrix} M_t^{-1} G_{1N}^t \right] (G_N^t)^{-1} \tag{6.30}$$

Substitution of (6.27), (6.29) and (6.30) into (6.24) leads to the formula coinciding with (6.23).

Thus, the procedure of the previous section requires the same number of iterations to get the solution as the standard one. On the other hand, for some classes of problems [for example, for the problems with block-diagonal $(G_2^T, G_3^T)^T$], each iteration of the former requires less computational effort, making this procedure preferable on the whole.

Note, in conclusion, that the right-hand eigenvectors V_t of the matrices A_t forming the bases

$$A_t + \epsilon B_t \triangleq \begin{pmatrix} G_2^t + \epsilon G_1^t \\ G_2^t \\ G_3^t \end{pmatrix}$$

are defined by (6.8). The matrix $(G_N^t)^{-1}$ changes from one iteration to the next one only if the index of the column leaving the basis is contained in the set of indices of the columns forming G_N^t; otherwise, $(G_N^t)^{-1}$ does not change. Note also that it is possible to reconstruct $(G_N^t)^{-1}$ from iteration to iteration when it does change. For block-diagonal $(G_2^T, G_3^T)^T$ the procedure has been proposed (although in quite different contexts) in various papers and is well-known as the compact inverse procedure (see, for example, [48, Chapter 5]).

CHAPTER 4

The Perturbation Method in Stochastic Programming

Optimization problems arising from stochastic systems are the most complicated ones, emphasizing the importance of suboptimal decision making.

In Section 4.1, as an introduction, one- and two-stage problems are considered. First, it is supposed that a probability density can be divided into a principal part and a perturbation. This corresponds to the hypothesis that some very rare random factors, e.g., damages, occur along with stable factors of uncertainty. Under this supposition, the one-stage stochastic program can be transformed to the standard perturbed program investigated in detail in Chapter 1. Another scheme of application for the perturbation method arises if perturbations are isolated in the functional description with a fixed probability distribution. Its efficacy is demonstrated using a model of a two-level control system in which the coordinating center may influence the low-level elements by the distribution of small resources only.

Section 4.2 is more important from the practical point of view. The optimal control problem is investigated here under the condition that the random disturbances are small. Quasilinear systems associated with a quadratic objective function are also considered in detail. The main result, which is valid for both classes of problems is the recommendation that the optimal solution to a reduced deterministic problem is used as a suboptimal control for the stochastic system. The situation where the reduced deterministic program has a nonunique solution is also of interest.

Section 4.3 is devoted to weakly controllable systems with random disturbances. A stochastic maximum principle is formulated, which allows one to construct a suboptimal control. If random factors are small, then this control, as a deterministic program for a stochastic system, attains the same efficacy as the suboptimal feedback constructed by means of solving the optimal deterministic problem. However, if the reduced trajectory is also stochastic, then a feedback can be found that is more efficient than any deterministic control. The way to construct such a feedback is given along with estimates of its efficiency. The results obtained are transferred to the case where the reduced problem allows the aggregation of the control.

In Sections 4.4–4.6 complicated multistage problems with state variable constraints are considered. Naturally, there is no chance of obtaining a general recommendation for such problems. The investigated problems have a particular character, but they are nonetheless of practical interest.

In Section 4.4, the operative control of industrial activity is analyzed under the assumption that raw material supplies are random but that the random deviations from their forecasted values are small. The main result can be interpreted as a statement of the suboptimality of a sliding planning principle. This principle recommends the choice of activity level at the nearest stage as part of the solution to the deterministic multistage optimization problem. The solution must be repeated at the beginning of each successive stage with new, corrected information about the stores of raw materials. Some formal estimates are obtained by a comparative analysis of dynamic programming relations corresponding to the stochastic and deterministic problems. More precise (in comparison with the sliding planning principle) suboptimal decision rules can be found if one uses a decomposition of the original dynamic programming equation into a number of independent equations, each of them corresponding to a one-dimensional inventory problem and having a simple solution.

The estimates obtained in Section 4.4 essentially depend on the number of stages, and they are not practically applicable to long-term planning. However, in Section 4.5, the estimates, which do not depend on this number, are given under the additional conditions of a linear objective and a discount factor. These estimates imply that sliding planning is suboptimal for an infinite-horizon problem. Moreover, some conditions are given for which fulfillment implies the optimality of the sliding planning principle, which ensures that all random trajectories are concentrated in an ϵ-vicinity of the turnpike solution to the deterministic problem.

The concluding section, 4.6, is close to Section 2.8 since the same hypothesis is used — that the controlled system is quasistatic. However, random disturbances are taken into account here. An original nonlinear control law is constructed, and the conditions of its optimality stated.

4.1. One- and two-stage problems

To begin with, we shall consider one-stage stochastic problems. Suppose a decision must be taken before the values of random factors are known, so the decision must be deterministic. If the decisionmaker estimates effects and expenditures using their mean values, he will find the optimal decision as a solution to the programs having the following structure

$$\max \{Ef(x,\omega) \,/\, Eg(x,\omega) \le 0\} \tag{1.1}$$

where E denotes the mathematical expectation with respect to random parameters ω.

It is not hard to see that the structure (1.1) is general enough. For example, if one needs to ensure that the expenditures do not exceed the admissible level of resources with probability $1 - \gamma$, i.e., the condition

$$p_r\{g(x,\omega) \leq 0\} \geq 1 - \gamma$$

must hold, then this condition can be rewritten in the form (1.1) by using the indicator function of the feasible set. Similarly, the problem to maximize the level, ξ, of effect achievable with a given probability

$$\max\ \{\xi \,/\, p_r\{f(x,\omega) \geq \xi\ ,\ g(x,\omega) \leq 0\} \geq 1 - \gamma\}$$

can be rewritten in the same ways too.

Suppose further that the distribution density, $p(\epsilon,\omega)$, of the random parameters is given as a linear function of a small parameter

$$p(\epsilon,\omega) = p_0(\omega) + \epsilon p_1(\omega) \tag{1.2}$$

where

$$p_0(\omega) \geq 0\ ,\ \int p_0(\omega)d\omega = 1 \tag{1.3}$$

$$p(\epsilon,\omega) \geq 0\ ,\ \int p_1(\omega)d\omega = 0 \tag{1.4}$$

In accordance with the general idea of the perturbation method, we shall suppose that averaging over the density $p_0(\omega)$ reduces problem (1.1) to a simpler one. Then the following method may be effective.

The program (1.1) may be written in the standard form

$$\max\ \{f_0(x) + \epsilon f_1(x) \,/\, g_0(x) + \epsilon g_1(x) \leq 0\} \tag{1.5}$$

$$f_0(x) \triangleq \int f(x,\omega)p_0(\omega)d\omega\ ,\ g_0(x) \triangleq \int g(x,\omega)p_0(\omega)d\omega \tag{1.6}$$

$$f_1(x) \triangleq \int f(x,\omega)p_1(\omega)d\omega\ ,\ g_1(x) \triangleq \int g(x,\omega)p_1(\omega)d\omega \tag{1.7}$$

Hence, to obtain a suboptimal solution, one can solve the simpler, reduced problem

$$\max\ \{f_0(x) \,/\, g_0(x) \leq 0\} \tag{1.8}$$

After that, if necessary, one solves the auxiliary problem

$$\max \{f_1(x) - \lambda^T(x) g_1(x) \,/\, x \in \Theta^*\} \tag{1.9}$$

where Θ^* is the solutions set to (1.8).

Naturally, the effectiveness of this scheme depends on the problem structure and on the properties of the densities $p_0(\omega), p_1(\omega)$. It can easily be shown that all the variants of the aggregation and decomposition schemes described in Chapter 2 could be developed here. For example, consider the linear program

$$\max \{Ec^T(\omega)x \,/\, E\sum_{r=1}^{n} A_r(\omega)x_r \le b \,,\, x_r \ge 0\} \tag{1.10}$$

If the activities separate out into groups such that the characteristics of all activities belonging to the some group have identical mean values with respect to $p_0(\omega)$, i.e.,

$$Ec_r(\omega) = c_i + \epsilon\sigma_r \,,\quad EA_r(\omega) = \bar{A}_i + \epsilon B_r \,,\quad r \in \Omega_i \,,$$

$$i = 1,\ldots,k \,,\quad \sum_{i=1}^{k} |\Omega_i| = n$$

then (1.10) is equivalent to a linear program, which is similar to the directly aggregatable and decomposable program (see Section 2.2).

As said earlier, the statement (1.1) of the problems corresponds to a situation of "here and now" type. Another situation (the "wait and see" type) is of interest, too. In this case, an uncertainty taking place at the moment of decision analysis partly (or even completely) vanishes up to the moment when the decision must be realized.

The simplest model of the "wait and see" situation is a one-stage problem of finding the decision schedule, $y(\omega)$, such that

$$f(y(\omega), \omega) \triangleq \max \{f(y, \omega) \,/\, g(y, \omega) \le 0\} \,,\quad \omega \in D \tag{1.11}$$

where D is a set of possible values of random parameters ω, and, after that, to estimate

$$Ef[y(\omega), \omega]$$

A more interesting scheme is well-known as that of two-stage stochastic programming (TSSP). Let the one-stage problem

$$f_2^*(x, \omega) \triangleq \max_y \{f_2(x, y, \omega) \,/\, g_2(x, y, \omega) \le 0\} \triangleq f_2(x, y(x, \omega), \omega) \tag{1.12}$$

have a solution $y(x, \omega)$ for any $x \in X$ and $\omega \in D$. Then the problem

$$f_1^* \triangleq \max_x \{Ef_1(x, f_2^*(x, \omega), \omega) \,/\, Eg_1(x, \omega) \leq 0\} \tag{1.13}$$

is called a TSSP problem, and problem (1.12) is called a problem of the second stage. Using perturbations of type (1.2) to select a simple reduced structure facilitates the solution to (1.13) but does not do similarly for the second state.

Another class of perturbations is of interest. Suppose that $p(\omega)$ does not depend on ϵ, but that the following representations are true:

$$\begin{aligned} &f_2(x, y, \omega) = f_{20}(x, y, \omega) + \epsilon\sigma_2(x, y, \omega) \\ &g_2(x, y, \omega) = g_{20}(x, y, \omega) + \epsilon\gamma_2(x, y, \omega) \\ &f_1(x, f_2^*(x, \omega), \omega) = f_{10}(x, f_2^*(x, \omega), \omega) + \epsilon\sigma_1(x, f_2^*(x, \omega), \omega) \\ &g_1(x, \omega) = g_{10}(x, \omega) + \epsilon\gamma_1(x, \omega) \end{aligned} \tag{1.14}$$

In this case, the perturbation method provides effective tools to solve both problems (1.13) and (1.12). Its scheme of application can be described in the following way.

To begin with, solve the reduced program

$$f_{20}^*(x, \omega) \triangleq \max_y \{f_{20}(x, \omega, y) \,/\, g_{20}(x, \omega, y) \leq 0\} \tag{1.15}$$

for all admissible x and ω. If the solution is not unique, then solve the auxiliary program

$$\Phi_2^*(x, \omega) \triangleq \max_y \{\sigma_2(x, y, \omega) - \lambda^T(x, y, \omega)\gamma_2(x, y, \omega) \,/\, y \in \Theta^*(x, \omega)\} \tag{1.16}$$

to find $y^*(x, \omega)$.

Here, $\Theta^*(x, \omega)$ is the set of solutions to (1.15), and $\lambda(x, y, \omega)$ is the Lagrange multipliers vector corresponding to $y \in \Theta^*(x, y, \omega)$. Using the results of Chapter 1, it is not hard to prove that if (a) $p_r\{\omega \in D\} = 1$ for some bounded and closed set D; (b) the functions entering the definition of the problems (1.12) and (1.13) are continuously differentiable; and (c) for all $x \in X$ and $\omega \in D$ the problem (1.15) satisfies the regularity conditions; then, uniformly over $X \times D$, the expansion

$$f_2^*(x, \omega) = f_{20}^*(x, \omega) + \epsilon\Phi_2^*(x, \omega) + o(\epsilon) \tag{1.17}$$

is valid.

Substituting (1.17) into (1.13), rewrite the TSSP problem in the form

$$f_1^* \triangleq \max \{\tilde{f}_{10}(x) + \epsilon\tilde{\sigma}_1(x) + o(\epsilon) \,/\, \tilde{g}_{10}(x) + \epsilon\tilde{\gamma}_{10}(x) \leq 0\} \tag{1.18}$$

where

$$\tilde{f}_{10}(x) \triangleq Ef_{10}(x, f_{20}^*(x,\omega),\omega) \;, \quad \tilde{g}_{10}(x) = Eg_{10}(x,\omega) \;,$$

$$\tilde{\sigma}_1(x) \triangleq E\{\sigma_1(x, f_{20}^*(x,\omega),\omega) + \frac{\partial f_{10}(x, f_{20}^*(x,\omega),\omega)}{\partial f_2}\Phi_2^*(x,\omega)\}$$

$$\tilde{\gamma}_1(x) \triangleq E\gamma_1(x,\omega)$$

Further, using the perturbation method, find an approximate solution to (1.18), i.e., determine the set Θ^* of solutions to

$$\max \{\tilde{f}_{10}(x) \,/\, \tilde{g}_{10}(x) \leq 0\} \tag{1.19}$$

and next solve the auxiliary problem

$$\max \{\tilde{\sigma}_1(x) - \lambda^T(x)\tilde{\gamma}_1(x) \,/\, x \in \Theta^*\} \tag{1.20}$$

One can show (see Chapter 1) that the solution $\bar{x}$ to the problem (1.20) and the decision rules $y^*(x,\omega)$ are feasible to within $O(\epsilon)$, with the necessary corrections being indicated. One can show also that they attain the optimal value of the objective function to within $o(\epsilon)$.

As an example, consider a hierarchical system consisting of subsystems that are not directly connected to each other, but can be controlled by the same "coordinator". Assume that each subsystem i ($i = 1,2,\ldots,k$) chooses a vector of activities $y^i(x,\omega)$ that maximizes its profit under any given realization of the random factors ω and any level, x, of the coordinator influence. For this purpose, one needs to solve

$$\begin{aligned} f_{2i}(x,\omega) &\triangleq \max_{y^i} \{(c_i^T + \epsilon\sigma_i^T(x,\omega))y^i \,/\, (A_i + \epsilon B_i(x,\omega))y^i = \\ &= b_i + \epsilon q_i(x,\omega) \;,\; y^i \geq 0\} \end{aligned} \tag{1.21}$$

where c_i ($n_i \times 1$), A_i ($m_i \times n_i$), $b_i(m_i \times 1)$ are constant matrices and $\sigma_i(\cdot)$, $B_i(\cdot)$, $g_i(\cdot)$ are matrix functions of appropriate dimensions.

The coordinator, on the basis of accessible information about the probability distribution of the random vector ω and the functions $f_{2i}(x,\omega)$ characterizing the behavior of the subsystems solves the problem of maximizing the expected system profit

$$\max \{E\sum_{i=1}^{k} f_{2i}(x,\omega) \,/\, Ax = b \;,\; x \geq 0\} \tag{1.22}$$

As follows from (1.21), both the level of coordinator influence and the intensity of random factors are assumed to be relatively small.

For $\epsilon = 0$ the problems (1.21) become the linear program

$$\max \{c_i^T y^i \,/\, A_i y^i = b \;,\; y^i \geq 0\} \tag{1.23}$$

Suppose the following conditions are fulfilled:

(1) A compact set D, such as $p_r\{\omega \in D\} = 1$ exists.
(2) The functions included in the definition of problem (1.21) are continuously differentiable.
(3) The solutions to the problems (1.23) and their dual problems are unique.
(4) The set $X = \{x \,/\, Ax = b \;,\; x \geq 0\}$ is bounded.

Then it is easy to prove that there exists $\epsilon_0 > 0$ such that, under $0 < \epsilon \leq \epsilon_0$, the solution of each i-subproblem (1.21) can be written in the form

$$y^i(x,\omega) = (A_{ib} + \epsilon B_{ib}(x,\omega))^{-1}(b_i + \epsilon q_i(x,\omega))$$
$$f_{2i}(x,\omega) = (c_i^T + \epsilon \sigma_i^T(x,\omega))(A_{ib} + \epsilon B_{ib})^{-1}(b_i + \epsilon q_i(x,\omega))$$

where A_{ib} is the optimal basic matrix of (1.23) and $B_{ib}(\cdot)$ is a corresponding submatrix of $B_i(\cdot)$. Problem (1.22) in this case will take the form

$$\begin{aligned} \max \; \{&\sum_{i=1}^{k} E[(c_i^T + \epsilon\sigma_i^T(x,\omega))(A_{ib} + \epsilon B_{ib}(x,\omega))^{-1}(b_i + \epsilon q_i(x,\omega))] \,/ \\ &Ax = b \;,\; x \geq 0\} \end{aligned} \tag{1.24}$$

Expanding the objective function (1.24) into a power series, we get

$$\begin{aligned} \max \; \{&c_i^T A_{ib}^{-1} b_i + \epsilon \sum_{i=1}^{k} [E\sigma_i^T(x,\omega) A_{ib}^{-1} b_i + c_i^T A_{ib}^{-1} E q_i(x,\omega) - \\ &- c_i^T A_{ib}^{-1} E B_{ib}(x,\omega) A_{ib}^{-1} b_i] + O(\epsilon^2) \,/\, Ax = b \;,\; x \geq 0\} \end{aligned} \tag{1.25}$$

Evidently, the solution $\bar{x}$ to the problem

$$\max \sum_{i=1}^{k} [E\sigma_i^T(x,\omega)A_{ib}^{-1}b_i + c_i^T A_{ib}^{-1} Eq_i(x,\omega) - \\ - c_i^T A_{ib}^{-1} EB_{ib}(x,\omega) A_{ib}^{-1} b_i] \,/\, Ax = b \ , \ x \geq 0\} \tag{1.26}$$

together with the decision rules, $y^i(\bar{x},\omega)$, attain the optimal value of the objective (1.25) to within $O(\epsilon^2)$.

In the particular case where $\sigma_i(x,\omega)$, $g_i(x,\omega)$, $B_{ib}(x,\omega)$ are linear functions of x and where the solutions to problem (1.26) and its dual are unique, we can prove that there exists $\epsilon_0 > 0$ such that, if $0 < \epsilon \leq \epsilon_0$, the plan $\bar{x}$ and the decision rules $y^i(\bar{x},\omega)$, $i = 1,\ldots,k$ are the unique solution to (1.21) and (1.22).

The simplicity of the described procedure is determined by the following factors: (1) the small influence of the random factors; (2) the weakness of the coordinator control (which leads, in particular, to the weak interconnections between the blocks); and (3) the linearity of the reduced structure. An effective procedure, based on the perturbation method, can also be developed in the case where at least one of the above-mentioned specific features is present. Such procedures will be examined later on as applied to general multi-stage problems.

4.2. Optimal control problems with small random perturbations

Consider some control problems associated with the optimization of dynamic systems under uncertainties. We shall adhere to the following formulation of these problems.

Let the state of a system at the moment t, $t = 0,1,\ldots,N$, be described by the n-vector of state variables $x^t(n \times 1)$, and let these variables behave according to the stochastic recurrence relation

$$x^{t+1} = f(x^t, u^t, \omega^t, t) \ , \ t = 0,1,\ldots,N-1,N \ , \ x^0 \text{ is given} \tag{2.1}$$

where ω^0, ω^1 ,..., ω^{N-1} are mutually independent random l-vectors. The controls u^t satisfy the inclusion

$$u^t \in U_t \tag{2.2}$$

and are sought in the form of a feedback law

$$u^t = u^t(x) \tag{2.3}$$

that attains the maximum value of the performance index, i.e., that solves

$$\max EG(x^N) \tag{2.4}$$

This formulation is based on the assumption that at each moment t the whole state vector x^t is observable and is undoubtedly classified as a multistage problem.

There is another form in which we can write (2.1)–(2.4). Let $B^{N-t}(x)$ $(t = 0,1,\ldots,N)$ be the Bellman function of the system, i.e., $B^{N-t}(x)$ is the optimal value of the performance index (2.4) corresponding to the case where the trajectory starts at moment t and initial state x. In accordance with the optimality principle (all parts of the optimal trajectories are also optimal), we write the equations

$$\begin{aligned} B^{N-(t-1)}(x) &\triangleq \max_{u\in U_{t-1}} EB^{N-t}(f(x,u,\omega^{t-1},t-1)) \ , \ t = 1,\ldots,N \\ B^0(x) &= G(x) \end{aligned} \tag{2.5}$$

At first glance, it seems that (2.5) contains an algorithm to find the solutions to (2.1)–(2.4). Indeed, beginning from stage N, we may consider the program

$$B^1(x) = \max_{u\in U_{N-1}} E\{G(f(x,u,\omega^{N-1},N-1))\} \tag{2.5$'$}$$

find $u^{N-1}(x)$, $B^1(x)$, and then solve the problem

$$B^2(x) = \max_{u\in U_{N-2}} E\{B^1(f(x,u,\omega^{N-2},N-2))\} \tag{2.5$''$}$$

then find $u^{N-2}(x)$, $B^2(x)$, etc.

Problems (2.5′) and (2.5″), however, are parametric, with the components of the n-vector x as the parameters. To construct the solution to such problems is not generally possible. That is why the equations can be used only for the inductive elucidation of some characteristics of Bellman functions (continuity, differentiability, etc.).

In accordance with the main subject of this book, we shall apply the perturbation method to obtain approximate solutions to (2.1)–(2.4). First, we shall consider a situation where the random perturbations are small and the perturbed system (2.1) can be described in the form

$$x^{t+1} = f(x^t,u^t,t) + \epsilon\gamma(x^t,u^t,\omega^t,t) \ , \ t = 0,1,\ldots,N-1 \ , \ x^0 \text{ is given} \tag{2.6}$$

where ϵ is a small parameter. It is natural to suppose here that the solutions to the problems (2.2)–(2.4) and (2.6) are located in some neighborhood of the solution to the deterministic reduced problem

$$x^{t+1} = f(x^t,u^t,t) \ , \ u^t \in U_t \ , \ t = 0,1,\ldots,N-1 \ ; \ \max G(x^N) \tag{2.7}$$

Relations (2.5) can be written as perturbed ones and, hence, suboptimal solutions can be found at each step by means of perturbation theory for the deterministic programs. We will not, however, investigate general procedures of that kind and will consider two particular cases where the perturbation method takes the simplest form.

4.2.1. Optimization of a system with linear quadratic structure

Consider the case where we seek the optimal control in the problem

$$\begin{aligned} &x^{t+1} = A_t x^t + D_t u^t + \epsilon\gamma(x^t, u^t, \omega^t, t) \ , \ t = 0,1,\ldots,N-1 \\ &\min E\frac{1}{2}\sum_{t=1}^{N}[(x^t)^T Q_t x^t + (u^{t-1})^T R_{t-1} u^{t-1}] \end{aligned} \tag{2.8}$$

where Q_t $(n \times n)$, R_{t-1} $(m \times m)$, $t = 1,\ldots,N$, are positive definite matrices, the controls being unconstrained. The reduced problem here is of the form

$$\begin{aligned} &x^{t+1} = A_t x^t + D_t u^t \ , \ t = 0,1,\ldots,N-1 \\ &\min \sum_{t=1}^{N}\frac{1}{2}[(x^t)^T Q_t x^t + (u^{t-1})^T R_{t-1} u^{t-1}] \end{aligned} \tag{2.9}$$

It is not hard to establish that the optimal feedback control for (2.9) is a linear function of the state variables

$$\bar{u}^t(x^t) = -K_t x^t \ , \ K_t \triangleq (R_t + D_t^T P_{t+1} D_t)^{-1} D_t^T P_{t+1} A_t \tag{2.10}$$

and that the corresponding Bellman function [representing, as in (2.5), the optimal value of the performance index when the motion starts from moment t and initial state x] is equal to

$$\bar{B}^{N-t}(x) = \frac{1}{2} x^T P_t x \tag{2.11}$$

where P_t, $t = 1,\ldots,N$, are positive definite matrices determined recursively by

$$\begin{aligned} P_{t-1} = \ & A_{t-1}^T P_t A_{t-1} - A_{t-1}^T P_t D_{t-1}(R_{t-1} + D_{t-1}^T P_t A_{t-1})^{-1} \times \\ & \times D_{t-1}^T P_t A_{t-1} + Q_{t-1} \ , \ P_N \triangleq Q_N \end{aligned} \tag{2.12}$$

The connection between the solutions to the perturbed problem and the reduced problem is given by Theorem 2.1, which also contains the proof of (2.10)–(2.12).

Theorem 2.1.

Suppose that the functions $\gamma(x, u, \omega, t)$, $t = 0,1,\ldots,N$, are smooth and that there exists a compact set D such that

$$p_r\{\omega^t \in D\} = 1 \quad , \quad t = 0,1,\ldots,N-1 \tag{2.13}$$

Then the optimal control $u^\tau(x)$ for the perturbed problem (2.8) and its Bellman function $B^{N-\tau}(x)$, $\tau = 0,1,\ldots,N-1$, may be written in the form

$$u^\tau(x) = \bar{u}^\tau(x) + \epsilon y_1^\tau(x) + \cdots + \epsilon^i y_i^\tau(x) + O(\epsilon^{i+1}) \tag{2.14}$$

$$B^{N-\tau}(x) = \bar{B}^{N-\tau}(x) + \epsilon \eta_1^{N-\tau}(x) + \cdots + \epsilon^{i+1} \eta_{i+1}^{N-\tau}(x) + O(\epsilon^{i+2}) \tag{2.15}$$

where i is any finite number. The evaluations are uniform for x belonging to any compact set in R^n, and the corrections can be calculated recursively according to the following relations

$$\begin{aligned} y_j^\tau(x) = {} & (R_\tau + D_\tau^T P_{\tau+1} D_\tau)^{-1} \psi_j(x, \bar{u}^\tau(x), y_1^\tau(x), \ldots, y_{j-1}^\tau(x)) \ , \\ & j = 1,\ldots,i \end{aligned} \tag{2.16}$$

$$\eta_j^{N-\tau}(x) = \varphi_j^{N-\tau}(x, \bar{u}^\tau(x), y_1^\tau(x), \ldots, y_{j-1}^\tau(x)) \ , \quad j = 1,\ldots,i+1 \tag{2.17}$$

(How to form the functions $\psi_j^\tau(\cdot)$, $\varphi_j^{N-\tau}(\cdot)$ is indicated in the proof.)

Proof. Let us write out Bellman's relations for the perturbed problem (2.8)

$$\begin{aligned} B^{N-(\tau-1)}(x) = {} & \max_u E\{B^{N-\tau}(A_{\tau-1}x + D_{\tau-1}u + \epsilon\gamma(x, u, \omega^{\tau-1}, \tau-1)) + \\ & + x^T Q_{\tau-1} x + u^T R_{\tau-1} u\} \ , \quad \tau = 1,2,\ldots,N \ , \\ & B^0(x) \triangleq x^T Q_N x \end{aligned} \tag{2.18}$$

By definition

$$B^0(x) = \bar{B}^0(x) = x^T Q_N x \tag{2.19}$$

and so we have (2.14) and (2.15) with

$$y_j^N(\cdot) \equiv 0 \;\;,\;\; \eta_j^N(\cdot) \equiv 0 \;\;,\;\; j = 1,2,\ldots \tag{2.20}$$

Further, let us suppose inductively that, for $\tau = t$, the functions $u^t(x)$, $B^{N-t}(x)$ may be written in the form

$$u^t(x) = \bar{u}^t(x) + \epsilon y^t(x\,,\epsilon) \tag{2.21}$$

$$B^{N-t}(x) = \bar{B}^{N-t}(x) + \epsilon\eta^{N-t}(x\,,\epsilon) \tag{2.22}$$

where $y^t(x\,,\epsilon)$, $\eta^{N-t}(x\,,\epsilon)$ are smooth functions.

For any finite i, let the following expansions be valid

$$y^t(x\,,\epsilon) = y_1^t(x) + \epsilon y_2^t(x) + \cdots + \epsilon^{i-1} y_i^t(x) + O(\epsilon^i)$$

$$\eta^{N-t}(x\,,\epsilon) = \eta_1^{N-t}(x) + \epsilon\eta_2^{N-t}(x) + \cdots + \epsilon^i \eta_{i+1}^{N-t}(x) + O(\epsilon^{i+1})$$

Suppose also that a way to expand the functions $\psi_i^t(\cdot)$, $\varphi_i^{N-t}(\cdot)$ has been indicated. Using (2.18) and (2.22) yields

$$\begin{aligned} B^{N-(t-1)}(x) &= \max_u E\{\bar{B}^{N-t}(A_{t-1}x + D_{t-1}u + \epsilon\gamma(x\,,u\,,\omega^{t-1}\,,t-1)) + \\ &\quad + \epsilon\eta^{N-t}(A_{t-1}x + D_{t-1}u + \epsilon\gamma(x\,,u\,,\omega^{t-1}\,,t-1)\,,\epsilon) + \\ &\quad + \frac{1}{2}x^T Q_{t-1}x + \frac{1}{2}u^T R_{t-1}u\} \\ &= \max_u\{\frac{1}{2}(x^T A_{t-1}^T + u^T D_{t-1}^T) P_t (A_{t-1}x + D_{t-1}u) + \\ &\quad + \frac{1}{2}x^T Q_{t-1}x + \frac{1}{2}u^T R_{t-1}u + \epsilon\beta^{t-1}(x\,,u\,,\epsilon)\} \end{aligned} \tag{2.23}$$

where

$$\begin{aligned} \beta^{t-1}(x\,,u\,,\epsilon) \triangleq\; & E\{\frac{1}{2}\gamma^t(x\,,u\,,\omega^{t-1}\,,t-1)P_t(A_{t-1}x + D_{t-1}u) + \\ & + \frac{1}{2}(x^T A_{t-1}^T + u^T D_{t-1}^T)P_t\gamma(x\,,u\,,\omega^{t-1}\,,t-1) + \\ & + \frac{1}{2}\epsilon\gamma^T(x\,,u\,,\omega^{t-1}\,,t-1)P_t\gamma(x\,,u\,,\omega^{t-1}\,,t-1) + \\ & + \eta^{N-t}(A_{t-1}x + D_{t-1}u + \epsilon\gamma(x\,,u\,,\omega^{t-1}\,,t-1)\,,\epsilon)\} \end{aligned}$$

The smoothness of the functions $\gamma(\cdot)$, $\eta^{N-t}(\cdot)$ and (2.13) lead to the smoothness of the function $\beta^{t-1}(\cdot)$. [Condition (2.13) can be replaced by the assumption of the existence and smoothness of the expectations of the functions determining $\beta^{t-1}(x, u, \epsilon)$.] Consequently, the optimal control $u^{t-1}(x)$ is the solution to the system

$$D_{t-1}^T P_t (A_{t-1} x + D_{t-1} u) + R_{t-1} u + \epsilon \frac{\partial \beta^{t-1}(x, u, \epsilon)}{\partial u} = 0 \tag{2.24}$$

We shall write it in the form

$$u^{t-1}(x) = \bar{u}^{t-1}(x) + \epsilon y^{t-1}(x, \epsilon) \tag{2.25}$$

Substitution (2.25) into (2.24) and taking (2.10) into account, we have that

$$\begin{aligned} &D_{t-1}^T P_t D_{t-1} y^{t-1}(x, \epsilon) + R_{t-1} y^{t-1}(x, \epsilon) + \\ &+ \frac{\partial \beta^{t-1}[x, \bar{u}^{t-1}(x) + \epsilon y^{t-1}(x, \epsilon), \epsilon]}{\partial u} = 0 \end{aligned} \tag{2.26}$$

The Jacobian of system (2.26) with respect to y is not equal to zero when $\epsilon = 0$. Hence, equations (2.26) define the function $y^{t-1}(x, \epsilon)$, which, for any i, admits the expansion

$$y^{t-1}(x, \epsilon) = y_1^{t-1}(x) + \epsilon y_2^{t-1}(x) + \cdots + \epsilon^{i-1} y_i^{t-1}(x) + O(\epsilon^i) \tag{2.27}$$

where

$$y_1^{t-1}(x) = (R_{t-1} + D_{t-1}^T P_t D_{t-1})^{-1} \left[- \frac{\partial \beta^{t-1}(x, \bar{u}^{t-1}(x), 0)}{\partial u} \right] \tag{2.28}$$

This proves (2.14) for $\tau = t - 1$ and, as follows from (2.28), gives the correspondence

$$\psi_1^{t-1}(x, \bar{u}^{t-1}(x)) \equiv - \frac{\partial \beta^{t-1}(x, \bar{u}^{t-1}(x), 0)}{\partial u}$$

The functions $\psi_i^{t-1}(\cdot)$, $i > 1$ can be found by substituting (2.27) into (2.26). The expansion (2.15) for $\tau = t - 1$ follows from (2.27), and the functions $\varphi_i^{N-(t-1)}(\cdot)$ are defined by the substitution of (2.27) into (2.23)

$$\varphi_1^{N-(t-1)}(x) = \beta^{t-1}(x, \bar{u}^{t-1}(x), 0), \ldots$$

Naturally, one can use only a finite number of terms in (2.14) to construct the control law, and it is of interest to evaluate the difference between the optimal value of the performance index and the one obtained using the truncated control with finite i

$$\tilde{u}_0^\tau \triangleq \bar{u}^\tau(x)$$
$$\tilde{u}_i^\tau(x,\epsilon) \triangleq \bar{u}^\tau(x) + \epsilon y_1^\tau(x) + \cdots + \epsilon^i y_i^\tau(x) \quad , \quad I = 1,2,\ldots$$

Let $F_i^{N-\tau}(x)$ be the value of the performance index under the condition that the motion is started from moment τ with initial state vector x and that the control $\tilde{u}_i^{\tau+l}(\cdot)$ is used at every stage $\tau + l$ ($l = 0,1,\ldots,N-1-\tau$), i.e., by definition

$$\begin{aligned} F_i^{N-(\tau-1)}(x) = \ & E\{F_i^{N-\tau}(A_{\tau-1}x + D_{\tau-1}\tilde{u}_i^{\tau-1}(x,\epsilon) + \\ & + \epsilon\gamma(x,\tilde{u}_i^{\tau-1}(x,\epsilon),\omega^{\tau-1},\tau-1)) + x^T Q_{\tau-1}x + \\ & + (\tilde{u}_i^{\tau-1}(x,\epsilon))^T R_{\tau-1}\tilde{u}_i^{\tau-1}(x,\epsilon)\} \ , \ F_i^0(x) \triangleq x^T Q_N x \end{aligned} \tag{2.29}$$

Theorem 2.2.

Let the assumptions of Theorem 2.1 be fulfilled. Then, uniformly in x belonging to any compact set in R^n,

$$B^{N-\tau}(x) = F_i^{N-\tau}(x) + O(\epsilon^{i+2}) \tag{2.29$'$}$$

Proof. Reasoning as we did above, let us suppose that (2.29′) is true for $\tau = N-1,\ldots,t$ [for $\tau = N$ this is the consequence of definitions (2.19) and (2.29)]. In accordance with our assumption and (2.29), we can write

$$\begin{aligned} F_i^{N-(t-1)}(x) = \ & E\{B^{N-t}(A_{t-1}x + D_{t-1}\tilde{u}_i^{t-1}(x,\epsilon) + \\ & + \epsilon\gamma(x,\tilde{u}_i^{t-1}(x,\epsilon),\omega^{t-1},t-1)) + x^T Q_{t-1}x + \\ & + (\tilde{u}_i^{t-1}(x,\epsilon))^T R_{t-1}\tilde{u}_i^{t-1}(x,\epsilon)\} + O(\epsilon^{i+2}) \end{aligned}$$

The right-hand side can be expanded into an asymptotic power series in ϵ, the coefficients of this expansion up to the coefficient at ϵ^{i+1} determined by formulae

(2.17). Hence, they will coincide with those obtained when expanding $B^{N-(t-1)}(x)$, and this implies that

$$F_i^{N-(t-1)}(x) = B^{N-(t-1)}(x) + O(\epsilon^{i+2})$$

As follows from (2.24), it is sufficient to calculate only the first i coefficients of the Bellman function $B^{N-t}(x)$ if we want to find the first i coefficients of $u^{t-1}(x)$. So, finding the suboptimal control that ensures $O(\epsilon^{i+2})$ proximity to the optimum requires the calculations of only the first i coefficients of (2.14) and (2.15).

4.2.2. Optimization of a perturbed linear system

Suppose that the perturbed problem is of the form

$$\begin{aligned} &x^{t+1} = A_t x^t + D_t u^t + \epsilon\gamma(x^t, u^t, t, \omega^t) \quad , \quad t = 0,1,\ldots,N-1 \\ &u^t \in U_t \quad , \quad t = 0,1,\ldots,N-1 \\ &\max E\{(c^N)^T x^N\} \end{aligned} \tag{2.30}$$

where U_t, $t = 0,1,\ldots,N-1$, are polyhedra. The reduced problem here is a linear program

$$\begin{aligned} &x^{t+1} = A_t x^t + D_t u^t \quad , \quad t = 0,1,\ldots,N-1 \\ &u^t \in U_t \quad , \quad t = 0,1,\ldots,N-1 \quad ; \quad \max\ (c^N)^T x \end{aligned} \tag{2.31}$$

It is not difficult to show (see the proof of Theorem 2.3, below) that the optimal feedback control $\bar{u}^t$ in (2.31) does not depend on the state vector x

$$u^t(x) = \bar{u}^t \tag{2.32}$$

and is the solution to the linear program

$$\max\ \{(\bar{\lambda}^{t+1})^T D_t u \,/\, u \in U_t\} \triangleq f_t^* \tag{2.33}$$

where $\bar{\lambda}^1,\ldots,\bar{\lambda}^N$ is the solution to the conjugate system

$$\bar{\lambda}^\tau = A_\tau^T \bar{\lambda}^{\tau+1} \quad , \quad \tau = N-1,\ldots,1 \quad ; \quad \bar{\lambda}^N \triangleq c^N \tag{2.34}$$

The Bellman functions corresponding to the reduced problem are linear functions of the state variables

$$\bar{B}^{N-t}(x) = (\bar{\lambda}^t)^T x + \sum_{\tau=t}^{N-1} f_\tau^* \quad , \quad t = 0,1,\ldots,N-1$$
$$\bar{B}^0(x) \triangleq (c^N)^T x \triangleq (\bar{\lambda}^N)^T x \tag{2.35}$$

Theorem 2.3.

Let the solution to the problem (2.33) and its dual be unique for all $t = 0,1,\ldots,N-1$. Let the functions $\gamma(x,u,t,\omega^t)$, $t = 0,1,\ldots,N-1$, be continuously differentiable and let there be† a compact set $D \subset R^l$ such that

$$p_r\{\omega^t \in D\} = 1 \quad , \quad t = 0,1,\ldots,N-1 \tag{2.36}$$

Then for any compact set $\Theta \subset R^n$ we can choose $\epsilon_\Theta > 0$ such that for $0 < \epsilon \leq \epsilon_\Theta$ the optimal control $u^\tau(x)$ and the Bellman functions $B^{N-\tau}(x)$ of the perturbed problem can be written in the form

$$u^\tau(x) = \bar{u}^\tau \quad , \quad \tau = 0,1,\ldots,N-1 \quad , \quad x \in \Theta \tag{2.37}$$

$$B^{N-\tau}(x) = \bar{B}^{N-\tau}(x) + \epsilon\eta^{N-\tau}(x,\epsilon) \quad , \quad \tau = 0,1,\ldots,N-1 \quad , \quad x \in \Theta \tag{2.38}$$

where $\eta^{N-\tau}(x,\epsilon)$ are continuously differentiable functions formed in the proof.

Proof. By definition,

$$\begin{aligned} B^{N-1}(x) &= \max_{u \in U_{N-1}} E\{(c^N)^T(A_{N-1}x + D_{N-1}u + \epsilon\gamma(x,u,\omega^{N-1},n-1))\} \triangleq \\ &\triangleq (\bar{\lambda}^{N-1})^T x + \max_u \{(\bar{\lambda}^N)^T D_{N-1}u + \epsilon\eta^{N-1}(x,u)\} \end{aligned} \tag{2.39}$$

where $\eta^{N-1}(x,u)$ is a continuously differentiable function defined as the expectation

$$\eta^{N-1}(x,u) \triangleq E(\bar{\lambda}^N)^T \gamma(x,u,\omega^{N-1},N-1) \tag{2.40}$$

† Similarly to Theorem 2.1, the assumption about constraints on the realizations of the random vectors ω^t $(t = 0,1,\ldots,N-1)$ can be replaced by an assumption about the existence and differentiability of the corresponding expectations [see relations (2.43) below].

Consider the problem

$$\max_{u \in U_{N-1}} \{(\bar{\lambda}^N)^T D_{N-1} u + \epsilon \eta^{N-1}(x, u)\} \tag{2.41}$$

to which the reduced one is given by (2.33). The uniqueness of the solutions to the former and its dual leads to the fulfillment of the regularity conditions from which, on the basis of the results of Section 1.3, we can conclude that

$$\lim_{\epsilon \to 0} u^{N-1}(\epsilon, x) = \bar{u}^{N-1} \quad , \quad \lim_{\epsilon \to 0} \mu^{N-1}(\epsilon, x) = \bar{\mu}^{N-1} \tag{2.42}$$

with $u^{N-1}(\epsilon, x)$ the solution to (2.41), $\mu^{N-1}(\epsilon, x)$ the corresponding Lagrange vector, and $\bar{\mu}^{N-1}$ the solution to the dual problem to (2.33) with $t = N-1$. It is not difficult to establish that the convergence in (2.42) is uniform for x in any compact set of R^n.

Consequently, for any compact set $\Theta \subset R^n$, we can select $\epsilon_{1\Theta} > 0$ such that $0 < \epsilon \le \epsilon_{1\Theta}$, $x \in \Theta$,

$$\bar{\mu}_r^{N-1} > 0 \Rightarrow \mu_r^{N-1}(\epsilon, x) > 0 \quad , \quad x \in \Theta$$

This means that, for $0 < \epsilon \le \epsilon_{1\Theta}$, $x \in \Theta$ the vertex of U, which is the solution to (2.33), is also the solution to (2.41);

$$u^{N-1}(\epsilon, x) = \bar{u}^{N-1} \Rightarrow$$
$$B^{N-1}(x) = (\bar{\lambda}^{N-1})^T x + f_{N-1}^* + \epsilon \eta^{N-1}(x, \bar{u}^{N-1}) = \bar{B}^{N-1}(x) + \epsilon \eta^{N-1}(x)$$

So the conditions (2.37) and (2.38) are proved for $\tau = N-1$. Let us suppose that they are true for $\tau = t$ and that the way to expand $\eta^{N-t}(x, \epsilon)$ has been indicated. Then, taking into account (2.35), we can write down

$$\begin{aligned} B^{N-(t-1)}(x) &= \max_{u \in U_{t-1}} E\{(\bar{\lambda}^t)^T (A_{t-1} x + D_{t-1} u + \epsilon \gamma(x, u, \omega^{t-1}, t-1)) + \\ &\quad + \epsilon \eta^{N-t}(A_{t-1} x + D_{t-1} u + \epsilon \gamma(x, u, \omega^{t-1}, t-1), \epsilon) + \\ &\quad + \sum_{\tau=t}^{N-1} f_\tau^* \qquad (2.43) \\ &= (\bar{\lambda}^{t-1})^T x + \max_{u \in U_{t-1}} \{(\bar{\lambda}^t)^T D_{t-1} u + \epsilon \eta^{N-(t-1)}(x, u, \epsilon)\} + \\ &\quad + \sum_{\tau=t}^{N-1} f_\tau^* \end{aligned}$$

where

$$\eta^{N-(t-1)}(x,u,\epsilon) \triangleq E\{(\bar{\lambda}^t)^T\gamma(x,u,\omega^{t-1},t-1) + \\ + \epsilon\eta^{N-t}(A_{t-1}x + D_{t-1}u + \epsilon\gamma(x,u,\omega^{t-1},t-1),\epsilon)\}$$

Investigating (2.43) by means similar to (2.39) and introducing the notation

$$\eta^{N-(t-1)}(x,\epsilon) \triangleq \eta^{N-(t-1)}(x,\bar{u}^{t-1},\epsilon)$$

we get the relations (2.37) and (2.38) for $\tau = t - 1$.

As follows from the proof, the functions $\eta^{N-\tau}(x,\epsilon)$ are connected by the difference equation

$$\eta^{N-(\tau-1)}(x,\epsilon) = E\{(\bar{\lambda}^\tau)^T\gamma(x,\bar{u}^{\tau-1},\omega^{\tau-1},\tau-1) + \\ + \epsilon\eta^{N-\tau}(A_{\tau-1}x + D_{\tau-1}\bar{u}^{\tau-1} + \\ + \epsilon\gamma(x,\bar{u}^{\tau-1},\omega^{\tau-1},\tau-1),\epsilon)\}, \\ \tau = 1,\ldots,N-1$$

$$\eta^1(x,\epsilon) \triangleq E\{(\bar{\lambda}^N)^T\gamma(x,\bar{u}^{N-1},\omega^{N-1},N-1)\}$$

and it is these functions that define the dependence of ϵ_Θ on the set Θ under consideration. In particular, if the functions $\gamma(x,u,\omega^{t-1},t-1)$ do not depend on x, there exists $\hat{\epsilon} > 0$ such that, when $0 < \epsilon \leq \hat{\epsilon}$, the relations (2.37) and (2.38) will be valid for all x from R^n. This $\hat{\epsilon}$ is defined by the stability factor of the optimal bases in the linear programs (2.34).

Both the examples given are characterized by the uniqueness of the solutions to the reduced problems (2.9) and (2.11). Moreover, it is easy to see that these problems satisfy the regularity conditions. That is, the results of Theorems 2.1–2.3 are similar to those of Section 1.4, where the same assumptions were made. We shall go further and obtain some results concerning problems (2.2)–(2.4) and (2.6) that establish some resemblance to the deterministic case in the general situation where the solution of the basic problem is not unique.

Let us consider the perturbed problem (2.2)–(2.4) and (2.6) and the reduced problem (2.2)–(2.4) and (2.7), supposing that the former has a set of optimal controls $\{u^\beta\}$ (see Section 2.8). For each u^t_β, $t = 1,\ldots,N-1$,

$$u^t_\beta \ (t = 0,1,\ldots,N-1) \in \{u^\beta\} \tag{2.44}$$

we can find the solutions x^t_β, λ^t_β, $t = 1,\ldots,N$ of the primal and the conjugate systems

$$x_\beta^{t+1} = f(x_\beta^t, u_\beta^t, t) \ , \ t = 0,1,\ldots,N-1 \ , \ x_\beta^0 \triangleq x^0 \tag{2.45}$$

$$\lambda_\beta^t = f_x^T(x_\beta^t, u_\beta^t, t)\lambda_\beta^{t+1} \ , \ t = N-1,\ldots,1 \ , \ \lambda_\beta^N = G_x'(x_\beta^N) \tag{2.46}$$

Let us define $\{u^{*0}, u^{*1},\ldots,u^{*N-1}\}$ as a solution to the problem

$$\Phi^* \triangleq \max_\beta E\{\sum_{\tau=0}^{N-1} (\lambda_b^{\tau+1})^T \gamma(x_\beta^\tau, u_\beta^\tau, \omega^\tau, \tau)\} \tag{2.47}$$

which is the auxiliary problem for a one-stage stochastic programming problem

$$\max \ \{G(x^N) \,/\, x^{t+1} = f(x^t, u^t, t) + \epsilon E\gamma(x^t, u^t, \omega^t, t) \ , \\ u^t \in U_t \ , \ t = 0,1,\ldots,N-1 \ , \ x^0 \text{ is given}\}$$

Theorem 2.4.

Let the sets U_t, $t = 0,1,\ldots,N-1$ be compact, and let the functions $f(\cdot)$, $f_x(\cdot)$, $G(\cdot)$, $G_x(\cdot)$, $\gamma(\cdot)$ be continuous. Suppose there exists a compact set such that

$$p_r\{\omega^t \in D\} = 1 \ , \ t = 0,1,\ldots,N-1 \tag{2.48}$$

Then

(1) When using the control $\{u^{*0}, u^{*1},\ldots, u^{*N-1}\}$ in (2.6) the value of the performance index is approximately equal to the optimal one given by the Bellman function to within $o(\epsilon)$

$$B^N(x^0) = \tilde{G}_\epsilon + o(\epsilon) \tag{2.49}$$

(2) $B^N(x^0)$ may be written in the form

$$B^N(x^0) = G^* + \epsilon\Phi^* + o(\epsilon) \tag{2.50}$$

where G^*, Φ^* are the optimal values of the objective functions (2.7) and (2.47).

(3) If, in addition to the above conditions, the solution $\{u^{*0}, u^{*1},\ldots, u^{*N-1}\}$ to (2.47) is unique, then[†]

[†] The notation $z_\epsilon(\omega) \xrightarrow[\epsilon \to 0]{p} z$ means the convergence of the random variable $z_\epsilon(\omega)$ to z in probability.

$$u_\epsilon^{*t}(\omega^0,\ldots,\omega^{t-1}) \xrightarrow[\epsilon\to 0]{p} u^{*t} \quad , \quad t = 0,1,\ldots,N-1 \tag{2.51}$$

where $u_\epsilon^{*t}(\omega^0,\ldots,\omega^{t-1})$, $t = 0,1,\ldots,N-1$, is the random optimal control generated by the optimal feedback control in (2.6).

Proof. It is not difficult to establish that

$$\tilde{x}_\epsilon^t = x^t + \epsilon x_1^t \quad , \quad t = 1,\ldots,N \quad , \quad \|x_1^t\| \le L = \text{constant} \tag{2.52}$$

where $\tilde{x}_\epsilon^t$, $t = 1,\ldots,N$, is the random trajectory of (2.6) corresponding to the control $\{u^{*0}, u^{*1},\ldots, u^{*N-1}\}$ and x^t, x_1^t, $t = 1,\ldots,N$, are the solutions to the difference equations

$$x^{t+1} = f(x^t, u^{*t}, t) \quad , \quad t = 0,1,\ldots,N-1 \quad , \quad x^0 \text{ is given} \tag{2.53}$$

$$\begin{aligned} x_1^{t+1} = {} & f_x(x^t, u^{*t}, t)x_1^t + \gamma(x^t, u^{*t}, \omega^t, t) + \nu(\epsilon) \ , \\ & x_1^0 \triangleq 0 \ ; \ \nu(\epsilon) \to 0 \ , \ \epsilon \to 0 \end{aligned} \tag{2.54}$$

Consequently, the value of the performance index is in the form of the expansion

$$\begin{aligned} \tilde{G}_\epsilon & \triangleq EG(\tilde{x}_\epsilon^N) = G(x^N) + \epsilon E\{(G_x(x^N))^T x_1^N\} + o(\epsilon) \triangleq \\ & \triangleq G^* + \epsilon E\{(G_x(x^N))^T x_1^N\} + o(\epsilon) \end{aligned} \tag{2.55}$$

Let λ^t, $t = 1,\ldots,N$, be the solution to the conjugate system (2.46) corresponding to u^{*t}, $t = 0,1,\ldots,N-1$. Then the coefficient of ϵ in (2.55) can be transformed in accordance with the relations

$$\begin{aligned} & E\{(G_x(x^N))^T x_1^N\} \triangleq E\{(\lambda^N)^T x_1^N\} = \\ & = E\{(\lambda^N)^T[f_x(x^{N-1}, u^{*N-1}, N-1)x_1^{N-1} + \\ & \quad + \gamma(x^{N-1}, u^{*N-1}, \omega^{N-1}, N-1)]\} + O(\nu(\epsilon)) = \\ & = E\{(\lambda^{N-1})^T x_1^{N-1} + (\lambda^N)^T \gamma(x^{N-1}, u^{*N-1}, \omega^{N-1}, N-1)\} + \\ & \quad + O(\nu(\epsilon)) = \cdots = \\ & = E\{(\lambda^1)^T x_1^1 + \sum_{\tau=1}^{N-1} (\lambda^{\tau+1})^T \gamma(x^\tau, u^{*\tau}, \omega^\tau, \tau)\} + O(\nu(\epsilon)) = \\ & = E \sum_{\tau=0}^{N-1} (\lambda^{\tau+1})^T \gamma(x^\tau, u^{*\tau}, \omega^\tau, \tau) + O(\nu(\epsilon)) = \Phi^* + O(\nu(\epsilon)) \ , \ \nu(\epsilon) \to 0 \\ & \qquad\qquad\qquad\qquad\qquad\qquad\qquad\qquad\qquad\qquad\qquad\qquad \epsilon \to 0 \end{aligned} \tag{2.55'}$$

and (2.55) can be rewritten in the following form

$$\begin{aligned}\tilde{G}_\epsilon &= G^* + \epsilon E \sum_{\tau=0}^{N-1} (\lambda^{\tau+1})^T \gamma(x^\tau, u^{*\tau}, \omega^\tau, \tau) + o(\epsilon) = \\ &= G^* + \epsilon \Phi^* + o(\epsilon)\end{aligned} \tag{2.56}$$

Similarly, we can easily obtain

$$x_\epsilon^{*t} = \bar{x}_\epsilon^t + \epsilon \bar{x}_{1\epsilon}^t \ , \ t = 1,\ldots,N \ , \ \|x_{1\epsilon}^t\| \le L = \text{constant} \tag{2.57}$$

$$\bar{x}_\epsilon^{t+1} = f(\bar{x}_\epsilon^t, u_\epsilon^{*t}, t) \ , \ t = 0,1,\ldots,N \ , \ x^0 \text{ is given} \tag{2.58}$$

$$\begin{aligned}\bar{x}_{1\epsilon}^{t+1} &= f_x(\bar{x}_\epsilon^t, u_\epsilon^{*t}, t)\bar{x}_{1\epsilon}^t + \gamma(\bar{x}_\epsilon^t, u_\epsilon^{*t}, \omega^t, t) + O(\nu(\epsilon)) \\ &\bar{x}_{1\epsilon}^0 = 0 \ , \ \nu(\epsilon) \underset{\epsilon \to 0}{\to} 0\end{aligned} \tag{2.59}$$

$$B^N(x^0) = E\{G(\bar{x}_\epsilon^N)\} + \epsilon E\{(G_x(\bar{x}_\epsilon^N)^T \bar{x}_{1\epsilon}^N\} + o(\epsilon) \tag{2.60}$$

where x_ϵ^{*t}, $t = 1,\ldots,N$ is the trajectory of (2.6) using the optimal control. Note that by definition,

$$u_\epsilon^{*t}(\omega^0,\ldots,\omega^{t-1}) = u_\epsilon^{*t}(x_\epsilon^{*t})$$

For the sake of brevity we shall write u_ϵ^{*t} instead of $u_\epsilon^{*t}(\omega^0,\ldots,\omega^{t-1})$.

By definition, we have that

$$B^N(x^0) \ge \tilde{G}_\epsilon \tag{2.61}$$

On the other hand,

$$G(\bar{x}_\epsilon^N) \le G^* \Rightarrow E\{G(\bar{x}_\epsilon^N)\} = G^* \tag{2.62}$$

Comparing (2.55) with (2.60) and considering (2.61) and (2.62) we can write

$$\lim_{\epsilon \to 0} E\{G(\bar{x}_\epsilon^N)\} = G^* \tag{2.63}$$

$$E\{G_x^T(\bar{x}_\epsilon^N)\bar{x}_{1\epsilon}^N\} \ge E\{G_x^T(x^N)x_1^N\} + O(\nu(\epsilon)) \ , \ \nu(\epsilon) \to 0 \ , \ \epsilon \to 0 \tag{2.64}$$

Let us now prove that (2.63) implies the existence of the function

$$u^t_{\beta\epsilon}(t = 0,1,\ldots,N-1) \in \{u^\beta\}$$

such that

$$\sum_{t=0}^{N-1} \|u^{*t}_\epsilon - u^t_{\beta\epsilon}\| \xrightarrow[\epsilon\to 0]{p} 0 \tag{2.65}$$

Suppose on the contrary, that there are positive numbers δ, γ and a subsequence $\{\epsilon'\}$ such that

$$p_r\{\inf_\beta \sum_{t=0}^{N-1} \|u^{*t}_{\epsilon'} - u^t_\beta\| \geq \delta / u^t_\beta(t = 0,1,\ldots,N-1) \in \{u^\beta\}\} \geq \gamma > 0 \ , \quad \epsilon' \to 0 \tag{2.66}$$

As is easily seen, any positive δ will have a corresponding δ_1 such that for all u^t $(t = 0,1,\ldots,N-1)$, satisfying the condition

$$\inf_\beta \{\sum_{t=0}^{N-1} \|u^t - u^t_\beta\| / u^t \in U_t \ , \ t = 0,1,\ldots,N-1 \ , \quad u^t_\beta(t = 0,1,\ldots,N-1) \in \{u^\beta\}\} \geq \delta$$

the following inequality will hold

$$G^* - G(x^N_u) \geq \delta_1 > 0$$

where x^t_u $(t = 1,\ldots,N)$ is the trajectory of (2.45) that corresponds to u^t, $t = 0,1,\ldots,N-1$. Hence, on the basis of (2.66), we write

$$E\{G^* - G(\bar{x}^N_\epsilon)\} \geq E\{(G^* - G(\bar{x}^N_\epsilon))\chi[\inf_\beta \sum_{t=0}^{N-1} \|u^{*t}_{\epsilon'} - u^t_\beta\| > \delta / \quad u^t_\beta(t = 0,1,\ldots,N-1) \in \{u^\beta\}]\} \geq \delta_1\gamma > 0$$

The former contradicts (2.63) and, consequently, proves (2.65). Here and later, $\chi(\omega)$ is referred to as the indicator function of event ω.

Let $u^t_{\beta\epsilon}(t = 0,1,\ldots,N-1)$ be the control belonging to $\{u^\beta\}$ and satisfying (2.65); $\hat{x}^t_\epsilon$, $\hat{\lambda}^t_\epsilon$, $t = 1,\ldots,N$, be the corresponding solutions to (2.45) and (2.46). Let $\hat{x}^t_{1\epsilon}$ be the solution to the system

$$\hat{x}^{t+1}_{1\epsilon} = f_x(\hat{x}^t_\epsilon, u^t_{\beta\epsilon}, t)\hat{x}^t_{1\epsilon} + \gamma(\hat{x}^t_\epsilon, u^t_{\beta\epsilon}, \omega^t, t) \ , \ \hat{x}^0_{1\epsilon} \triangleq 0 \tag{2.67}$$

According to (2.65), we have

$$\begin{gathered}\|\hat{x}^N_\epsilon - \bar{x}^N_\epsilon\| \underset{\epsilon\to 0}{\overset{p}{\longrightarrow}} 0 \ , \ \|\hat{x}^N_{1\epsilon} - \bar{x}^N_{1\epsilon}\| \underset{\epsilon\to 0}{\overset{p}{\longrightarrow}} 0 \Rightarrow \\ |E\{(G_x(\bar{x}^N_\epsilon))^T\bar{x}^N_{1\epsilon}\} - E\{G_x(\hat{x}^N_\epsilon))^T\hat{x}^N_{1\epsilon}\}| \underset{\epsilon\to 0}{\longrightarrow} 0\end{gathered} \tag{2.68}$$

Similarly to (2.55′)

$$E\{(G_x(\hat{x}^N_\epsilon))^T\hat{x}^N_{1\epsilon}\} = E\{\sum_{\tau=0}^{N-1}(\hat{\lambda}^{\tau+1}_\epsilon)^T\gamma(\hat{x}^\tau_\epsilon, u^\tau_{\beta\epsilon}, \omega^\tau, \tau)\} \Rightarrow \tag{2.69}$$

$$\begin{aligned}E\{(G_x(\bar{x}^N_\epsilon))^T\bar{x}^N_{1\epsilon}\} = \ & E\{\sum_{\tau=0}^{N-1}(\hat{\lambda}^{\tau+1}_\epsilon)^T\gamma(\hat{x}^\tau_\epsilon, u^\tau_{\beta\epsilon}, \omega^\tau, \tau)\} + \\ & + O(\nu(\epsilon)), \nu(\epsilon) \to 0 \\ & \qquad\qquad\qquad \epsilon \to 0\end{aligned} \tag{2.70}$$

Let us substitute (2.70) and (2.55′) into (2.64):

$$E\{\sum_{\tau=0}^{N-1}(\hat{\lambda}^{\tau+1}_\epsilon)\gamma(\hat{x}^\tau_\epsilon, u^\tau_{\beta\epsilon}, \omega^\tau, \tau)\} \geq \Phi^* + O(\nu(\epsilon)) \ , \ \underset{\epsilon \to 0}{\nu(\epsilon) \to 0}$$

However, by definition,

$$\begin{gathered}E\{\sum_{\tau=0}^{N-1}(\hat{\lambda}^{\tau+1}_\epsilon)^T\gamma(\hat{x}^\tau_\epsilon, u^\tau_{\beta\epsilon}, \omega^\tau, \tau)\} \leq \Phi^* \Rightarrow \\ E\{\sum_{\tau=0}^{N-1}(\hat{\lambda}^{\tau+1}_\epsilon)^T\gamma(\hat{x}^\tau_\epsilon, u^\tau_{\beta\epsilon}, \omega^\tau, \tau)\} = \Phi^* + O(\nu(\epsilon)) \ , \ \nu(\epsilon) \to 0 \ , \ \epsilon \to 0\end{gathered} \tag{2.71}$$

This together with (2.70) and (2.60) leads to the first and second results of the theorem. In the case of the uniqueness of the auxiliary problem solution, by (2.71)

$$u^\tau_{\beta\epsilon} \underset{\epsilon\to 0}{\overset{p}{\longrightarrow}} u^{*\tau} \ , \ \tau = 0,1,\ldots,N-1$$

and so (2.65) is equivalent to (2.51).

4.3. Discrete dynamic systems with weak or aggregatable controls. An asymptotic stochastic maximum principle

Consider the problem

$$x^{t+1} = f(x^t, t) + \epsilon\gamma(x^t, t, u^t, \omega^t) \ , \ t = 0,1,\ldots,N-1 \ , \ x^0 \text{ is given} \tag{3.1}$$

$$u_t \in U_t \ , \ t = 0,1,\ldots,N-1 \tag{3.2}$$

$$\max EG(x^N) \tag{3.3}$$

which is of the class (2.2)–(2.4) and (2.6). The peculiarity of this problem is that the reduced system

$$\bar{x}^{t+1} = f(\bar{x}^t, t) \ , \ t = 0,1,\ldots,N-1 \ , \ x^0 \text{ is given} \tag{3.4}$$

is not only deterministic but also uncontrolled. Stated formally, any control u^t, $t = 0,1,\ldots,N-1$, satisfying (3.2) is an optimal one for the basic problem. The conjugate system takes the form

$$\bar{\lambda}^t = f_x^T(\bar{x}^t, t)\bar{\lambda}^{t+1} \ , \ t = 0,1,\ldots,N-1 \ , \ \bar{\lambda}^N \triangleq G_x(\bar{x}^N) \tag{3.5}$$

and the auxiliary problem (2.47) decomposes into N subproblems

$$\max_{u^t \in U_t} \{(\bar{\lambda}^{t+1})^T E\gamma(\bar{x}^t, t, u^t, \omega^t)\} \ , \ t = 0,1,\ldots,N-1 \tag{3.6}$$

We shall now formulate some results that make the evaluation of Theorem 2.4 more precise.

Theorem 3.1.

Let the functions $f(\cdot)$, $f_x(\cdot)$, $\gamma(\cdot)$, $G(\cdot)$, $G_x(\cdot)$ satisfy the Lipschitz conditions with respect to x within some neighborhood of the reduced trajectory $\bar{x}^t$, $t = 0,\ldots,N-1$ uniformly with respect to the other variables.

Assume that the sets U_t, $t = 0,1,\ldots,N-1$ are compact and that there exists a compact set $D \subset R^l$ such as

$$p_r\{\omega^t \in D\} = 1 \ , \ t = 0,1,\ldots,N-1 \tag{3.7}$$

Then:

(1) When using the control $\{u^{*0}, u^{*1}, \ldots, u^{*N-1}\}$, which is determined at each moment as the solution to (3.6), the value $\tilde{G}_\epsilon$ of the performance index attained is approximately equal to the optimal one

$$B^N(x^0) = \tilde{G}_\epsilon + O(\epsilon^2) \tag{3.8}$$

(2) For the Bellman function $B^N(x^0)$ of (3.4) and (3.3), the following expression is valid:

$$B^N(x^0) = G(x^N) + \epsilon \sum_{\tau=1}^{N} \max_{u^{\tau-1} \in U_{\tau-1}} \{(\bar{\lambda}^\tau)^T E\gamma(\bar{x}^{\tau-1}, \tau-1, u^{\tau-1}, \omega^{\tau-1})\} + O(\epsilon^2) \tag{3.9}$$

(3) If the solutions to the problems (3.7) are unique, then the following limits exist:

$$u_\epsilon^{*t}(\omega^0, \omega^1, \ldots, \omega^{t-1}) \xrightarrow[\epsilon \to 0]{p} u^{*t} \quad , \quad t = 0,1,\ldots,N-1 \tag{3.10}$$

where $u_\epsilon^{*t}(\omega^0, \omega^1, \ldots, \omega^{t-1})$ is the optimal random control in (3.1) that is generated by the optimal feedback control for the random trajectory of (3.1).

We see that here auxiliary problem (3.6) takes the form of some stochastic maximum principle: at each moment t, a control maximizing an approximate Hamiltonian of the system is sought. However, as distinct from the strict maximum principle that gives the necessary optimality conditions, we obtain in (3.6), asymptotic sufficient conditions for optimality. We do not assume here the existence of the solutions in (3.1)–(3.3) and assert that the indicated proximity to the optimum is attained when using the solutions to (3.6).

The important feature of the suboptimal control $\{u^{*0}, u^{*1}, \ldots, u^{*N-1}\}$ is that it does not depend on the phase coordinates, and, consequently, can be calculated *a priori* without any further observations of the phase vector's realizations. (Controls of this type are called programmed ones.) Of course, this effect occurs only if the influence of random factors is small. Now we shall examine a different situation where the perturbed problem is of the form

$$x^{\tau+1} = f(x^\tau, \tau, \omega^\tau) + \epsilon\gamma(x^\tau, \tau, u^\tau, \omega^\tau) \quad , \quad u^\tau \in U_\tau \quad , \quad \tau = 0,1,\ldots,N-1 \tag{3.11}$$

$$\max EG(x^N) \tag{3.12}$$

The reduced problem for (3.11) is uncontrolled [as it was in (3.4)] but at the same time is essentially dependent on random parameters

$$x^{\tau+1} = f(x^\tau, \tau, \omega^\tau) \quad , \quad \tau = 0,1,\ldots,N-1 \tag{3.13}$$

Both the perturbed and the reduced problem are conveniently presented here in the form of Bellman's equations:

$$\begin{aligned} B^{N-(\tau-1)}(x) = & \max_{u^{\tau-1}\in U_{\tau-1}} EB^{N-\tau}[f(x,\tau-1,\omega^{\tau-1}) + \\ & + \epsilon\gamma(x^{\tau-1},\tau-1,u^{\tau-1},\omega^{\tau-1})] \ , \\ & \tau = 1,2,\ldots,N-1 \ , \ B^0(x) \triangleq G(x) \end{aligned} \tag{3.12'}$$

$$\begin{aligned} \bar{B}^{N-(\tau-1)}(x) = & E\bar{B}^{N-\tau}(f(x,\tau-1,\omega^{\tau-1})) \ , \ \tau = 1,2,\ldots,N-1 \\ & \bar{B}^0(x) \triangleq G(x) \end{aligned} \tag{3.13'}$$

Assume that the reduced Bellman functions are differentiable in x, i.e., that there exist the functions

$$\lambda^\tau(x) \triangleq \nabla \bar{B}^{N-\tau}(x) \ , \ \tau = 0,1,\ldots,N \tag{3.14}$$

By differentiating (3.13′), we shall obtain the relations

$$\lambda^\tau(x) = E\{f_x^T(x,\tau,\omega^\tau)\lambda^{\tau+1}(f(x,\tau,\omega^\tau))\} \tag{3.15}$$

representing a stochastic variant of the conjugate system (3.5).

Similarly to (2.47), we can write down the auxiliary problem

$$\begin{aligned} & \max_{u^\tau\in U_\tau} E\{(\lambda^{\tau+1}(f(x,\tau,\omega^\tau)))^T\gamma(x,\tau,u^\tau,\omega^\tau)\} \triangleq \Phi_\tau^*(x) \ , \\ & \tau = 0,1,\ldots,N-1 \end{aligned} \tag{3.16}$$

which is also in the form of a maximum principle but, unlike (2.47), generates controls dependent on the phase coordinates. Consider the following.

Theorem 3.2.

Let the sets U_τ, $\tau = 0,1,\ldots,N-1$ be bounded and closed, let the functions $f(\cdot)$, $\gamma(\cdot)$, $G(\cdot)$ be continuous; and let compact sets D, P_0 exist such that

$$p_r\{\omega^\tau \in D\} = 1 \ , \ \tau = 0,1,\ldots,N-1 \ ; \ p_r\{x^0 \in P_0\} = 1$$

where x^0 is a random initial vector of phase coordinates. Let us also suppose that, for any compact set P in R^n,† we can identify a constant k_p such that for $\bar{x} \in P$, $\bar{\bar{x}} \in P$

$$\|\lambda^\tau(\bar{x}) - \lambda^\tau(\bar{\bar{x}})\| \le k_p \|\bar{x} - \bar{\bar{x}}\| \tag{3.17}$$

$$\|f(\bar{x},\tau,\omega^\tau) - f(\bar{\bar{x}},\tau,\omega^\tau)\| \le k_p \|\bar{x} - \bar{\bar{x}}\| \tag{3.18}$$

$$\|\gamma(\bar{x},\tau,u^\tau,\omega^\tau) - \gamma(\bar{\bar{x}},\tau,u^\tau,\omega^\tau)\| \le k_p \|\bar{x} - \bar{\bar{x}}\| \tag{3.19}$$

Then by substituting in (3.11) the control

$$u^{*0}(\tilde{x}^0), u^{*1}(\tilde{x}^1), \ldots, u^{*N-1}(\tilde{x}^{N-1}) \tag{3.20}$$

where $u^{*\tau}(x)$, $\tau = 0,1,\ldots,N-1$ are the solutions to (3.16), we get a performance index value that is optimal to within $O(\epsilon^2)$. Uniformly for x in any compact set in R^n,

$$\Phi_\tau^*(x) - \Phi_\tau(x, u_\epsilon^{*\tau}(x)) = O(\epsilon) \tag{3.21}$$

where

$$\Phi_\tau(x,u) \triangleq E\{(\lambda^{\tau+1}(f(x,\tau,\omega^\tau)))^T \gamma(x,\tau,u,\omega^\tau)\}$$

and $u_\epsilon^{*\tau}(x)$ is the strict optimal feedback control in (3.11). In the case where for given x and τ the solution to (3.16) is unique, the following limit exists:

$$\lim_{\epsilon \to 0} u_\epsilon^{*\tau}(x) = u^{*\tau}(x) \tag{3.22}$$

The proof of the theorem is based on the following result.

Lemma 3.1. Let the conditions of Theorem 3.2 be fulfilled. Then the representations

$$B^{N-\tau}(x) = \bar{B}^{N-\tau}(x) + \epsilon\eta^{N-\tau}(x) + O(\epsilon^2) \ , \ \tau = 0,1,\ldots,N-1 \tag{3.23}$$

are valid uniformly for x in any compact set in R^n, where $\eta^{N-\tau}(x)$ satisfy Lipshitz conditions.

† The functions $\lambda^\tau(x)$ will satisfy (3.17) if, for example, the functions $f_x(x,\tau,\omega^\tau)$, $\nabla G(x)$ satisfy Lipshitz conditions with respect to x in every compact set in R^n.

Proof. Let us note first that, from (3.17)–(3.19), it follows that for $\bar{x} \in P$, $\bar{\bar{x}} \in P$ (P is a compact set) we can identify a constant $\hat{k}_p$ such that

$$|\Phi_\tau(\bar{x}, u) - \Phi_\tau(\bar{\bar{x}}, u)| \leq \hat{k}_p \|\bar{x} - \bar{\bar{x}}\| \Rightarrow \tag{3.24}$$

$$\begin{cases} \Phi_\tau(\bar{x}, u^\tau) \leq \Phi_\tau(\bar{\bar{x}}, u^\tau) + \hat{k}_p \|\bar{x} - \bar{\bar{x}}\| \\ \Phi_\tau(\bar{\bar{x}}, u^\tau) \leq \Phi_\tau(\bar{x}, u^\tau) + \hat{k}_p \|\bar{x} - \bar{\bar{x}}\| \end{cases} \Rightarrow$$

$$\begin{cases} \Phi_\tau^*(\bar{x}) \triangleq \max\limits_{u^\tau \in U_\tau} \Phi_\tau(\bar{x}, u^\tau) \leq \max\limits_{u^\tau \in U_\tau} \Phi_\tau(\bar{\bar{x}}, u^\tau) + \hat{k}_p \|\bar{x} - \bar{\bar{x}}\| \\ \Phi_\tau^*(\bar{\bar{x}}) \triangleq \max\limits_{u^\tau \in U_\tau} \Phi_\tau(\bar{\bar{x}}, u^\tau) \leq \max\limits_{u^\tau \in U_\tau} \Phi_\tau(\bar{x}, u^\tau) + \hat{k}_p \|\bar{x} - \bar{\bar{x}}\| \end{cases} \Rightarrow$$

$$|\Phi_\tau^*(\bar{x}) - \Phi_\tau^*(\bar{\bar{x}})| \leq \hat{k}_p \|\bar{x} - \bar{\bar{x}}\|$$

By definition,

$$B^0(x) = \bar{B}^0(x) = G(x)$$

Let us suppose that (3.23) is valid for $\tau = t$

$$B^{N-t}(x) = \bar{B}^{N-t}(x) + \epsilon\eta^{N-t}(x) + O(\epsilon^2) \tag{3.25}$$

and that $\eta^{N-t}(x)$ satisfies Lipshitz conditions.

Taking into account (3.25), we can write

$$\begin{aligned} B^{N-(t-1)}(x) = \max_{u^{t-1} \in U_{t-1}} &\{E\bar{B}^{N-t}(f(x, t-1, \omega^{t-1}) + \epsilon\gamma(x, t-1, u^{t-1}, \omega,)) + \\ &+ \epsilon\eta^{N-t}(f(x, t-1, \omega^{t-1}) + \epsilon\gamma(x, t-1, u^{t-1}, \omega^{t-1}))\} + O(\epsilon^2) \end{aligned} \tag{3.26}$$

$$\begin{aligned} &E\bar{B}^{N-t}(f(x, t-1, \omega^{t-1}) + \epsilon\gamma(x, t-1, u^{t-1}, \omega^{t-1})) = \\ &= \bar{B}^{N-(t-1)}(x) + \epsilon\Phi_{t-1}(x, u^{t-1}) + O(\epsilon^2) \end{aligned} \tag{3.27}$$

with the estimate $O(\epsilon^2)$ being uniform for x in any compact set in R^n.

We also have that

$$\begin{aligned} &\eta^{N-t}(f(x, t-1, \omega^{t-1}) + \epsilon\gamma(x, t-1, u^{t-1}, \omega^{t-1})) = \\ &\eta^{N-t}(f(x, t-1, \omega^{t-1})) + O(\epsilon) \Rightarrow \end{aligned} \tag{3.28}$$

$$B^{N-(t-1)}(x) = \bar{B}^{N-(t-1)}(x) + \epsilon(\Phi_{t-1}^*(x) + E\eta^{N-t}(f(x, t-1, \omega^{t-1})) + O(\epsilon^2) \tag{3.29}$$

From which, using the notation

$$\eta^{N-(t-1)}(x) = \Phi_{t-1}^{*}(x) + E\eta^{N-t}(f(x,t-1,\omega^{t-1})) \tag{3.30}$$

we get (3.23) with $\tau = t-1$.

Proof of Theorem 3.2. It is easy to see that the value $\tilde{J}$ of the performance index obtained using the control (3.20) may be written in the form

$$\tilde{J} = \int p(x^0)F^N(x^0)\mathrm{d}x^0$$

where $p(x^0)$ is the probability distribution density of the random vector x^0 of initial conditions, and where

$$\begin{gathered} F^{N-(\tau-1)}(x) = EF^{N-\tau}(f(x,\tau-1,\omega^{\tau-1}) + \epsilon\gamma(x,\tau-1,u^{*\tau-1}(x),\omega^{\tau-1})) \ , \\ \tau = N,N-1,\ldots,1 \ , \ F^0(x) \triangleq G(x) \end{gathered} \tag{3.31}$$

On the other hand, the optimal index J^* is given by

$$J^* = \int p(x^0)B^N(x^0)\mathrm{d}x^0$$

Consequently the first assertion will be proved if

$$B^{N-\tau}(x) = F^{N-\tau}(x) + O(\epsilon^2) \ , \ \tau = 0,1,\ldots,N \tag{3.32}$$

is shown to be true.

As above, we have that

$$B^0(x) = F^0(x) = G(x)$$

and suppose that uniformly for x in any compact set in R^n

$$\begin{gathered} B^{N-t}(x) = F^{N-t}(x) + O(\epsilon^2) \Rightarrow \\ F^{N-(t-1)}(x) = EF^{N-t}(f(x,\tau-1,\omega^{\tau-1}) + \epsilon\gamma(x,\tau-1,u^{*\tau-1}(x),\omega^{\tau-1})) = \\ = EB^{N-t}(f(x,\tau-1,\omega^{\tau-1}) + \epsilon\gamma(x,\tau-1,u^{*\tau-1}(x),\omega^{\tau-1})) + O(\epsilon^2) \end{gathered} \tag{3.33}$$

Using (3.23) yields

$$\begin{gathered} F^{N-(t-1)}(x) = E\bar{B}^{N-t}(f(x,\tau-1,\omega^{\tau-1}) + \epsilon\gamma(x,\tau-1,u^{*\tau-1}(x),\omega^{\tau-1})) + \\ + \epsilon\eta^{N-t}(f(x,\tau-1,\omega^{\tau-1}) + \epsilon\gamma(x,\tau-1,u^{*\tau-1}(x),\omega^{\tau-1})) + O(\epsilon^2) \Rightarrow \\ F^{N-(t-1)}(x) = \bar{B}^{N-(t-1)}(x) + \epsilon\eta^{N-(t-1)}(x) + O(\epsilon^2) \end{gathered}$$

which together with (3.23), establishes (3.32) for $\tau = t-1$. To prove (3.21), let us note that, by definition,

$$\begin{gathered} EB^{N-(\tau+1)}(f(x,\tau,\omega^\tau) + \epsilon\gamma(x,\tau,u_\epsilon^{*\tau}(x),\omega^\tau)) \geq \\ \geq EB^{N-(\tau+1)}(f(x,\tau,\omega^\tau) + \epsilon\gamma(x,\tau,u^{*\tau}(x),\omega^\tau)) \end{gathered} \tag{3.34}$$

Similarly to (3.29),

$$EB^{N-(\tau+1)}(f(x,\tau,\omega^\tau) + \epsilon\gamma(x,\tau,u^\tau,\omega^\tau)) = \bar{B}^{N-\tau}(x) + \epsilon\Phi_\tau(x,u^\tau) + \\ + \epsilon E\eta^{N-(\tau+1)}(f(x,\tau,\omega^\tau)) + O(\epsilon^2) \quad , \quad \forall\, u^\tau \in U_\tau$$

The substitution of the former relation into (3.34) yields the inequality

$$\Phi_\tau(x, u_\epsilon^{*\tau}(x)) \geq \Phi_\tau^*(x) + O(\epsilon)$$

which is fulfilled uniformly for any compact set in R^n and is equivalent to (3.21). The relation (3.22) is the obvious consequence of the above.

We now want to connect these results to the more general situation where the perturbed problem is of the form

$$\begin{aligned} &x^{t+1} = f(x^t, t, \psi^t(u^t), \omega^t) + \epsilon\gamma(x^t, t, u^t, \omega^t) \\ &t = 0,1,\ldots,N-1 \quad , \quad p(x^0) \text{ is given} \\ &u^t \in U_t \quad , \quad \psi^t(u^t) \in \tilde{U}_t \quad , \quad t = 0,1,\ldots,N-1 \quad ; \quad \max\, EG(x^N) \end{aligned} \tag{3.35}$$

The reduced problem for (3.35) admits the aggregation of controls and is written in the form

$$\begin{aligned} &x^{t+1} = f(x^t, t, \psi^t, \omega^t) \quad , \quad t = 0,1,\ldots,N-1 \quad , \quad p(x^0) \text{ is given} \\ &\psi^t \in \tilde{U}_t \quad , \quad t = 0,1,\ldots,N-1 \quad ; \quad \max\, EG(x^N) \end{aligned} \tag{3.36}$$

Suppose that the optimal feedback control $\xi^t(x)$ has been computed in (3.36) and that its Bellman functions $\bar{B}^{N-t}(x)$, $t = 0,1,\ldots,N-1$ are differentiable. Let us introduce the auxiliary problem as

$$\begin{aligned} &\max\{E\{(\lambda^{t+1}(f(x,t,\xi^t(x),\omega^t))^T\gamma(x,t,u^t,\omega^t)\} \,/\, u^t \in U_t\,, \\ &\varphi^t(u^t) = \xi^t(x)\} \quad , \quad \lambda^t(x) \triangleq \nabla\bar{B}^{N-t}(x) \quad , \quad t = 0,1,\ldots,N-1 \end{aligned} \tag{3.37}$$

Theorem 3.3.

Let the conditions of Theorem 3.2 be satisfied. Then the control $u^{*t}(x)$ ($t = 0,1,\ldots,N-1$) as obtained from (3.37) yields a value of the performance index equal to the optimal value to within $o(\epsilon)$. If, in addition, for given x and t, the solution to (3.37) is unique, the following limit exists

$$\lim_{\epsilon \to 0} u_\epsilon^{*t}(x) = u^{*t}(x)$$

where $u_\epsilon^{*t}(x)$ is the optimal control in (3.35).

The proof of the theorem is similar to the proof of Theorem 3.2.

4.4. Sliding planning and suboptimal decomposition of operative control in a production system

The most general and complex problems of stochastic programming are multistage ones where the feasible sets appear to be dependent on the state variables. The general description of the perturbation procedure is given by an inductive (with respect to the number of stages) extension of the procedure presented for the two-stage problems in Section 4.1.

Without going into detail concerning such general constructions, we shall concentrate on a particular problem of this class that is of considerable interest from the practical point of view. We shall interpret it in terms of optimizing the operative control in a production system under uncertainty about raw material supplies. Let the problem be to develop a policy that maximizes the objective function

$$B^N(x) \triangleq \max E\{\sum_{\tau=1}^{N} [(c^\tau)^T x(\tau) - h^\tau(x(\tau - 1))]\} \tag{4.1}$$

such that

$$x(t) \in \Theta_t(s(t-1)) \ , \ \Theta_t(s) \triangleq \{x \, / \, A_t x \leq s \ , \ D_t x \leq d^t \ , \ x \geq 0\} \tag{4.2}$$

where $s(t)$ is the sequence of m-vectors such that

$$s(t) = s(t-1) - A_t x(t) + q(t) \ , \ t = 1,\ldots,N-1 \ , \ s(0) \triangleq s \geq 0 \tag{4.3}$$

and $q(t)$ (q^t) are random nonnegative mutually independent m-vectors.

Interpreting (4.1)–(4.3) as a problem of optimal planning (scheduling), we shall associate the n-vector $x(t)$ of industrial activities at stage t; $s(t-1)$ with the stock levels of raw materials at the beginning of this stage, and $q(t)$ with the supplies of raw materials coming from outside.

According to this interpretation, the relations (4.3) mean that the level $s(t)$ at the beginning of stage $t+1$ is formed from both the unspent remainders left over from the preceding stage and new deliveries, where A_t, $t = 1,\ldots,N$, are matrices of input–output coefficients. The conditions (4.2) represent natural constraints on production capacities — in particular, according to these restrictions, one cannot spend more raw materials than there are on hand. The performance index (4.1) is the sum of the profit $(c^\tau)^T x(\tau)$ at each stage minus storage expenses.

The functions $B^{N-t}(s)$ defined in (4.1) satisfy the equations

$$\begin{aligned} B^{N-t}(x) = \ & \max \{(c^{t+1})^T x - h^{t+1}(s) + EB^{N-t-1}(s - A_{t+1}x + q^{t+1}) \, / \\ & x \in \Theta_{t+1}(s)\} \ , \ t = 0,1,\ldots,N-1 \ , \ B^0(s) \triangleq 0 \end{aligned} \tag{4.4}$$

which are equivalent to the original problem. In order to use the perturbation method, let us suppose first, for the sake of simplicity, that the variance of q_i^t is time independent and, second, that

$$Eq_i^t = \bar{q}_i^t \ , \ E(q_i^t - \bar{q}_i^t)^2 = \epsilon^2 D_i^t \ , \ i = 1,2,\ldots,m \tag{4.5}$$

i.e., random deviations from the forecasted supplies are supposed to be small. However, this "smallness" is treated in a different sense from Section 4.2.

The reduced problem can be presented here in the two equivalent forms:

$$\begin{aligned} \bar{B}^{N-t}(s) = \ & \max \{(c^{t+1})^T x - h^{t+1}(s) + \bar{B}^{N-t-1}(s - A_{t+1}x + \bar{q}^{t+1}) \, / \\ & x \in \Theta_{t+1}(s)\} \ , \ t = 0,1,\ldots,N-1 \ , \ \bar{B}^0(s) \triangleq 0 \end{aligned} \tag{4.6}$$

$$\begin{aligned} \bar{B}^{N-t}(s) \triangleq \ & \max \sum_{t+1}^{N} \{(c^\tau)^T x^\tau - h^\tau(s^{\tau-1})) \, / \, x^\tau \in \Theta_\tau(s^{\tau-1}) \ , \\ & s^\tau = s^{\tau-1} - A_\tau x^\tau + \bar{q}^\tau \ , \tau = t+1,\ldots,N \ , \ s^t \triangleq s\} \end{aligned} \tag{4.7}$$

where we write $\bar{q}^t$ instead of q^t in (4.4) and (4.1)–(4.3) and where the operations of expectation are omitted.

The simplest and most widely used approach of developing decision rules for problems of type (4.1)–(4.3) is based on the sliding planning principle. The main feature of this principle may be characterized as follows:

(1) Some dynamic multistage model is constructed to describe the inventory and production processes.
(2) One assumes the availability of correct information about the levels of resources at the current stage and of a statistical prediction concerning successive stages.
(3) The level of industrial activity at the current stage is determined as part of the solution to the deterministic multistage optimization problem.
(4) The solution is repeated at the beginning of each successive stage with newly corrected information if the real state of the system at the end of this stage differs from that predicted.

In the case we are examining, according to the sliding planning principle, the deciding rules that would be optimal if at subsequent stages $t+1,\ldots,N-1$ the random supplies $q^{t+1},\ldots,q^{N-1}$ were equal to their expected values, are those recommended for selection. The advantage of the procedure is that it does not require any statistical characteristics of the deviations $q^t - \bar{q}^t$ at each stage. We use the current random vector of state coordinates as initial conditions and solve the deterministic program with expectations instead of random factors. On the other hand, it is obvious that such decision rules may turn out to be "suboptimal" only under the condition that deviations $q^t - \bar{q}^t$ are, in a certain sense, small.

We shall now substantiate their suboptimality, basing our reasoning on hypothesis (4.3). Formally, the sliding planning procedure consists of replacing, at every stage t, the optimal decision rules

$$\begin{aligned} x_\epsilon^{t+1}(s) = \ & \arg\max \{(c^{t+1})^T x + EB^{N-t-1}(s - A_{t+1}x + q^{t+1}) / \\ & x \in \Theta_{t+1}(s)\} \end{aligned} \tag{4.8}$$

by rules defined by

$$\bar{x}^{t+1}(s) = \arg\max \{(c^{t+1})^T x + \bar{B}^{N-t-1}(s - A_{t+1}x + \bar{q}^{t+1}) / x \in \Theta_{t+1}(s)\} \tag{4.9}$$

Designating $F^{N-t}(s)$ the value of the objective function (4.1) obtained using the sliding planning principle under the condition that the process is started from the moment t, we can write

$$\begin{aligned} F^{N-t}(s) \triangleq \ & (c^{t+1})^T \bar{x}^{t+1}(s) - h^{t+1}(s) + \\ & + EF^{N-t-1}(s - A_{t+1}\bar{x}^{t+1}(s) + q^{t+1}) \ , \\ & t = 0,1,\ldots,N-1 \ , \quad F^0(s) \triangleq 0 \end{aligned} \tag{4.10}$$

Theorem 4.1.

Let us suppose that the functions $h^t(s)$, $t = 1,\ldots,N$, satisfy Lipschitz conditions in R_m^+

$$|h^t(s') - h^t(s'')| \le k_h \|s' - s''\| \ , \quad s \in R_m^+ \ , \quad s'' \in R_m^+ \tag{4.11}$$

and that the following relations are true

$$\begin{aligned} & 0 \in \Theta_s(s) \ , \quad \forall s \in R_m^+ \ , \quad t = 1,2,\ldots,N \\ & \max\{\|x\| / x \in \Theta_t(0)\} \le L = \text{constant} \ , \quad t = 1,2,\ldots,N \end{aligned} \tag{4.12}$$

Then the functions $B^{N-t}(s)$, $\bar{B}^{N-t}(s)$, $t = 0,\ldots,N-1$, satisfy Lipschitz conditions in R_m^+ with some constant k and the following are valid

$$|B^{N-t}(s) - \bar{B}^{N-t}(s)| \leq k(N-t)\sqrt{\textstyle\sum_{i=1}^{m} D_i}\,\epsilon \ , \ \forall\, s \in R_m^+ \quad t = 0,1,\ldots,N-1 \tag{4.14}$$

$$|F^{N-t}(s) - \bar{B}^{N-t}(s)| \leq k(N-t)\sqrt{\textstyle\sum_{i=1}^{m} D_i}\,\epsilon \ , \ \forall\, s \in R_m^+ \quad t = 0,1,\ldots,N-1 \tag{4.15}$$

The proof of the theorem is given in [134] and based on results in [97].

We can propose a more accurate procedure for solving (4.1)–(4.3) if the representations

$$h^t(s) = \sum_{i=1}^{m} h_i^t(s_i) \ , \ s \triangleq \{s_i\} \ , \ i = 1,\ldots,m \tag{4.16}$$

are possible. Let us rewrite (4.4) in the form

$$\begin{aligned} B^{N-t}(s) = \ & \max\,\{-h^{t+1}(s) + EB^{N-t-1}(s - b + q^{t+1}) + \\ & + \max\,\{(c^{t+1})^T x \,/\, A_{t+1}x = b \ , \ D_{t+1}x \leq d^{t+1} \ , \ x \geq 0\} \,/ \\ & b \leq s \ , \ b \in U_{t+1}\} \ , \ t = 0,1,\ldots,N-1 \ , \ B^0(s) \triangleq 0 \end{aligned} \tag{4.17}$$

where U_{t+1} are sets of $b(m \times 1)$ such that the internal problems have solutions.

The former are linear problems. Let us assume for now that the solutions of the problems dual to them are unique

$$\lambda^{t+1}(b) = \lambda^{t+1} \ , \ \mu^{t+1}(b) = \mu^{t+1} \tag{4.18}$$

and that the sets U_{t+1} are written as

$$U_{t+1} = \{b \,/\, \check{b}^t \leq b \leq \hat{b}^t\} \tag{4.19}$$

Then, by induction, it is easy to establish that

$$B^{N-t}(s) = \sum_{i=1}^{m} B_i^{N-t}(s_i) \ , \ s \triangleq \{s_i\} \ , \ i = 1,2,\ldots,m \tag{4.20}$$

where

$$B_i^{N-t}(s_i) = \max_{b_i} \{\lambda_i^{t+1} b_i - h_i^{t+1}(s_i) + EB_i^{N-t-1}(s_i - b_i + q_i^{t+1}) / \\ b_i \le s_i \, , \, \check{b}_i^t \le b_i \le \hat{b}_i^t\} + \frac{(\mu^{t+1})^t d^{t+1}}{m} \, , \, B_i^0(s_i) = 0 \; ; \tag{4.21}$$

$$\lambda^{t+1} \triangleq \{\lambda_i^{t+1}\} \, , \, \check{b}^t \triangleq \{\check{b}_i^t\} \, , \, \hat{b}^t = \{\hat{b}_i^t\} \, , \, i = 1,\ldots,m \tag{4.22}$$

Of course, in the general case, assumptions (4.18) and (4.19) cannot be true and, consequently, accurate expansions (4.20) cannot be made for all $s \in R_m^+$. However, when hypothesis (4.5) proves true, one may expect that, in most cases, random trajectories are in a neighborhood of the reduced optimal deterministic trajectory such that (4.18) is quite acceptable.

We shall now formulate some results substantiating (4.22) for subsets of R_m^+ that are characterized by special regularity conditions on the reduced optimal trajectories.

Theorem 4.2.

Let the conditions of Theorem 4.1 and (4.16) be satisfied. Let us suppose also that the following three regularity conditions are true:

(1) When $t = 0$, $s = s^0$, the solution $\bar{x}^1, \bar{x}^2, \ldots, \bar{x}^N$ to the program (4.7) is unique, which means, in particular, that the solutions

$$x^{l+1}(b^{l+1}) \triangleq M_{l+1} b^{l+1} + V_{l+1} = \bar{x}^{l+1} \, , \, l = 0,1,\ldots,N-1 \tag{4.23}$$

to the programs

$$\max \{(c^{l+1})^T x \,/\, A_{l+1} x = b^{l+1} \, , \, D_{l+1} x \le d^{l+1} \, , \, x \ge 0\} \, , \\ b^{l+1} \triangleq A_{l+1} \bar{x}^{l+1} \, , \, l = 0,1,\ldots,N-1 \tag{4.24}$$

are unique, where $M_{l+1}(n \times m)$, $V_{l+1}(n \times 1)$ are defined by the optimal basic matrices in (4.24).

(2) There exist positive vectors $\hat{\Delta}^l(m \times 1)$ such that $t = 1$, $s \in \{s \,/\, s^l - \hat{\Delta}^l \le s \le s^l + \hat{\Delta}^l\}$, the solutions to the programs (4.7) are unique. [Note that (1) and (2) are automatically true if $h_i^{t+1}(s_i)$ are strictly convex.]

(3) The solutions $(\lambda^{l+2}, \mu^{l+1})$, $l = 0,1,\ldots,N-1$ to the programs dual to (4.24) are unique.

Then there exist positive numbers ϵ_0, $\tilde{k}$, Δ_i^t and numbers $\check{b}_i^{t+1}, \hat{b}_i^{t+1}$, $t = 0,1,\ldots,N-1$, $i = 1,\ldots,m$, such that for $0 < \epsilon \le \epsilon_0$, $t = 0,1,\ldots,N-1$

$$|B^{N-t}(s) - \sum_{i=1}^{m} B_i^{N-t}(s_i)| \leq \tilde{k}\epsilon^2 \quad , \quad s \in \Omega^t \quad , \quad s \triangleq \{s_i\}$$
$$\Omega^t \triangleq \{s \,/\, s^t - \Delta^t \leq s \leq s^t + \Delta^t \,, s \geq 0\} \quad , \quad s^t \triangleq s^0 - \sum_{\tau=1}^{t} (A_\tau \bar{x}^\tau - \bar{q}^\tau) \tag{4.25}$$

where each $B_i^{N-t}(s_i)$ satisfies Lipschitz conditions in R^1 and is defined by the equation

$$B_i^{N-t}(s_i) = \max_{b_i} \{\lambda_i^{t+1} b_i + EB_i^{N-t-1}(s_i + q_i^{t+1} - b_i) \,/$$
$$\check{b}_i^t \leq b_i \leq s_i\} + \frac{(\mu^{t+1})^T d^{t+1}}{m} \quad , \text{ for } s_i \geq \check{b}_i \ , \tag{4.26}$$
$$B_i^{N-t}(s_i) = B_i^{N-t}(\check{b}_i) \quad , \text{ for } s_i < \check{b}_i$$

with initial conditions $B_i^0(s_i) = 0$, $i = 1,\ldots,m$, $\lambda^{T+1} \triangleq \{\lambda_i^{t+1}\}$.

Let us denote by $\Phi^{N-t}(s)$ the value of the performance index (4.1) corresponding to different initial conditions and obtained with the decision rules

$$\tilde{x}^{t+1}(s) \triangleq (M_{t+1}\tilde{b}^{t+1}(s) + V_{t+1})\alpha_{t+1}(s) \in \Theta_{t+1}(s) \tag{4.27}$$

where M_{t+1}, V_{t+1} are as in (4.23),

$$\alpha_{t+1}(s) \triangleq \min\,\{1 \,, \min_i \{\frac{s_i}{\check{b}_i^{t+1}} \,/\, \check{b}_i^{t+1} > 0\}\}$$

and where the components of $\tilde{b}^{t+1}(s)$ are the solutions to (4.26)

$$\tilde{b}^{t+1}(s) \triangleq \{\tilde{b}_i^{t+1}(s_i)\} \quad , \quad s\{s_i\} \quad , \quad i = 1,\ldots,m$$
$$\tilde{b}_i^{t+1}(s_i) \triangleq \arg\max_{b_i}\{\lambda_i^{t+1} b_i + EB_i^{N-t}(s_i - b_i + q_i^{t+1}) \,/\, \check{b}_i^{t+1} \leq b_i \leq \check{b}_i^{t+1} \,, \tag{4.28}$$
$$b_i \leq s_i\} \,, \text{for } s_i \geq \check{b}_i^{t+1} \,;\, \tilde{b}_i^{t+1}(s_i) = \tilde{b}_i^{t+1}(\check{b}_i^{t+1}) \,, \text{for } s_i < \check{b}_i^{t+1}$$

The functions $\Phi^{N-t}(s)$ are obviously connected by the equations

$$\Phi^{N-t}(s) \triangleq (c^{t+1})^T \tilde{x}^{t+1}(s) - h^{t+1}(s) +$$
$$+ E\Phi^{N-t-1}(s - A_{t+1}\tilde{x}^{t+1}(s) + q^{t+1}) \quad , \quad \Phi^0(s) \triangleq 0$$

Then for $0 < \epsilon \leq \epsilon_0$, $t = 0,1,\ldots,N-1$

$$|\Phi^{N-t}(s) - \sum_{i=1}^{m} B_i^{N-t}(s_i)| \leq \tilde{k}\epsilon^2 \quad , \quad s \in \Omega^t \tag{4.29}$$

Replacing (4.5) by the stronger supposition

$$p_r\{q^t \in [\bar{q}^t - \epsilon p^t \; , \; \bar{q}^t + \epsilon p^t]\} = 1 \quad , \quad p^t \geq 0 \tag{4.30}$$

we can improve the results.

Theorem 4.3.

Let the conditions of Theorem 4.2 be satisfied with (4.30) instead of (4.5). Then there exists $\bar{\epsilon}_0$ such that, for $0 < \epsilon \leq \bar{\epsilon}_0$, $t = 0,1,\ldots,N-1$

$$B^{N-t}(s) = \Phi^{N-t}(s) = \sum_{i=1}^{m} B_i^{N-t}(s_i) \quad , \quad s \in \Omega^t \tag{4.31}$$

where the $B_i^{N-t}(s_i)$, Ω^t are as in Theorem 4.2

The values of Δ_i^t, $\check{b}_i^{t+1}$, $\hat{b}_i^{t+1}$ defining the functions $B_i^{N-t}(s_i)$, $F^{N-t}(s)$ and the domains Ω^t depend on the sizes of cubes that may be constructed inside the stability domains of the optimal basis in the linear programs (4.24) and their duals (see [134]).

4.5. Sliding planning on an infinite horizon

In the previous section, we got some results evaluating the "suboptimality" of the sliding planning procedure for the problem (4.1)–(4.3). The estimates, however, essentially depended on the length N of the planning interval. Here we shall formulate results that substantiate the procedure in the case when $N \to \infty$. The proofs are given in detail in [134].

Let us consider the problem

$$B^N(s) \triangleq \max E\{\sum_{\tau=1}^{N} \alpha^{\tau-1}[c^T x(\tau) - h^T s(\tau-1)]\} \tag{5.1}$$

$$s(t) = s(t-1) - Ax(t) + q(t) \quad , \quad t = 1,2,\ldots,N-1 \quad , \quad s(0) \triangleq s \geq 0 \tag{5.2}$$

$$Eq_i(t) = \bar{q}_i > 0 \quad , \quad E\{(q_i(t) - \bar{q}_i)^2\} = \epsilon^2 D_i \quad , \quad i = 1,\ldots,m \tag{5.3}$$

$$x(t) \in \Theta(s(t-1)) \quad , \quad \Theta(s) \triangleq \{x \,/\, Ax \leq s \; , \; Dx \leq d \; , \; x \geq 0\} \tag{5.4}$$

that is a particular case of (4.1)–(4.3) with

$$\begin{aligned} &c^t = \alpha^{t-1}c \ , \ A_t = A \ , \ D_t = D \ , \ d^t = d \\ &h^t(s) = \alpha^{t-1}h^T s \end{aligned} \tag{5.5}$$

where

$$0 < \alpha < 1$$

is a discount factor.

Denoting by $B^N(s)$ the optimal value of the performance index (5.1) in an N-stage problem with initial conditions $s(0) = s$, we can write

$$\begin{aligned} B^N(s) = \ & \max \{c^T x - h^T s + \alpha E B^{N-1}(s - Ax + q) \,/\, x \in \Theta(s)\} \ , \\ & N = 1,2,\ldots, \ \ B^0(s) \triangleq 0 \end{aligned} \tag{5.6}$$

The reduced Bellman equations take the form

$$\begin{aligned} \bar{B}^N(s) = \ & \max \{c^T x - h^T s + \alpha \bar{B}^{N-1}(s - Ax + \bar{q} \,/\, x \in \Theta(s)\} \ , \\ & N = 1,2,\ldots, \ \ \bar{B}^0(s) \triangleq 0 \end{aligned} \tag{5.7}$$

where $\bar{B}^N(s)$ is the optimal value of the linear program

$$\begin{aligned} \bar{B}^N(s) = \ & \max \{ \sum_{\tau=1}^{N} \alpha^{\tau-1}[c^T x^\tau - h^T s^{\tau-1}] \,/\, x^\tau \in \Theta(s^{\tau-1}) \ , \\ & s^\tau = s^{\tau-1} - Ax^\tau + \bar{q} \ , \ \tau = 1,2,\ldots,N-1 \ , \ S^0 \triangleq S\} \end{aligned} \tag{5.8}$$

Interpreting (5.1)–(5.4) as a problem of optimal planning (see the previous section), it makes sense to suppose that

$$\begin{aligned} &A \geq 0 \ , \ D \geq 0 \ , \ d \geq 0 \ , \ h \geq 0 \ , \\ &\max \{\|x\| \,/\, x \in \Theta(0)\} \leq L = \text{constant} \end{aligned} \tag{5.9}$$

which leads to the inclusion

$$0 \in \Theta(s) \ , \ \forall \, s \in R_m^+ \tag{5.10}$$

By virtue of (5.9) and Lemma 4.1, for any finite N ($N = 1,2,\ldots$), the functions $B^N(s)$, $\bar{B}^N(s)$ satisfy Lipschitz conditions in R_m^+. However, it is possible here to prove an essentially stronger result.

Lemma 5.1. Let the relations (5.9) take place. Then the functions $\bar{B}^N(s)$, $N = 1,2,\ldots$ satisfy Lipschitz conditions in R_m^+, with constant k independent of N. This lemma enables one to establish a uniform (with respect to $N = 1,2,\ldots$) neighborhood of strictly optimal decision rules

$$y_\epsilon^N(s) = \arg\max \{c^T x - h^T s + \alpha E B^{N-1}(s - Ax + q) \,/\, x \in \Theta(s)\} \tag{5.11}$$

and the decision rules we obtain using the sliding planning principle

$$\bar{y}^N(s) = \arg\max \{c^T x - h^T s + \alpha \bar{B}^{N-1}(s - Ax + \bar{q}) \,/\, x \in \Theta(s)\} \tag{5.12}$$

Note that we write $y(\cdot)$ instead of $x(\cdot)$ since a different indexation is used here to that of (4.8). In (5.11) $y_\epsilon^N(s)$ is the decision rule at the first stage of the N-stage problem and in (4.8) $x_\epsilon^N(s)$ is the decision rule for stage N.

To formulate the results, let us introduce, as above, the functions

$$\begin{aligned} F^N(s) = {} & c^T \bar{y}^N(s) - h^T s + \alpha E F^{N-1}(s - A\bar{y}^N(s) + q) \;, \\ & N = 1,2,\ldots, \quad F^0(s) \triangleq 0 \end{aligned} \tag{5.13}$$

being the values of the performance index under the condition that the decision rules (5.12) are used.

Theorem 5.1.

Let the relations (5.9) be satisfied. Then for all $N = 1,2,\ldots$

$$|B^N(s) - \bar{B}^N(s)| \le \epsilon \frac{k\alpha}{1-\alpha} \sqrt{\sum_i D_i} \;, \quad \forall\, s \in R_m^+ \tag{5.14}$$

$$|F^N(s) - \bar{B}^N(s)| \le \epsilon \frac{k\alpha}{1-\alpha} \sqrt{\sum_i D_i} \;, \quad \forall\, s \in R_m^+ \tag{5.15}$$

where k is the Lipschitz constant of the functions $\bar{B}^N(s)$, $N = 1,2,\ldots$.

There exist continuous functions on R_m^+, $\bar{B}^\infty(s)$, $B^\infty(s)$

$$\bar{B}^\infty(s) = \lim_{N\to\infty} \bar{B}^N(s) \;, \quad B^\infty(s) = \lim_{N\to\infty} B^N(s) \tag{5.16}$$

$$|B^\infty(s) - \bar{B}^\infty(s)| \le \frac{\epsilon k\alpha}{1-\alpha} \sqrt{\sum_i D_i} \tag{5.17}$$

Let us note that when proving the theorem the assumption of small random factors is not used. So we can rewrite (5.14) and (5.15) in the form

$$|B^N(s) - \bar{B}^N(s)| \leq \frac{k\alpha}{1-\alpha}\sqrt{\sum_1^m E(q_i - \bar{q}_i)^2} \ , \ s \in R_m^+ \ , \ N = 1,2,\dots \tag{5.18}$$

$$|F^N(s) - \bar{B}^N(s)| \leq \frac{k\alpha}{1-\alpha}\sqrt{\sum_1^m E(q_i - \bar{q}_i)^2} \ , \ s \in R_m^+ \ , \ N = 1,2,\dots \tag{5.19}$$

The functions $B^\infty(s)$, $\bar{B}^\infty(s)$ defined in (5.16) are the Bellman functions of (5.1)–(5.4) and (5.8) with $N = \infty$. Letting N tend to ∞ in (5.6) and (5.7), it is easy to establish that $B^\infty(s)$, $\bar{B}^\infty(s)$ are the solutions to the functional equations

$$B^\infty(s) = \max\{c^Tx - h^Ts + \alpha EB^\infty(s - Ax + q) \,/\, x \in \Theta(s)\} \tag{5.20}$$

$$\bar{B}^\infty(s) = \max\{c^Tx - h^Ts + \alpha\bar{B}^\infty(s - Ax + q) \,/\, x \in \Theta(s)\} \tag{5.21}$$

Let us also define the stationary decision rules

$$y_\epsilon^\infty = \arg\max\{c^Tx + \alpha EB^\infty(s - Ax + q) \,/\, x \in \Theta(s)\} \tag{5.22}$$

$$\bar{y}^\infty = \arg\max\{c^Tx + \alpha\bar{B}^\infty(s - Ax + q) \,/\, x \in \Theta(s)\} \tag{5.23}$$

Now we wish to adduce similar results to those of Theorems 4.2 and 4.3. For problems with a linear reduced structure, "sliding planning" appears to be as effective for some subsets in R_m^+ as the decomposition procedure described in the previous section. Such subsets are constructed in a neighborhood of a regular trajectory characterized by the following statements in which, for simplicity, we shall assume that $h = 0$.

Lemma 5.2. Let the relations (5.9) be fulfilled and let $h = 0$. Let us suppose in addition that

(1) The solution (ν^*, μ^*) to the program

$$\check{\Phi} \triangleq \min\{\nu^T\bar{q} + \mu^Td \,/\, (\nu,\mu) \geq 0 \ , \nu^TA + \mu^TD \geq c^T\} \tag{5.24}$$

is unique and $\nu^* > 0$.

(2) The solution x^* to the program dual to (5.24)

$$\check{\Phi} = \max\{c^Tx \,/\, x \in \Theta(\bar{q})\} \tag{5.25}$$

is unique.

(3) For $N = N^*$ and $s = s^0$ the solution $\bar{x}^1, \bar{x}^2, \ldots, \bar{x}^{N^*}$ to the reduced problem

$$\bar{B}^N(s) = \max \{ \sum_{\tau=1}^{N} \alpha^{\tau-1} c^T x^\tau \, / \, Dx^t \leq d \, , \, \sum_{\tau=1}^{t} Ax^\tau \leq s + (t-1)\bar{q} \, , \\ x^t \geq 0 \, , \, t = 1,\ldots,N\} \tag{5.26}$$

and the solution $(\bar{\lambda}_1, \bar{\mu}_1), (\bar{\lambda}_2, \bar{\mu}_2), \ldots, (\bar{\lambda}_{N^*}, \bar{\mu}_{N^*})$ to its dual

$$\bar{B}^N(s) = \min \{ \sum_{\tau=1}^{N^*} [\lambda_\tau^T (s + (\tau - 1)\bar{q}) + \mu_\tau^T d] \, / \, \sum_{\tau=t}^{N} \lambda_\tau^T A + \\ + \mu_t^T D \geq \alpha^{t-1} c^T \, , \, (\lambda_t, \mu_t) \geq 0 \, , \, t = 1,2,\ldots,N\} \tag{5.27}$$

are unique, with the former satisfying the inequality

$$\frac{\bar{\lambda}_{N^*}}{\alpha^{N^*-1}} \geq \nu^* \tag{5.28}$$

Then for $N \geq N^*$

$$\bar{B}^N(s^0) = \bar{B}^{N^*}(s^0) + \frac{\alpha^{N^*} - \alpha^N}{1 - \alpha} \check{\Phi} \tag{5.29}$$

$$\bar{B}^\infty(s^0) = \bar{B}^{N^*}(s^0) + \frac{\alpha^{N^*}}{1 - \alpha} \check{\Phi} \tag{5.30}$$

$$\bar{y}^N(s^0) = \bar{y}^\infty(s^0) = \bar{y}^{N^*}(s^0) \tag{5.31}$$

When (5.28) holds, the solution $x^1(s^0), \ldots, x^{N^*}(s^0), x^{N^*+1}(s^0), \ldots, x^N(s^0)$ to (5.26) for any $N \geq N^*$, $s = s^0$ satisfies the condition

$$x^{N^*+1}(s^0) = x^*, \ldots, x^N(s^0) = x^*$$

i.e., it is most profitable to spend at every stage (beginning from N^*) all the delivered raw materials $\bar{q}$, choosing the activity level maximizing income (5.25) at each stage.

As they are valid for $s = s^0$, the conditions of the theorem will be true for any s within some neighborhood of s^0. Using this fact, it is possible to prove the following theorems.

Theorem 5.2.

Let $\bar{x}^1, \bar{x}^2, \ldots, \bar{x}^{N^*}$ be the unique solution to (5.26) for s^0, $N = N^*$ and, with the given s^0 and N^*, let the conditions of Lemma 5.2 be satisfied. Let us suppose also that, for

$$s = s^l \triangleq s^0 - \sum_{\tau=1}^{l} Ax^\tau + l\bar{q}$$

and $N = N^* - l$, $l = 1,2,\ldots,N^*-1$, the solutions to problem (5.27) are unique.

Then under the condition that

$$E(q_i - \bar{q}_i)^2 = \epsilon^2 D_i \quad , \quad i = 1,2,\ldots,m \tag{5.32}$$

there exist positive vectors Δ^l, Δ and constants ϵ_0, k_1 such that, for $0 < \epsilon \le \epsilon_0$,

$$s \in \Omega^l \triangleq \{s \,/\, s^l - \Delta^l \le s \le s^l + \Delta^l \,,\, s \ge 0\}$$
$$N \ge N^* - l \quad , \quad l = 0,1,\ldots,N^*-1$$

and

$$s \in \Omega \triangleq \{s \,/\, \bar{q} - \Delta \le s \le \bar{q} + \Delta \,,\, s \ge 0\}$$

the estimates

$$|B^N(s) - \bar{B}^N(s)| \le k_1\epsilon^2 \quad , \quad |B^\infty(s) - \bar{B}^\infty(s)| \le k_1\epsilon^2 \tag{5.33}$$

$$|F^N(s) - \bar{B}^N(s)| \le k_1\epsilon^2 \tag{5.34}$$

are valid.

Theorem 5.3.

Let the conditions of Theorem 5.2 be satisfied with

$$p_r\{q \in [\bar{q} - \epsilon p \,,\, \bar{q} + \epsilon p]\} = 1 \quad , \quad p \ge 0 \tag{5.35}$$

instead of (5.33). Then there exist positive Δ^l, Δ, ϵ_0 such that, for

$$0 < \epsilon \leq \epsilon_0 \ , \ s \in \Omega^l \triangleq \{s \, / \, s^l - \Delta^l \leq s \leq s^l + \Delta^l \ , s \geq 0\} \ , \ N \geq N^* - l$$

and for

$$s \in \Omega \triangleq \{s \, / \, q - \Delta \leq s \leq q + \Delta \ , s \geq 0\} \ , \ N \geq 1$$

the following relations will be valid

$$B^N(s) = \bar{B}^N(s) = F^N(s) \ , \ B^\infty(s) = \bar{B}^\infty(s) \tag{5.36}$$

4.6. Control of weakly dynamic systems under random disturbances

In Section 2.8 we considered the control problem for a technological process described as a weakly dynamic system whose state variables are essentially constrained. The control laws developed seem simple and attractive from the practical point of view. However, it was noted that the impossibility of predicting the real system reaction precludes the direct use of these laws. Let us now show that efficient prototypes of such laws (decision rules) can be developed by perturbation theory, and that these laws will be robust with respect to stochastic disturbances.

To describe the process, consider the same relations as in Section 2.8:

$$x^{t+1} = \epsilon f(x^t, u^t) + \varphi(u^t, \omega^t) \ , \ t = 0,1,\ldots,N-1 \ , \ x^0 = x_0 \tag{6.1}$$

But now suppose that the disturbances ω^t are unknown at the moment t, when the control action u^t must be realized. It is supposed also that

(1) The values x^τ, $\tau \leq t$; u^τ, $\tau \leq t - 1$ are known.
(2) The value ω^{t-1} can be found by means of the relations

$$x^t = \epsilon f(x^{t-1}, u^{t-1}) + \varphi(u^{t-1}, \omega^{t-1}) \tag{6.2}$$

(3) The sequence $\{\omega^t \ , t = 0,1,\ldots\}$ is stochastic, the increments

$$\varsigma^t \triangleq \omega^t - \omega^{t-1} \ , \ t = 1,2,\ldots \ ; \ \omega^0 = 0 \tag{6.3}$$

being mutually independent bounded random variables

$$\varsigma^t = (\varsigma_i^t) \ , \ |\varsigma_i^t| \leq \Delta_i \tag{6.4}$$

so that, for all t, $t = 0,1,\ldots,N-1$, $\omega^t \in D$ where D is a compact set.

A control is constructed as a decision rule

$$u^t = u[\tilde{x}^{(t)}] \ , \ \tilde{x}^{(t)} \triangleq (x^0,\ldots,x^t) \tag{6.5}$$

such that the stage variable constraints

$$x^{t+1} \leq \bar{x} \tag{6.6}$$

are not violated by any disturbances that satisfy (6.3) and (6.4). Moreover, u^t must belong to the given polyhedron U. The goal of the choice of u^t from the described class of decision rules is maximization of the expectation of the following function

$$\sum_{t=0}^{N-1} g(x^{t+1}, u^t) \tag{6.7}$$

Constraint (6.6) can be written in the form

$$\epsilon f_i(x^t, u^t) + \varphi_i(u^t, \omega^{t-1} + \varsigma^t) \leq \bar{x} \tag{6.8}$$

In so far as (6.8) must be fulfilled for all possible ς^t, they can be rewritten in the equivalent, deterministic form

$$\epsilon f_i(x^t, u^t) + \max_{\varsigma} \{\varphi_i(u^t, \omega^{t-1} + \varsigma^t) \,/\, |\varsigma_i^t| \leq \Delta_i] \leq \bar{x} \tag{6.9}$$

First, let us consider the properties of the reduced problem ($\epsilon = 0$). The choice of control u^t is bounded by the domain

$$U_d(\omega^{t-1}) \triangleq \{u \,/\, u \in U \ , \ \max_{\varsigma} [\varphi_i(u, \omega^{t-1} + \varsigma) \,/\, |\varsigma_i| \leq \Delta_i] \leq \bar{x}\} \tag{6.10}$$

Suppose that

$$U_d(\omega^{t-1}) = \emptyset \ , \ t = 0,1,\ldots,N \tag{6.11}$$

for all feasible ω^t. Then the recursive relations of dynamic programming

$$\begin{aligned} B^\tau(x,\omega) = \ & \max \{E_\varsigma g[\varphi(u, \omega + \varsigma)] + E_\varsigma B^{\tau-1}[\varphi(u, \omega + \varsigma), \omega + \varsigma] \,/ \\ & u \in U_d(\omega)\} \ ; \ B^0(x,\omega) = 0 \end{aligned} \tag{6.12}$$

are valid where $B^N(x_0, 0)$ is equal to the optimal objective value. It is clear from (6.12) that

$$B^\tau(x, \omega) = B^\tau(\omega) \tag{6.13}$$

and so, consequently, only the first term of the maximized expression is dependent on u. Thus the optimal control is computed for each step as the solution to the program

$$\max \{E_\varsigma g[\varphi(u, \omega + \varsigma), u] \,/\, u \in U_d(\omega)\} \tag{6.14}$$

as a function of ω. More precisely

$$u^t = u_o^t = u_0(\omega^{t-1}) \tag{6.15}$$

where the function $u_o(\omega)$ characterizes the dependence of the solution to problem (6.14) on the parameter ω. (This dependence is universal in the stationary case, but one must look for the solution to different problems for each t in a nonstationary case.)

Since the solvability of the equation

$$x^t = \varphi(u^{t-1}, \omega^{t-1}) \tag{6.16}$$

with respect to ω^{t-1} has been assumed, one can reconstruct (6.15) in the form of the feedback that uses measurements of the state variables x^t. In particular we have, for the linear system

$$\varphi(u, \omega) = Bu + \omega \tag{6.17}$$

an explicit form of the feedback.

This case is of practical importance even if the performance index is linear only with respect to the state variables. In this situation the control law does not depend on the detailed probabilistic characteristics of the process and is determined by the size of $\Delta = \{\Delta_i\}$ only. In what follows, we shall investigate in detail the completely linear problem where

$$\sum_{t=0}^{N-1} g(x^{t+1}, u^t) = \sum_{t=0}^{N-1} (c^T x^{t+1} + d^T u^t) \tag{6.18}$$

$$x^{t+1} = \epsilon A x^t + B u^t + \omega^t \tag{6.19}$$

Here, the synthesis of the reduced control is transformed to the solution of the linear program

$$\max\, [(c^T B + d^T) u \,/\, Bu + \omega + \Delta \leq \bar{x} \,,\, u \in U] \tag{6.20}$$

with ω a parameter which takes the value

$$\omega = x^t - Bu^{t-1} \tag{6.21}$$

while the control u^t is computed.

If u is a scalar, $\underline{u} \leq u \leq \bar{u}$ and $B > 0$, then we have the explicit formula

$$u^t = \min\, \{\bar{u} \,,\, \min_i b_i^{-1}(\bar{x}_i - x_i^t + b_i u^{t-1} - \Delta_i)\} \tag{6.22}$$

or, for $\bar{u} = \infty$,

$$u^t - u^{t-1} = \min_i \frac{\bar{x}_i - \Delta_i - x_i^t}{b_i} = -\max_i \frac{x_i^t - (\bar{x}_i - \Delta_i)}{b_i} \tag{6.23}$$

The law of (6.23) may be called the integral feedback with respect to deviations from the nearest boundary (Δ_i plays role of an "insurance level"). The developed laws are optimal for the reduced problem. However, we cannot ensure that they are feasible for the perturbed one. As in Section 2.8 one can construct feasible laws by solving, instead of (6.20) and (6.21), the program

$$\max\, [(c^T B + d^T) u \,/\, Bu + \tilde{\omega} + \Delta + \epsilon Ax \leq \bar{x} \,,\, u \in U] \tag{6.24}$$

where $\tilde{\omega}$ is a parameter taking the value

$$\tilde{\omega} = \omega^{t-1} + \epsilon Ax^t = x^t - Bu^{t-1} - \epsilon A(x^{t-1} - x^t) \tag{6.25}$$

at the moment t.

Proximity estimates of this scheme are contained in the following theorems given without the proof [134].

Theorem 6.1.

Let there exist $\hat{u} \in U$ such that

$$B\hat{u} + \omega + \Delta < \bar{x}$$

for all $\omega \in D$. Also, let the solutions to the linear programs

$$\max \left[(c^T B + d^T)u \,/\, Bu + \omega^{t-1} + \Delta \leq \bar{x} \,,\, u \in U\right] \;,\; t = 1,\ldots,N \tag{6.26}$$

be unique as well as the solutions to their duals. Then there exists an ϵ_0 such that the optimal control coincides with the solution to (6.24) over $0 < \epsilon \leq \epsilon_0$.

Theorem 6.2.

Let $U = E^1$, $B^T c + d > 0$, $B > 0$. Then there exists ϵ_0 such that, under $0 < \epsilon \leq \epsilon_0$, the optimal control is given by the formulae

$$\begin{aligned} u^t - u^{t-1} &= -\max_i \frac{1}{b_i}[x_i^t - (\bar{x}_i - \Delta_i) + \epsilon a_i (x_i^t - x_i^{t-1})] \\ u^0 &= -\max_i \frac{1}{b_i}[\omega^0 - (\bar{x}_i - \Delta_i) + \epsilon a_i x_i^0] \end{aligned} \tag{6.27}$$

where the a_i are the rows of the matrix A.

CHAPTER 5

Suboptimal Linear Regulator Design

This chapter opens the concluding part of the book, which is devoted to suboptimal control laws for dynamic objects. Similar problems were considered in preceding chapters; however, only recurrence relations or difference equations were used as a mathematical description of the objects' behavior over a finite time set. (An exception is Section 4.5, where a denumerable set of moments was considered.) Further to this, we shall take a system of differential equations as a model that is more applicable to engineering.

The fact that control solutions are characterized here by functions given over some interval of time complicates the mathematical apparatus. Therefore, this chapter is devoted to the simplest but possibly the most practically important problem of optimal control theory, which is well-known as the linear quadratic (LQ) problem: find a control for a linear system that minimizes a performance index characterizing a quadratic norm of deviations from a desired state and control. It can be proved that the optimal control law is a linear feedback based on measurements of all the state variables. If all conditions are invariant to translations in time, the optimal feedback gains are constants, which can be found easily as soon as a basic quadratic matrix equation is solved.

Thus, the LQ problem can be considered the foundation of linear control theory, e.g., as in [121]. Moreover, the principal elements of the mathematical apparatus of the LQ problem can be used to solve the more complex LQG problem of how to find an optimal control law for the linear system excited by random perturbations, the measurements being distorted by random Gaussian noise.

It would seem that one does not need to look for suboptimal solutions since the optimal one is available. However, this is not true owing to other considerations.

In the first place, the perturbation method allows one to obtain simpler control designs which are, as a rule, more reliable. On the other hand, pure computational difficulties appear if the basic quadratic matrix equation (the so-called algebraic Riccati equation) is large, and the perturbation method shows a way to avoid these. Last, using the perturbation method, we can investigate the robustness of optimal solutions with respect to small model errors. This especially relates to errors in estimating the order of the differential equations describing the system. This is the well-known small time constants problem of classical control theory. This aspect is of general interest because it is generally impossible to give an adequate description of a real object with a finite number of ordinary differential equations.

Some basic facts concerning the LQ problem are given in Section 5.1. After that, the perturbation method is applied to regular problems in which the condition of complete controllability is fulfilled for $\epsilon = 0$. A recursive algorithm for the optimal linear regulator, which is based on an ϵ-power expansion of the solution to the Riccati equation, is given. If a finite number, m, of terms of this expansion is used to construct a suboptimal control law, then the suboptimality can be estimated to a $O(\epsilon^{2m})$-error in the objective value. Moreover, by means of the duality theorem, explicit estimates can be obtained for the efficiency of a zero-order approximation. It is shown that this approximation admits a suboptimal decomposed control for weakly coupled dynamic systems.

Degeneracy effects are investigated in Section 5.2, and these are apparently of most interest. To begin with, primal degeneracy is discussed, i.e., the loss of complete controllability. By a nonsingular transformation of the state variables, one can separate out a weakly controllable subsystem such that an independent suboptimization scheme is applicable to another part of the system that is strongly controllable. If the weakly controllable subsystem, for $\epsilon = 0$, is stable, then this decomposition scheme is justified by the asymptotic expansion of the matrix solving the Riccati equation.

The critical situation, in which the stability conditions are determined by small terms, is of more importance. This is particularly relevant to the case of zero eigenvalues of the matrix of a reduced weakly controllable subsystem. It is shown that the dynamics of such a system is divisible into "slow" and "fast" processes, and that its investigation is a key to the analysis of the problem of "small time constraints". If the "fast" subsystem is stable and weakly controllable, then one can use a suboptimal feedback, which is based only on the "slow" variable measurements, and this corresponds to the heuristic of neglecting small time constants in the original model. Otherwise, it is necessary to expend part of the control energy on a preliminary stabilization of the "fast" subsystem. This decomposition is described as a zero-order approximation of the perturbation method, with efficiency estimates being given. Moreover, a recursive procedure admitting approximations of higher order is given.

Dual degeneracy arising with a violation of the complete observability condition is discussed, too. Here a suboptimal aggregation scheme is effective, and the dimensionality decreases with the "depth" of the degeneracy.

Two examples of applications are given in Section 5.3. These are:

(1) A flow shop control problem, in which a two-level scheme is justified. The upper level is an inventory control, and the lower level is the technological control of separate units.

(2) An ecological system control with decomposition into two levels: the aggregated level aiming to stabilize a number of "macropopulations" and the detailed level controlling the structures of the macropopulations.

The case of purely imaginary eigenvalues of the weakly controllable subsystem matrix (for $\epsilon = 0$) is considered in Section 5.4. This case is of importance for the control of oscillatory systems. Initially, a simpler dynamics is discussed corresponding to the absence of the strongly controllable part. This can be interpreted as the control of the oscillatory system, which is close to a conservative one. A suboptimal control law is constructed here in an explicit form. Each oscillatory mode is controlled independently, the feedback based on measurements of the generalized velocities only. After that, the general situation is considered. It is proved that a decomposition scheme can be constructed that includes the independent control of the oscillatory modes. A simple example of a mechanical system is given, and the construction of the suboptimal control demonstrated for a large interval of parameter values.

The concluding section is devoted to the LQG problem. The main result is a justification of a simple, practical heuristic approach. It is proved that one cannot take into account fast damping modes if an ϵ-optimal control is constructed. This allows one to simplify the numerical procedure, which is necessary to obtain the suboptimal feedback. Moreover, the structure of the control itself is also simplified so that the computational effort, which must be realized on-line, can be reduced.

5.1. The LQ problem. Suboptimal decomposition

Let us consider the following optimal control problem (LQ problem). Find a vector-function $u = u(t)$, $u \in E^m$ such that the solution $x(t)$, $t \in [0, \infty]$, to the system

$$\dot{x} = Ax + Bu \quad , \quad x(0) = x_0 \quad , \quad y = cx \tag{1.1}$$

minimizes the functional

$$J\{u\} = \frac{1}{2} \int_0^\infty (y^T Q y + u^T R u) \mathrm{d}t \tag{1.2}$$

over the class of square integrable functions $u(t)$, the corresponding trajectories $x(t)$ satisfying the condition

$$\lim_{t \to \infty} x(t) = 0 \tag{1.3}$$

It is well known (see, e.g., [45]) that the problem has a unique solution under some constraints on the constant matrices A, B, C, Q, R and, moreover, that this solution can be written in the form of a linear feedback control that admits the analytical design of linear regulators.

Let us first state some definitions.

Definition 1.1. A pair of matrices $A(n \times n)$, $B(n \times m)$ is completely controllable if and only if

$$\text{rank}\ \{B,AB,\ldots,A^{n-1}B\} = n \tag{1.4}$$

The system (1.1) is completely controllable if and only if A, B is completely controllable. A pair of matrices $A(n \times n)$, $C(m \times n)$ is completely observable if and only if

$$\text{rank}\ \{C^T,A^TC^T,\ldots,(A^T)^{n-1}C^t\} = n \tag{1.4$'$}$$

so that the pair A^T, C^T is completely controllable.

Definition 1.2. A matrix A is stable if all eigenvalues of A have negative real parts. The distance from the imaginary axis to the nearest eigenvalue of a stable matrix A is the stability rate of A.

The following result contains the principal elements of solution to the LQ problem. (Henceforth, the notation $Q > 0$ means that a symmetric matrix Q is positive definite.)

Theorem 1.1.

Let $Q > 0$, $R > 0$ and let the pairs $(A\,,B)$, (A^T,C^T) be completely controllable. Then, for any $x_0 \in E^n$ there exists the unique optimal control $u_{\text{opt}}(t)$ defined by the formula

$$u_{\text{opt}}(t) = -K_{\text{opt}}x_{\text{opt}}(t) \tag{1.5}$$

where

$$K_{\text{opt}} = R^{-1}B^TP_{\text{opt}} \tag{1.6}$$

and P_{opt} is the unique positive definite symmetric solution to the matrix quadratic equation

$$PA + A^TP - PBR^{-1}B^TP + C^TQC = 0 \tag{1.7}$$

The matrix

$$H \triangleq A - BK_{\text{opt}}$$

is stable and hence the system (1.1) with the feedback (1.6)

$$\dot{x} = (A - BK_{\text{opt}})x$$

is asymptotically stable. The minimal value of the performance index (1.2) is completely determined by the matrix P_{opt}

$$J\{u_{\text{opt}}\} = \frac{1}{2}x_0^T P_{\text{opt}} x_0 \tag{1.8}$$

for given initial x_0.

Proof of Theorem 1.1, which can be found in many textbooks (see, e.g., [45]), is based on the following important result. (Note that the first part of Lemma 1.1 is well known as the Liapunov Lemma.)

Lemma 1.1. Let K be a constant $(m \times n)$ matrix. Let

$$H = A - BK \ ; \ \tilde{Q} = C^T QC + K^T RK > 0 \tag{1.9}$$

Then

(1) The solution $M > 0$ to the linear matrix equation

$$MH + H^T M + \tilde{Q} = 0 \tag{1.10}$$

exists if and only if H is stable.

(2) If

$$u = -Kx \tag{1.11}$$

is used as a control law in (1.1) then

$$J(u) = \frac{1}{2}x_0^T M x_0 \tag{1.12}$$

Proof of the lemma is simple. Let H be stable. Then there exists the integral

$$M = \int_0^\infty e^{H^T t}\tilde{Q}e^{Ht}\mathrm{d}t$$

Integrating by parts, we obtain

$$\begin{aligned} H^T M &= \int_0^\infty H^T e^{H^T t} \tilde{Q} e^{Ht} \mathrm{d}t \\ &= e^{H^T t} \tilde{Q} e^{Ht} |_0^\infty - \int_0^\infty e^{H^T t} \tilde{Q} e^{Ht} H \mathrm{d}t \\ &= -\tilde{Q} - MH \end{aligned} \tag{1.13}$$

i.e., M is the solution to (1.10). The positive definiteness of M follows from (1.13) immediately.

Suppose now that $M > 0$ is a solution to (1.10). Consider equation (1.1) taking into account the control (1.11). It has the form

$$\dot{x} = Hx \quad , \quad x(0) = x_0 \tag{1.14}$$

Introduce the Liapunov function

$$V(x) = \frac{1}{2} x^T M x$$

which is a positive definite quadratic form. Its derivative, by (1.14), is negative definite since

$$\dot{V} = \frac{1}{2}[Hx]^T Mx + \frac{1}{2} x^T MHx = -\frac{1}{2} x^T \tilde{Q} x < 0$$

Hence, the system (1.14) is asymptotically stable and H is a stable matrix.

Further, we have

$$J\{u(x)\} = \frac{1}{2} \int_0^\infty x^T [C^T QC + K^T RK] x \mathrm{d}t \tag{1.15}$$

where

$$x(t) = e^{Ht} x_0$$

so that

$$J\{u(x)\} = \frac{1}{2} x_0^T \int_0^\infty e^{H^T t} \tilde{Q} e^{Ht} \mathrm{d}t x_0 = \frac{1}{2} x_0^T M x_0 \tag{1.16}$$

QED.

Theorem 1.1 admits the optimal control once the solution to (1.7) is found. However, this purely algebraic problem is difficult enough if the dimension n of the state vector x is large. Therefore, the perturbation method may be useful in this theoretically simple optimal control problem if, of course, one can choose a reduced problem allowing some direct decomposition or aggregation scheme.

Let the original problem be written in the form

$$J\{u_{\text{opt}}\} \triangleq \min \left\{ \frac{1}{2} \int_0^\infty [y^T Q y + u^T R u] \mathrm{d}t \right\} \tag{1.17}$$

such that

$$\begin{aligned} &y = c^T(\epsilon)x \ , \quad \dot{x} = A(\epsilon)x + B(\epsilon)u \ , \quad x(0) = x_0 \\ &A(\epsilon) = A_0 + \epsilon A_1 \ , \quad B(\epsilon) = B_0 + \epsilon B_1 \ , \quad C(\epsilon) = C_0 + \epsilon C_1 \end{aligned} \tag{1.18}$$

for some sufficiently small ϵ. Associate with (1.17) the reduced problem

$$\begin{aligned} &J_0\{u^0_{\text{opt}}\} \triangleq \min \left\{ \frac{1}{2} \int_0^\infty [y^T Q y + u^T R u] \mathrm{d}t \right\} , \\ &y = c_0^T x \ , \quad \dot{x} = A_0 x + B_0 u \ , \quad x(0) = x_0 \end{aligned} \tag{1.19}$$

Theorem 1.2.

Let $Q > 0$, $R > 0$ and let the pairs $\{A_0, B_0\}$, $\{A_0^T, C_0^T\}$ be completely controllable. Then there exists $\epsilon' > 0$ such that, for $0 \leq \epsilon \leq \epsilon'$, the solution to (1.17) and (1.18) is given by the formula

$$u_{\text{opt}}(x, \epsilon) = -K_{\text{opt}}(\epsilon)x \ , \quad K_{\text{opt}}(\epsilon) = R^{-1}B^T(\epsilon)P_{\text{opt}}(\epsilon) \tag{1.20}$$

where $P_{\text{opt}}(\epsilon)$ is the unique positive definite solution to the equation

$$\begin{aligned} \Phi(\epsilon, P) \triangleq\ & P(\epsilon)A(\epsilon) + A^T(\epsilon)P(\epsilon) - P(\epsilon)B(\epsilon)R^{-1}B^T(\epsilon)P(\epsilon) + \\ & + C^T(\epsilon)QC(\epsilon) \\ =\ & 0 \end{aligned} \tag{1.21}$$

$P_{\text{opt}}(\epsilon)$, $K_{\text{opt}}(\epsilon)$ are analytical functions of ϵ and $P_{\text{opt}}(0) = P_0$, $K_{\text{opt}}(0) = K_0$ where P_0, K_0 are determined by the reduced problem solution.

Proof. It is evident that

$$\Phi(0, P_0) = 0$$

Write down the Jacobian of $\Phi(\epsilon, P)$ at the point $(0, P_0)$ using the matrix differentiation rules

$$\frac{\partial(AX)}{\partial X} = A \otimes E_r \ ; \ \frac{\partial(XA)}{\partial X} = E_l \otimes A^T$$

if X is $(l \times r)$-matrix. Here $\otimes$ denotes the Kroneker product, E_s is the $(s \times s)$ identity matrix. We have

$$J_\Phi = |E_n \otimes H_0^T + H_0^T \otimes E_n| \ ,$$
$$H_0 = A_0 - B_0 K_0$$

By virtue of known properties of Kroneker product (see, e.g., [5]), all the eigenvalues of the matrix $A \otimes E + E \otimes B$ coincide with sums of eigenvalues of A and B. Hence, the Jacobian J_Φ is not equal to zero since H_0 is stable. The implicit function theorem gives the desired result. QED.

Note that the coefficients P_i, $i = 1,2,\ldots$ in the expansion

$$P_{\text{opt}}(\epsilon) = \sum_{i=0}^{\infty} P_i \epsilon^i \tag{1.22}$$

can be found recursively from the linear equations of Liapunov type

$$P_i H_0 + H_0^T P_i + \tilde{Q}_i = 0 \tag{1.23}$$

Determinants of all systems coincide with J_Φ and are non-zero. One can suppose that it is sufficient in practice to find either a zero-order approximation or some small number of terms in (1.22).

Theorem 1.3.

Let

$$u = u_m(x) = -K_m x$$

where

$$K_m = R^{-1}B^T(\epsilon)\bar{P}_m \ , \ \bar{P}_m = \sum_{i=0}^{m-1} P_i\epsilon^i$$

be used as a control law in the system (1.17). Then there exist $\bar{\epsilon}$, q such that

$$J\{u_m(x)\} - J\{u_{\text{opt}}(x)\} \leq q\epsilon^{2m} \tag{1.24}$$

for $0 \leq \epsilon \leq \bar{\epsilon}$.

Proof is based on the following lemma.

Lemma 1.2. Let the control

$$u = \bar{u}(x) = -\bar{K}x$$

be used in (1.17) and let the stability rate of the matrix

$$\bar{H} \triangleq A(\epsilon) - B(\epsilon)\bar{K}$$

be not less than α for $0 \leq \epsilon \leq \epsilon'$. Then

$$J\{\bar{u}(x)\} - J\{u_{\text{opt}}(x)\} \leq \frac{c_\Delta}{\alpha}\|\Delta\|^2\,\|x_0\|^2 \tag{1.25}$$

where $\Delta = \bar{K} - K_{\text{opt}}$ and c_Δ is independent of ϵ.

Proof of Lemma 1.2. By (1.12)

$$J\{\bar{u}(x)\} = \frac{1}{2}x_0^T M x_0$$

where M is the solution to the linear equation

$$\begin{aligned} &M\bar{H} + \bar{H}^T M + \hat{Q}(\epsilon) = 0 \\ &\hat{Q}(\epsilon) = C^T(\epsilon)QC(\epsilon) + \bar{K}^T R\bar{K} \end{aligned} \tag{1.26}$$

At the same time

$$J\{u_{\text{opt}}(x)\} = \frac{1}{2}x_0^T P_{\text{opt}} x_0$$

where P_{opt} satisfies, by (1.20) and (1.21), the equation

$$\begin{aligned} &P_{\text{opt}}H + H^T P_{\text{opt}} + C^T(\epsilon)QC(\epsilon) + K_{\text{opt}}^T R K_{\text{opt}} = 0 \\ &H \triangleq A(\epsilon) - B(\epsilon)K_{\text{opt}} = \bar{H} + B(\epsilon)\Delta \end{aligned} \tag{1.27}$$

Let $\gamma \triangleq P_{\text{opt}} - M$. We can present (1.26) and (1.27) in the form

$$P_{\text{opt}}\bar{H} + \bar{H}^T P_{\text{opt}} - \gamma\bar{H} - \bar{H}^T\gamma + C^T(\epsilon)QC(\epsilon) + (K_{\text{opt}} + \Delta)^T R(K_{\text{opt}} + \Delta) = 0$$
$$P_{\text{opt}}\bar{H} + \bar{H}^T P_{\text{opt}} + P_{\text{opt}}B(\epsilon)\Delta + \Delta B^T(\epsilon)P_{\text{opt}} + C^T(\epsilon)QC(\epsilon) + K_{\text{opt}}^T R K_{\text{opt}} = 0$$

Subtracting we obtain

$$\gamma\bar{H} + \bar{H}^T\gamma + P_{\text{opt}}B(\epsilon)\Delta + \Delta B^T(\epsilon)P_{\text{opt}} - K_{\text{opt}}^T R\Delta - \Delta^T R K_{\text{opt}} - \Delta^T R\Delta = 0$$

or, taking into account (1.20),

$$\gamma\bar{H} + \bar{H}^T\gamma - \Delta^T R\Delta = 0 \tag{1.28}$$

As in (1.13) the representation

$$\gamma = -\int_0^\infty e^{\bar{H}^T t}\Delta^T R\Delta e^{\bar{H}t}dt$$

is valid. Hence,

$$\|\gamma\| \le \|\Delta\|^2 \frac{1}{\alpha} \cdot \text{constant} \tag{1.29}$$

since

$$\int_0^\infty \|e^{\bar{H}t}\|dt \le \frac{1}{\alpha} \cdot \text{constant} \tag{1.30}$$

Inequality (1.29) yields the desired result.

Proof of Theorem 1.3. Since the stability rate of H_0 is some positive value that does not depend on ϵ, the stability rate of the matrix H_m,

$$H_m \triangleq A(\epsilon) - B(\epsilon)K_m$$

is bounded from below by some value $\alpha_m > 0$ in a neighborhood of $\epsilon = 0$ by virtue of the continuous dependence of eigenvalues on the parameter ϵ.

Further, one can use Lemma 1.2 to construct the inequalities

$$\begin{aligned} J\{u_m(x)\} - J\{u_{\text{opt}}(x)\} &\le \frac{c_\Delta}{\alpha_m}\|K_m - K_{\text{opt}}\|^2 \\ &\le \frac{c_\Delta}{\alpha_m}\|R^{-1}\|^2 \cdot \|B^T(\epsilon)\|^2 \cdot \|\bar{P}_m - P_{\text{opt}}\|^2 \\ &\le q\epsilon^{2m} \end{aligned}$$

since

$$\|\bar{P}_m - P_{\text{opt}}\| \le \text{constant } \epsilon^m$$

for small ϵ. QED.

To obtain explicit estimates of the efficiency of approximate solutions, one can use the duality theorem [91].

Theorem 1.4.

Let the pairs $\{A, B\}$, $\{A^T, C^T\}$ be completely controllable and let:

$$J_r \triangleq \min_u \left\{ \frac{1}{2}\int_0^\infty (y^TQy + u^TRu)dt \triangleq J_r(u) \right\} \tag{1.31}$$

such that $y = Cx$, $\dot{x} = Ax + Bu$, $x(0) = x_0$ and

$$J_d \triangleq \min_y \left\{ \frac{1}{2}\int_0^\infty (u^TR^{-1}u + y^TQ^{-1}y)dt = J_d(y) \right\} \tag{1.32}$$

such that $u = B^Tx$, $\dot{x} = -A^Tx + C^Ty$, $x(0) = x_0$. Then

$$J_r = \frac{1}{2}x_0^TMx_0 \ , \ J_d = \frac{1}{2}x_0^TNx_0$$

where $M > 0$, $N > 0$ are the solutions to the equations

$$MA + A^TM - MBR^{-1}B^TM + C^TQC = 0$$
$$-NA^T - AN - NC^TQCN + B^TR^{-1}B = 0$$

and

$$M = N^{-1} \tag{1.33}$$

Proof. The representations J_r, J_d take place by Theorem 1.1. Pre- and post-multiply by N^{-1} the equation (1.32). Then the result is the consequence of the uniqueness of the solution to the matrix quadratic equation.

Corollary. Let the conditions of Theorem 1.4 be satisfied and let M_0 be chosen such that $A - BK_0$ is stable if $K_0 = R^{-1}B^TM_0$. Let $1/2x_0^TM_1x_0$ be the value of (1.31) with the control $u = -K_0x$. Let $N_0 > 0$ be such that the matrix $-A^T - C^T\bar{K}_0$ is stable if $\bar{K}_0 = QCN_0$. Let $1/2x_0^TN_1x_0$ be the value of (1.32) with the control $y = -\bar{K}_0x$. Then

$$x_0^TN_1^{-1}x_0 \leq x_0^TMx_0 \leq x_0^TM_1x_0 \ , \ \forall\, x_0 \tag{1.34}$$

Proof is self-evident.

The bilateral estimate (1.34) of the optimal value shows a way to obtain an explicit estimate of zero-order approximation. In particular, consider problem (1.31) where

$$A = A(\epsilon) = A_0 + \epsilon A_1$$

and where all other parameters (B, C, Q, R) are independent of ϵ, the controllability and observability conditions being satisfied for $0 \leq \epsilon \leq \epsilon'$. (This holds if the conditions are true for $\epsilon = 0$.) Then the following result [87] gives the desired estimate.

Theorem 1.5.

Let $u = u_{opt}(x, \epsilon)$ be the solution to the perturbed problem; let $u^0(x)$ be the reduced solution. Let $J\{u, \epsilon\}$ be the objective value of the perturbed problem with the control u. Then there exists $\epsilon_1 < 0$ such that, for $0 \leq \epsilon \leq \epsilon_1$, the following inequality is valid

$$(1 + \epsilon\sigma_m)J\{u^0(x), 0\} \leq J\{u_{opt}(x, \epsilon), \epsilon\} \leq (1 - \epsilon\sigma_M)^{-1}J\{u^0(x), 0\} \tag{1.35}$$

where σ_m, σ_M are the minimal and maximal eigenvalues of the matrix

$$q \triangleq [K_0^T R K_0 + C^T Q C]^{-1}(P_0 A_1 + A_1^T P_0) \tag{1.36}$$

and

$$\epsilon_1^{-1} = \max\{|\sigma_m|, |\sigma_M|\} \tag{1.37}$$

Proof. By Theorem 1.1 we have that

$$u^0(x) = -K_0 x \ , \ K_0 = R^{-1}B^T P_0 \tag{1.38}$$

$$P_0 A_0 + A_0^T P_0 - P_0 B R^{-1} B^T P_0 + C^T Q C = 0 \tag{1.39}$$

$$J\{u^0, 0\} = \frac{1}{2} x_0^T P_0 x_0 \tag{1.40}$$

Let us take $V(x) = \frac{1}{2} x^T P_0 x$ as a Liapunov function of the perturbed system controlled by (1.38)

$$\dot{x} = A(\epsilon)x - BK_0 x \ , \ x(0) = x_0 \tag{1.41}$$

Then, as in the proof of Lemma 1.1, we have that system (1.41) is stable if the condition

$$\tilde{Q}_0 \triangleq C^T Q C + K_0^T R K_0 > \epsilon(P_0 A_1 + A_1^T P_0) \tag{1.42}$$

is satisfied. The objective value is bounded and equal to

$$J\{u_0, \epsilon\} = \frac{1}{2} x_0^T M_1 x_0 \tag{1.43}$$

where M_1 is the solution to the linear system

$$\begin{aligned} &M_1 H_0(\epsilon) + H_0^T(\epsilon) M_1 + \tilde{Q}_0 = 0 \ , \\ &H_0(\epsilon) \triangleq A(\epsilon) - BK_0 \end{aligned} \tag{1.44}$$

We look for the solution to (1.44) in the form

$$M_1 = \lambda_1 P_0 - \delta \ , \ \lambda_1 > 0 \tag{1.45}$$

Then the matrix δ and the scalar λ must satisfy the equation

$$\delta H_0(\epsilon) + H_0^T(\epsilon)\delta - \tilde{Q}_0 - \lambda_1[P_0 H_0(\epsilon) + H_0^T(\epsilon)P_0] = 0$$

Again, by Lemma 1.1 and the stability of $H_0(\epsilon)$,

$$\tilde{Q}_0 + \lambda_1[P_0 H_0(\epsilon) + H_0^T(\epsilon)P_0] \leq 0 \Rightarrow \delta \geq 0$$

Recall that

$$P_0 H_0 + H_0^T P_0 + \tilde{Q}_0 = 0$$

Then we have

$$(\lambda_1 - 1)\tilde{Q}_0 \geq \epsilon\lambda_1(P_0 A_1 + A_1^T P_0) \tag{1.46}$$

Thus, if (1.46) is true, then

$$P_{\text{opt}}(\epsilon) \leq M_1 \leq \lambda_1 P_0 \tag{1.47}$$

Applying a similar procedure to the dual problem and taking into account (1.33) we have that the condition

$$(\lambda_2 - 1)\tilde{Q}_0 \geq -\epsilon\lambda_2(P_0 A_1 + A_1^T P_0) \tag{1.48}$$

implies

$$N(\epsilon) \leq \lambda_2 N_0 \ , \ \lambda_2 > 0 \tag{1.49}$$

where

$$N^{-1}(\epsilon) = P_{\text{opt}}(\epsilon) \ , \ N_0^{-1} = P_0$$

Thus, the choice of λ_1, λ_2 according to (1.46) and (1.48) ensures that

$$\lambda_2^{-1} P_0 \leq P_{\text{opt}}(\epsilon) \leq M_1 \leq \lambda_1 P_0 \tag{1.50}$$

To improve the estimates, one needs to find minimal admissible values of λ_1, λ_2.

Rewrite (1.42), (1.46) and (1.48) in the form

$$\alpha - \frac{1}{\epsilon}\tilde{Q}_0 < 0 \ , \ \alpha - \frac{1}{\epsilon}(1 - \frac{1}{\lambda_1})\tilde{Q}_0 \leq 0 \ , \ \alpha - \frac{1}{\epsilon}(\frac{1}{\lambda_2} - 1)\tilde{Q}_0 \geq 0 \quad (1.51)$$

where

$$\alpha = P_0 A_1 + A_1^T P_0$$

Recall that the matrices α, $\tilde{Q}_0$ are symmetric and that $\tilde{Q}_0 > 0$. The relations (1.51) are equivalent to the inequalities for the corresponding quadratic forms

$$\begin{aligned} &x^T \alpha x - \frac{1}{\epsilon} x^T \tilde{Q}_0 x < 0 \ , \ x^T \alpha x - \frac{1}{\epsilon}(1 - \frac{1}{\lambda_1}) x^T \tilde{Q}_0 x \leq 0 \\ &x^T \alpha x - \frac{1}{\epsilon}(\frac{1}{\lambda_2} - 1) x^T Q_0 x \geq 0 \end{aligned} \quad (1.52)$$

for any $x \neq 0$.

It is known (see, e.g., [20]) that there exists a nonsingular matrix z such that

$$z^T \alpha z = \text{diag}\ \{\sigma_i\} \ , \ z^T \tilde{Q}_0 z = E$$

where σ_i are the real eigenvalues of the matrix $q \triangleq \tilde{Q}_0^{-1}\alpha$.

Applying the transformation $x = z\xi$ to (1.52), we obtain

$$\begin{aligned} &\sum_i(\sigma_i - \frac{1}{\epsilon})\xi_i^2 < 0 \ , \ \sum_i[\sigma_i - \frac{1}{\epsilon}(1 - \frac{1}{\lambda_1})]\ \xi_i^2 \leq 0 \\ &\sum_i[\sigma_i - \frac{1}{\epsilon}(\frac{1}{\lambda_2} - 1)]\xi_i^2 \geq 0 \end{aligned} \quad (1.53)$$

Let

$$\sigma_m = \min \sigma_i \ , \ \sigma_M = \max_i \sigma_i$$

If

$$\lambda_1 = \frac{1}{1 - \epsilon\sigma_M} \ ; \ \lambda_2 = \frac{1}{1 + \epsilon\sigma_m} \quad (1.54)$$

then the second and the third inequalities in (1.53) are true. The first inequality in (1.53) holds if $\epsilon \leq \epsilon_1$ where ϵ_1 satisfies (1.37).

The proof is complete since

$$J\{u_{\text{opt}},\epsilon\} = \frac{1}{2}x_0^T P_{\text{opt}}(\epsilon)x_0$$
$$J\{u^0,0\} = \frac{1}{2}x_0^T P_0 x_0 \;,\; J\{u^0,\epsilon\} = \frac{1}{2}x_0^T M_1 x_0 \tag{1.55}$$

QED.

Note that (1.35) can be rewritten in the form

$$\frac{J\{u^0,\epsilon\} - J\{u_{\text{opt}},\epsilon\}}{J\{u_{\text{opt}},\epsilon\}} \le \epsilon \frac{\sigma_M - \sigma_m + \sigma_M\sigma_m\epsilon}{1 - \epsilon(\sigma_M - \sigma_m + \sigma_M\sigma_m)} \tag{1.56}$$

which gives a direct estimate of the reduced control efficiency in the perturbed problem.

It is natural to apply the obtained results to optimal control problems for weakly connected systems, i.e., to problems of type (1.1) and (1.2) where

$$A = \text{diag}\,\{A_0^s\} + \epsilon A_1 \;,\; B = \text{diag}\,\{B^s\}$$
$$R = \text{diag}\,\{R^s\} \;,\; C^TQC = \text{diag}\,\{\bar{Q}^s\} \tag{1.57}$$

Let $\bar{Q}^s$, $R^s > 0$ and the pairs $\{A_0^s, B^s\}$, $s = 1,\ldots,l$ be completely controllable. Then the reduced problem is decomposable into l independent problems

$$J^s(u^s) \triangleq \frac{1}{2}\int_0^\infty [(x^s)^T\bar{Q}^s x^s + (u^s)^T R^s u^s]dt \tag{1.58}$$

such that

$$\dot{x}^s = A_0^s x^s + B^s u^s \;,\; x^s(0) = x_0^s \;,\; s = 1,\ldots,l \tag{1.59}$$

The solution to each subproblem may be written in the standard form

$$u_0^s = -K_0^s x^s \;,\; K_0^s = (R^s)^{-1}(B^s)^T P_0^s$$

where the matrices P_0^s are the solutions to the equations

$$P_0^s A_0^s + (A_0^s)^T P_0^s - P_0^s B^s (R^s)^{-1} (B^s)^T P_0^s + \bar{Q}^s = 0$$

By Theorem 1.3 the decomposed solution is equal to the optimal one to within $O(\epsilon^2)$.

To obtain the explicit estimate one needs to compute the eigenvalue of the matrix q defined by (1.36). Here that matrix has a simplified form

$$q = \text{diag}\,\{(\tilde{Q}^{(s)})^{-1}\}(P_0 A_1 + A_1^T P_0) \ , \ P_0 = \text{diag}\,\{P_0^s\} \ ,$$
$$\tilde{Q}^{(s)} = \bar{Q}^{(s)} + (K_0^s)^T R^s K_0^s \ , \ s = 1,\ldots,l$$

In the simplest case of a dichotomous decomposition, the reduced problem can be isolated such that

$$A_1 = \begin{Bmatrix} 0 & A_{12} \\ A_{21} & 0 \end{Bmatrix}$$

where A_{12}, A_{21} specify cross connections.

It is not hard to see that

$$q = \begin{Bmatrix} 0 & q_{12} \\ q_{21} & 0 \end{Bmatrix}$$

where

$$q_{12} = (\tilde{Q}^{(1)})^{-1}\alpha_1 \ , \ q_{21} = (\tilde{Q}^{(2)})^{-1}\alpha_1^T \ , \ \alpha_1 = P_0^{(1)} A_{12} + A_{21}^T P_0^{(2)}$$

To compute σ_m, σ_M one can use the well-known extremal properties

$$\sigma_m = \min \frac{\xi^T \alpha \xi}{\xi^T \tilde{Q}_0 \xi} \ , \ \sigma_M = \max \frac{\xi^T \alpha \xi}{\xi^T \tilde{Q}_0 \xi} \tag{1.60}$$

In the dichotomous case this gives

$$\sigma_m = \min \frac{2\xi_1^T \alpha_1 \xi_2}{\xi_1^T \tilde{Q}^{(1)} \xi_1 + \xi_2^T \tilde{Q}^{(2)} \xi_2} \ , \ \sigma_M = \max \frac{2\xi_1^T \alpha_1 \xi_2}{\xi_1^T \tilde{Q}^{(1)} \xi_1 + \xi_2^T \tilde{Q}^{(2)} \xi_2}$$

5.2. Loss of controllability, singularity, and suboptimal aggregation

Definition. The LQ problem

$$\min \frac{1}{2}\int_0^\infty [y^T Qy + u^T Ru]dt \quad , \quad Q > 0 \quad , \quad R > 0 \tag{2.1}$$

such that

$$y = Cx \quad , \quad \dot{x} = A(\epsilon)x + B(\epsilon)u \quad , \quad x(0) = x_0 \tag{2.2}$$

is degenerate if, for $\epsilon = 0$, either the controllability or observability conditions do not hold. It is supposed that both these conditions are true for $0 < \epsilon \leq \epsilon'$.

At first sight, the analysis of the degenerate case is practically useless since the set of completely controllable pairs of matrices is everywhere dence (see, e.g., [45]). That is, one can choose a completely controllable and observable system in a parametric neighborhood of the original system. However, as underlined above, the analysis of degenerate cases allows one to exploit the most effective possibilities of the perturbation method.

We previously considered a basic transformation. It is known [148] that the system (2.1) can be transformed by some nonsingular linear transformation to the form

$$\dot{x}^i = \sum_{j=1}^{4} A^{ij}x^j + B^i u \quad , \quad i = 1,2,3,4 \quad ; \quad y = \sum_{j=1}^{4} C^j x^j$$

where

$$A^{23} = 0 \quad ; \quad A^{ij} = 0 \quad , \quad j < i \quad ; \quad B^3 = 0 \quad ; \quad B^4 = 0 \quad ; \quad C^1 = 0 \quad ; \quad C^3 = 0$$

The components (x^1, x^2) are completely controllable, the components (x^2, x^4) are completely observable. Vice versa, the components (x^3, x^4) are uncontrollable, the components (x^1, x^3) are unobservable. The transformation can be realized in different ways, but the dimensions of the parts are invariant and determined only by the rank of the matrices

$$A_u \triangleq \{B, AB, \ldots, A^{n-1}B\} \quad ; \quad A_y \triangleq \{C^T, A^T C^T, \ldots, (A^T)^{n-1} C^T\}$$

so that

$$\dim x^1 + \dim x^2 = \operatorname{rank} A_u$$
$$\dim x^2 + \dim x^4 = \operatorname{rank} A_y$$

The characteristic determinant of the system does not depend on the transformation

$$\Delta(\lambda) = |\lambda E - A| = \prod_{i=1}^{4} |\lambda E - A^{ii}|$$

where E are the identity matrices of appropriate dimension.

Lemma 2.1. Let the eigenvalues of the matrices A^{ii}, $i = 1,2,3,4$ be distinct. Then there exists a nonsingular transformation $x \to \xi$ such that the system (2.1) takes the form

$$\dot{\xi}^i = \Lambda^i \xi^i + D^i u \ , \quad i = 1,2,3,4 \ ; \quad \xi = (\xi^i \ , i = 1,2,3,4)$$
$$y = \sum_{i=1}^{4} \sigma^i \xi^i$$

where $D^3 = 0$; $D^4 = 0$; $\sigma^1 = 0$, $\sigma^3 = 0$.

Proof. Let

$$\operatorname{rank} A_u = \bar{n} = n - k \ , \quad k > 0 \tag{2.3}$$

Let us introduce the $(n \times \bar{n})$-matrix W_R. The columns of W_R are the vectors of a basis for the linear subspace spanned by the columns of A_u. Let them be normed such that

$$W_R^T W_R = E_{\bar{n}}$$

Let us also construct the $(k \times n)$ matrix V_L. The columns of V_L constitute a basis for the linear subspace, orthogonal to the subspace mentioned above, so that

$$V_L^T W_R = 0 \ , \quad V_L^T V_L = E_k$$

If the eigenvalues of the matrices

$$\Lambda_1 \triangleq W_R^T A W_R \ , \quad \Lambda_2 \triangleq V_L^T A V_L$$

are distinct, then there exists a unique solution β to the matrix equation

$$-\beta\Lambda_1^T + \Lambda_2^T\beta = -V_L^T A^T W_R$$

Let

$$W_L \triangleq W_R + V_L\beta \; ; \; V_R \triangleq -W_R\beta^T + V_L$$

Let us show that the following relations are true

$$W_L^T W_R = E_{\bar{n}} \; ; \; V_L^T V_R = E_k \; ; \; V_L^T W_R = 0 \; ; \; W_L^T V_R = 0 \tag{2.4}$$

$$V_L^T B = 0 \; , \; V_L^T A W_R = 0 \; ; \; W_L^T A V_R = 0 \tag{2.5}$$

The first four relations (2.4) and (2.5) are direct consequences of the definitions. Further, since the subspace spanned by the columns of A_u is invariant with respect to A, there exists a matrix γ such that

$$A W_R = W_R\gamma \Rightarrow V_L^T A W_R = V_L^T W_R \gamma = 0$$

Direct computation shows that

$$\begin{aligned} W_L^T A V_R &= (W_R^T + \beta^T V_L^T) A(-W_R\beta^T + V_L) \\ &= -W_R^T A W_R \beta^T + W_R^T A V_L + \beta^T V_L^T A V_L - \beta^T V_L^T A W_R \beta^T \\ &= -\Lambda_1\beta^T + \beta^T\Lambda_2 + W_R^T A V_L \end{aligned}$$

but by the definition of β the last expression is equal to zero.

If we take

$$x = W_R\xi_1 + V_R\xi_2 \tag{2.6}$$

or

$$\xi_1 = W_L^T x \; , \; \xi_2 = V_L^T x \tag{2.7}$$

then, by (2.1)

$$W_R\dot{\xi}_1 + V_R\dot{\xi}_2 = A W_R\xi_1 + A V_R\xi_2 + Bu$$

and, after premultiplication by V_L^T, W_L^T we obtain

$$\dot{\xi}_1 = \Lambda_1 \xi_1 + D_1 u \quad , \quad \dot{\xi}_2 = \Lambda_2 \xi_2$$

where

$$D_1 \triangleq W_L^T B$$

Thus, the procedure of system decomposition into completely controllable and completely uncontrollable parts is realized. Similarly, one can realize a decomposition into completely observable and completely unobservable parts by choosing W_R with columns constituting the basis of the linear subspace spanned by the columns of A_y and by constructing V_L similarly by means of the orthogonal subspace basis. To obtain the form mentioned in the lemma, it is sufficient to realize both types of decomposition successively. QED.

Let the matrices in (2.2) $A(\epsilon)$, $B(\epsilon)$ have the form

$$A(\epsilon) = A_0 + \epsilon A_1 \quad ; \quad B(\epsilon) = B_0 + \epsilon B_1$$

and let

$$\text{rank } A_u^0 = n - k \tag{2.3'}$$

where

$$A_u^0 = \{B_0, A_0 B_0, \ldots, A_0^{n-1} B_0\}$$

Construct the matrices W_R, V_R, W_L, V_L, associated with A_0, B_0, assuming that the lemma conditions are fulfilled.

The transformation (2.6) and (2.7) applied to (2.2) gives

$$\begin{aligned} \dot{\xi}_1 &= \Lambda_1 \xi_1 + D_1 u + \epsilon(\Lambda_{11} \xi_1 + \Lambda_{12} \xi_2 + D_{11} u) \\ \dot{\xi}_2 &= \Lambda_2 \xi_2 + \epsilon(\Lambda_{21} \xi_1 + \Lambda_{22} \xi_2 + D_{21} u) \end{aligned} \tag{2.8}$$

where

$$\begin{aligned} \Lambda_1 &= W_L^T A_0 W_R \ , & \Lambda_2 &= V_L^T A_0 V_R \ , & D_1 &= W_L^T B_0 \\ \Lambda_{11} &= W_L^T A_1 W_R \ , & \Lambda_{12} &= W_L^T A_1 V_R \ , & D_{11} &= W_L^T B_1 \\ \Lambda_{21} &= V_L^T A_1 W_R \ , & \Lambda_{22} &= V_L^T A_1 V_R \ , & D_{21} &= V_L^T B_1 \end{aligned} \tag{2.9}$$

Let us rewrite the objective

$$J\{u\} = \frac{1}{2}\int_0^\infty (\xi^T \bar{Q}(\epsilon)\xi + u^T R u)dt \tag{2.10}$$

where

$$\begin{aligned}
\bar{Q}(\epsilon) &\triangleq \begin{Bmatrix} \overset{n-k}{\bar{Q}_{11}(\epsilon)} & \overset{k}{\bar{Q}_{12}(\epsilon)} \\ \bar{Q}_{21}(\epsilon) & \bar{Q}_{22}(\epsilon) \end{Bmatrix} \\
\bar{Q}_{11}(\epsilon) &= W_R^T C^T(\epsilon) Q C(\epsilon) W_R \\
\bar{Q}_{22}(\epsilon) &= V_R^T C^T(\epsilon) Q C(\epsilon) V_R \\
\bar{Q}_{12}(\epsilon) &= \bar{Q}_{21}^T(\epsilon) = W_R^T C^T(\epsilon) Q C(\epsilon) V_R
\end{aligned} \tag{2.11}$$

Let us discuss the results of the transformation. The behavior of $\xi_2(t)$ is determined, for $\epsilon = 0$, by the initial values $\xi_2(0)$ only and does not depend on either the control or the behavior of any other subsystem that is completely controllable for $\epsilon = 0$.

Correspondingly, the perturbed system where ϵ is small but nonzero contains the weakly controllable subsystem. The difference in the control action scales leads us to suppose that the optimization procedure can be simplified. In fact, let us suppose that the matrix Λ_2 is stable. Then small perturbations do not change the essential behavior of the weakly controlled subsystem. Hence, an independent procedure for the completely controlled subsystem is possible, ensuring the stability of the system on the whole.

The components of ξ_1 are linear combinations of the original state variables $x = (x_j\,, j = 1,\ldots,n)$, i.e., one can consider ξ_1 as the vector of aggregated variables. Then the procedure mentioned above can be interpreted as a suboptimal aggregation.

To justify this approach we write down the basic matrix quadratic equation of (1.7) type for the problem (2.8) and (2.10). Write $P(\epsilon)$ in the block form

$$P(\epsilon) = \begin{Bmatrix} \overset{n-k}{P_{11}(\epsilon)} & \overset{k}{P_{12}(\epsilon)} \\ P_{21}(\epsilon) & P_{22}(\epsilon) \end{Bmatrix}$$

By (1.7) we have that

$$\begin{aligned}
&\begin{Bmatrix} P_{11}(\epsilon) & P_{12}(\epsilon) \\ P_{21}(\epsilon) & P_{22}(\epsilon) \end{Bmatrix} \begin{Bmatrix} \Lambda_1 + \epsilon\Lambda_{11} & \epsilon\Lambda_{12} \\ \epsilon\Lambda_{21} & \Lambda_2 + \epsilon\Lambda_{22} \end{Bmatrix} + \\
&+ \begin{Bmatrix} (\Lambda_1 + \epsilon\Lambda_{11})^T & \epsilon\Lambda_{21}^T \\ \epsilon\Lambda_{12}^T & \Lambda_2^T + \epsilon\Lambda_{22}^T \end{Bmatrix} \begin{Bmatrix} P_{11}(\epsilon) & P_{12}(\epsilon) \\ P_{21}(\epsilon) & P_{22}(\epsilon) \end{Bmatrix} - \\
&- \begin{Bmatrix} P_{11}(\epsilon) & P_{12}(\epsilon) \\ P_{21}(\epsilon) & P_{22}(\epsilon) \end{Bmatrix} \begin{Bmatrix} D_1 + \epsilon D_{11} \\ \epsilon D_{21} \end{Bmatrix} R^1\{(D_1 + \epsilon D_{11})^T \epsilon D_{21}^T\} \times \\
&\times \begin{Bmatrix} P_{11}(\epsilon) & P_{12}(\epsilon) \\ P_{21}(\epsilon) & P_{22}(\epsilon) \end{Bmatrix} + \begin{Bmatrix} \bar{Q}_{11}(\epsilon) & \bar{Q}_{12}(\epsilon) \\ \bar{Q}_{21}(\epsilon) & \bar{Q}_{22}(\epsilon) \end{Bmatrix} = 0
\end{aligned} \tag{2.12}$$

or

$$\begin{aligned}
&P_{11}(\epsilon)\Lambda_1 + \Lambda_1^T P_{11}(\epsilon) - P_{11}(\epsilon) D_1 R^{-1} D_1^T P_{11}(\epsilon) + \bar{Q}_{11}(0) = \\
&= \epsilon\varphi_{11}(P(\epsilon), \epsilon) \\
&P_{12}(\epsilon)\Lambda_2 + \Lambda_1^T P_{12}(\epsilon) - P_{11}(\epsilon) D_1 R^{-1} D_1^T P_{12}(\epsilon) + \bar{Q}_{12}(0) = \\
&= \epsilon\varphi_{12}(P(\epsilon), \epsilon) \\
&P_{22}(\epsilon)\Lambda_2 + \Lambda_2^T P_{22}(\epsilon) - P_{12}^T(\epsilon) D_1 R^{-1} D_1^T P_{12}(\epsilon) + \bar{Q}_{22}(0) = \\
&= \epsilon\varphi_{22}(P(\epsilon), \epsilon)
\end{aligned} \tag{2.12'}$$

The equations (2.12′) give an implicit description of the matrix-functions $P_{11}(\epsilon)$, $P_{22}(\epsilon)$, $P_{12}(\epsilon) = P_{21}^T(\epsilon)$. By the controllability of $\{\Lambda_1, D_1\}$ there exists a unique, positive definite solution to the reduced subsystem

$$P_{11}\Lambda_1 + \Lambda_1^T P_{11} - P_{11} D_1 R^{-1} D_1^T P_{11} + Q_{11}(0) = 0 \tag{2.13}$$

if $\bar{Q}_{11}(0) \triangleq \bar{Q}_{11} > 0$.

Introduce the matrix H_1

$$H_1 \triangleq \Lambda_1 - D_1 R^{-1} D_1^T P_{11}(0) \tag{2.14}$$

which determines the behavior of the aggregated subsystem if the control u^0 is used

$$u^0 \triangleq -R^{-1}\{D_1^T, 0\}\begin{Bmatrix} P_{11}(0) & P_{12}(0) \\ P_{21}(0) & P_{22}(0) \end{Bmatrix}\begin{Bmatrix} \xi_1 \\ \xi_2 \end{Bmatrix} = $$
$$= -R^{-1}D_1^T P_{11}(0)\xi_1 - R^{-1}D_1^T P_{12}(0)\xi_2 \tag{2.15}$$

The matrix H_1 is stable. If Λ_2 is also stable, then the equations

$$\begin{aligned} &P_{12}\Lambda_{12} + H_1^T P_{12} + \bar{Q}_{12}(0) = 0 \\ &P_{22}\Lambda_2 + \Lambda_2^T P_{22} - P_{12}^T D_1 R^{-1} D_1^T P_{12} + \bar{Q}_{22}(0) = 0 \end{aligned} \tag{2.16}$$

are solvable in succession. In fact, the first one is linear, and its determinants is equal to

$$|E_{n-k} \otimes \Lambda_2^T + H_1^T \otimes E_k| \neq 0$$

The second equation is linear with respect to P_{22} and its determinant is also nonzero.

Thus the computation of a zero-order approximation consists of solution of the aggregated problem

$$\begin{aligned} &\min \ \{\frac{1}{2}\int_0^\infty [\xi_1^T \bar{Q}_{11}\xi_1 + u^t R u]dt \,/\, \dot{\xi}_1 = \Lambda_1\xi_1 + D_1 u \ ; \\ &\qquad \xi_1(0) = W_L^T x_0 = \xi_{10}\} = \frac{1}{2}\xi_{10}^T P_{11}(0)\xi_{10} \end{aligned} \tag{2.17}$$

and the solutions of the linear matrix equations.

The Jacobian of the system (2.16) at the point $\{P(\epsilon) = P(0), \epsilon = 0\}$ may be written in the form

$$|E_{n-k} \otimes \Lambda_2^T + H_1^T \otimes E_k| \cdot |\Lambda_2^T \otimes E_k + E_k \otimes \Lambda_2^T|$$

and is nonzero. Hence, the implicit function theorem is applicable. This remark establishes the proof of the following result.

Theorem 2.1.

Let $\bar{Q}_{11} > 0$. Then there exists ϵ' such that, for $0 \leq \epsilon \leq \epsilon'$, the optimal value of the perturbed system be written in the form

$$J\{u_{\text{opt}}\} = \frac{1}{2} x_0^T [W_L P_{11}(\epsilon) W_L^T + 2 W_L P_{12}(\epsilon) V_L^T + V_L P_{22}(\epsilon) V_L^T] x_0 \tag{2.18}$$

where $P_{11}(\epsilon)$, $P_{12}(\epsilon)$, $P_{22}(\epsilon)$ are analytical functions of ϵ. $P_{11}(0)$ can be calculated from the solution to the aggregated problem (2.17). To calculate $P_{12}(0)$, $P_{22}(0)$ as well as the coefficients of ϵ^l ($l > 1$) in the power expansion of $P_{11}(\epsilon)$, $P_{12}(\epsilon)$, $P_{22}(\epsilon)$ one needs to solve the linear equations.

Corollary. If, in addition, $Q_{12}(0) = 0$, $Q_{22}(0) = 0$, i.e., the objective is weakly dependent on ξ_2, then the aggregated control law

$$u_1^0 = -R^{-1} D_1^T P_{11}(0) \xi_1 \tag{2.19}$$

is $O(\epsilon)$-optimal.

Note that the control constructed with the first m terms of the expansion $P(\epsilon)$ is optimal to within $O(\epsilon^{2m})$. This can be proved as in Theorem 1.3. Note also that the behavior of the reduced system is close to that of the perturbed system. This is stipulated by the preservation of the stabilizability condition [121] for $\epsilon = 0$.

Singular cases where this condition is not fulfilled are also of interest. Naturally, this relates only to the situation where some eigenvalues of Λ_2 are situated on the imaginary axes, and where others have negative real parts, since the perturbation method is not effective if Λ_2 is unstable.

Let us consider in detail the case $\Lambda_2 = 0$, which is equivalent to

$$V_L^T A(0) = 0 \tag{2.20}$$

The last relation means that columns of V_L are the left eigenvectors of the matrix A_0 corresponding to the k-multiple semi-simple zero eigenvalue [36].

The system (2.8) takes the form

$$\begin{aligned} \dot{\xi}_1 &= \Lambda_1 \xi_1 + D_1 u + \epsilon[\Lambda_{11}\xi_1 + \Lambda_{12}\xi_2 + D_{11} u] \\ \dot{\xi}_2 &= \epsilon[\Lambda_{21}\xi_1 + \Lambda_{22}\xi_2 + D_{21} u] \end{aligned} \tag{2.21}$$

If we introduce the "slow" time scale $\tau = \epsilon t$, we obtain

$$\begin{aligned} \epsilon \frac{d\xi_1}{d\tau} &= \Lambda_1 \xi_1 + D_1 u + \epsilon[\Lambda_{11}\xi_1 + \Lambda_{12}\xi_2 + D_{11} u] \\ \frac{d\xi_1}{d\tau} &= \Lambda_{21}\xi_1 + \Lambda_{22}\xi_2 + D_{21} u \end{aligned} \tag{2.22}$$

with the objective in the form

$$J\{u\} = \frac{1}{\epsilon}\frac{1}{2}\int_0^\infty [\xi^T \bar{Q} \xi + u^T R u] d\tau \triangleq \frac{1}{\epsilon} \bar{J}(\epsilon) \tag{2.23}$$

One can see that the variables ξ_2 change more slowly than ξ_1. If $\epsilon = 0$, the order of the system decreases since the first subsystem is described by an algebraic relation which can be used to eliminate the variables ξ_1. This class of problems is of special practical interest and is useful, moreover, for the interpretation of control theoretic results. In fact, it is impossible to establish *a priori* the complexity of the mathematical model, which is necessary for a "proper" description of the real control object. However, the methods and algorithms for optimal control are based on the hypothesis of a fixed model order. It is not clear whether a control law that is optimal for the given model is necessarily effective for the real object or even for other models that describe the real object more accurately. Therefore, the problem of "insensitivity" to change in the model order, which is well-known in classical control theory [35, 50] as "a small time constant problem", is a very real one. A path to its solution may be based on the physically reasonable hypothesis that increasing the model order takes into account progressively faster processes. A classical frequency domain approach to linear feedback system design shows that the "proper" order depends on the feedback gain necessary for some desired precision, i.e., on the desired value of a performance index. It is inappropriate to introduce subsystems with small time constants affecting the transfer function only outside the desired frequency band since that is equivalent to the exclusion of fast damping processes. On the other hand, accounting for processes having strongly different speeds in one and the same model provides essential difficulties for numerical analysis (see [75, 111] and other works on the theory of "stiff" systems). Evidently these difficulties are greater in the associated optimization problems.

Thus, the singular optimal control problem described by equation (2.21) or (2.22) is not an "exotic" one, but it is of special interest. Singular perturbed problems of optimal control were investigated by many authors (see, e.g., [101–104]) who did not connect these problems with violation of the controllability condition. The original description was

$$\begin{aligned} \epsilon\frac{dx^{(1)}}{d\tau} &= a_{11}x^{(1)} + a_{12}x^{(2)} + b_1 u \\ \frac{dx^{(2)}}{d\tau} &= a_{21}x^{(1)} + a_{22}x^{(2)} + b_2 u \end{aligned} \tag{2.22'}$$

Evidently, the system (2.22) can be written in the form (2.22′). On the other hand, the system (2.22′) can be transformed into (2.22) by a suitable choice of basis. In fact, introducing the "fast" time scale $t = \tau/\epsilon$, one can write (2.22′) in the form (2.22) with

$$A_0 = \begin{pmatrix} a_{11} & a_{12} \\ 0 & 0 \end{pmatrix}, \quad A_1 = \begin{pmatrix} 0 & 0 \\ a_{21} & a_{22} \end{pmatrix}, \quad B_0 = \begin{pmatrix} b_1 \\ 0 \end{pmatrix}, \quad B_1 = \begin{pmatrix} 0 \\ b_2 \end{pmatrix}$$

After that the standard transformation (2.6) and (2.7) gives the desired result if we adopt

$$V_L = \begin{pmatrix} 0 \\ E_{n_2} \end{pmatrix} \ , \quad V_R = \begin{pmatrix} -a_{11}^{-1}a_{12} \\ E_{n_2} \end{pmatrix} \ , \quad W_R = \begin{pmatrix} E_{n_1} \\ 0 \end{pmatrix}$$

where n_1, n_2 are respectively the dimensions of the "fast" subsystem of (2.22′) and the "slow" subsystem. Thus (2.22) and (2.22′) are equivalent.

To begin with, we shall describe the main heuristic notion. One supposes that the control includes two components. The first component is based on measurements of the "fast" subsystem state and ensures the optimal jump to quasistationary behavior. It is natural to choose this component as in the previous regular case. The second component can be constructed as a feedback of the "slow" state and ensures the optimal change to quasistationary behavior. It is admissible to suppose that the stabilization of the "fast" subsystem by the "fast" feedback provides a "stiff" relation between the state variables and the "slow" control. One can isolate the case of $D_1 = 0$, i.e., the case where the system on the whole is weakly controllable. Then the first stage is not realizable but is useless anyway if the "fast" subsystem is stable.

Generally speaking this situation is impossible if the decomposition is realized as described above, i.e., if the first subsystem is completely controllable for $\epsilon = 0$. However, if such a part is absent, one can try to divide the weakly controllable subsystem into a stable subsystem and a neutral one. The case $D_1 = 0$ can be interpreted in this way.

Let us now formulate the algorithm that realizes the heuristic approach mentioned above.

(1) Take $u = u_1 + u_2$. Put $u_1 = 0$ if Λ_1 is stable and $D_1 = 0$ otherwise.
(2) Choose u_1 as a solution to the "fast" LQ subproblem

$$\begin{aligned} \min \ &\{\frac{1}{2}\int_0^\infty (\xi_1^T\bar{Q}_{11}(0)\xi_1 + u_1^TRu_1)\mathrm{d}\tau \,/\, \epsilon\frac{\mathrm{d}\xi_1}{\mathrm{d}\tau} = \Lambda_1\xi_1 + D_1u_1 \ ; \\ &\xi_1(0) = \xi_{10}\} \end{aligned} \tag{2.24}$$

or, in the "fast" time scale,

$$\begin{aligned} \min \ &\{\frac{1}{2}\int_0^\infty (\xi_1^T\bar{Q}_{11}(0)\xi_1 + u_1^TRu_1)\mathrm{d}t \,/\, \dot{\xi}_1 = \Lambda_1\xi_1 + D_1u_1 \ ; \\ &\xi_1(0) = \xi_{10}\} \end{aligned} \tag{2.24′}$$

The solution has the form

$$u_1 = -K_1\xi_1 \ , \quad K_1 = R^{-1}D_1^TP_{11}^0 \tag{2.25}$$

where P_{11}^0 is the positive definite solution to the quadratic equation

$$P_{11}\Lambda_1 + \Lambda_1^T P_{11} - P_{11}D_1R^{-1}D_1^TP_{11} + \bar{Q}_{11}(0) = 0 \tag{2.26}$$

the matrix

$$H_1 \triangleq \Lambda_1 - D_1K_1 \tag{2.27}$$

being stable.

(3) Given (2.25), rewrite (2.22) and (2.23) in the form

$$\begin{cases} \epsilon\dfrac{d\xi_1}{d\tau} = H_1\xi_1 + D_1u_2 + O(\epsilon) \\ \dfrac{d\xi_2}{d\tau} = (\Lambda_{21} - D_{21}K_1)\xi_1 + \Lambda_{22}\xi_2 + D_{21}u_2 \end{cases} \tag{2.28}$$

$$\bar{J}(u) = \frac{1}{2}\int_0^\infty [\frac{1}{2}\xi^T\bar{Q}(\epsilon)\xi + \frac{1}{2}(u_2 - K_1\xi_1)^TR(u_2 - K_1\xi_1)]d\tau \tag{2.29}$$

Put $\epsilon = 0$. Using the "stiff" relation

$$\xi_1 = -H_1^{-1}D_1u_2 \tag{2.30}$$

isolate the "slow" subsystem

$$\frac{d\xi_2}{d\tau} = \Lambda_{22}\xi_2 + [D_{21} - (\Lambda_{21} - D_{21}K_1)H_1^{-1}D_1]u_2 \tag{2.31}$$

with the objective

$$\begin{aligned} \bar{J} = \frac{1}{2}\int_0^\infty [&\xi_2^T\bar{Q}_{22}(0)\xi_2 + u_2^TD_1^T(H_1^T)^{-1}\bar{Q}_{11}(0)H_1^{-1}D_1u_2 + \\ &+ u_2^T(E_m + K_1H_1^{-1}D_1)^TR(E_m + K_1H_1^{-1}D_1)u_2 - \\ &- u_2^TD_1^T(H_1^T)^{-1}Q_{12}(0)\xi_2 - \xi_2^TQ_{12}^T(0)H_1^{-1}D_1u_2]d\tau \end{aligned} \tag{2.32}$$

(4) Introducing the auxiliary control function

$$\begin{aligned} \tilde{u} &\triangleq u_2 - L\xi_2 \\ L &\triangleq R^{-1}D_1^T(H_1^T)^{-1}\bar{Q}_{12}(0) \end{aligned} \tag{2.33}$$

obtain the problem in the standard form

$$J = \frac{1}{2}\int_0^\infty [\xi_2^T \tilde{Q}\xi_2 + \tilde{u}^T R\tilde{u}]\mathrm{d}\tau \to \min \tag{2.34}$$

$$\frac{\mathrm{d}\xi_2}{\mathrm{d}\tau} = \tilde{M}\xi_2 + \tilde{D}\tilde{u} \tag{2.35}$$

where

$$\begin{aligned} \tilde{M} &= \Lambda_{22} + \tilde{D}L \\ \tilde{D} &= D_{21} - (\Lambda_{21} - D_{21}K_1)H_1^{-1}D_1 \\ \tilde{Q} &= \bar{Q}_{22}(0) - L^T RL \end{aligned} \tag{2.36}$$

(5) Find the solution $\tilde{P}^0$ to the equation

$$\tilde{P}\tilde{M} + \tilde{M}^T\tilde{P} - \tilde{P}\tilde{D}R^{-1}\tilde{D}^T\tilde{P} + \tilde{Q} = 0 \tag{2.37}$$

and the minimizing function

$$\tilde{u} = -\tilde{K}\xi_2 \ , \quad \tilde{K} = R^{-1}\tilde{D}^T\tilde{P}^0 \tag{2.38}$$

Thus, the heuristic algorithm gives the following control law

$$u^0 = u_1 + u_2 = -K_1\xi_1 - K_2\xi_2 \ , \quad K_2 \triangleq \tilde{K} - L \tag{2.39}$$

if equations (2.26) and (2.37) are solvable.

The result can be rewritten in the original variables

$$\begin{aligned} u^0 &= -K^0 x = -R^{-1}G^T x \\ G &\triangleq W_L P_{11}^0 W_L^T B_0 + V_L \tilde{P}^0 V_L^T B_1 + V_L P_{21}^0 W_L^T B_0 \\ P_{21}^0 &\triangleq -[\bar{Q}_{21}(0) + \tilde{P}^0(\Lambda_{21} - D_{21}K_1)]H_1^{-1} \end{aligned} \tag{2.40}$$

where the notations (2.25) and (2.27) are used.

Let us pass to a justification of the algorithm. This can be done by using either the original description (2.1) and (2.2) or the transformed ones (2.21) and (2.10) and (2.22) and (2.23). This is because these descriptions are equivalent if the basic hypothesis concerning the existence of a matrix V_L such that

$$V_L^T B(0) = 0 \quad , \quad V_L^T A(0) = 0 \tag{2.41}$$

is true.

The form (2.21) and (2.10) is methodologically preferable to prove the results, but it is more effective to have the resultant expressions in the original language.

Theorem 2.2.

Let the pair $\tilde{M}$, $\tilde{D}$ be completely controllable and let $\tilde{Q} > 0$. Let $\{A(\epsilon), B(\epsilon)\}$, $\{A^T(\epsilon), C^T(\epsilon)\}$ be completely controllable over $0 < \epsilon \leq \epsilon'$, $\epsilon' > 0$.

(1) Moreover, let the pair $\{\Lambda_1, D_1\}$ be completely controllable[†] and let $\bar{Q}_{11}(0) > 0$. Then there exists ϵ'' such that, over $0 < \epsilon \leq \epsilon''$,

(a) The control u^0 ensures the stability of the system (2.2).
(b) The optimal control to (2.1) and (2.2) may be written in the form

$$U_{\text{opt}} = u^0 + O(\epsilon)x = -[K^0 + O(\epsilon)]x = -R^{-1}[G + \epsilon G_1 + \cdots]x \tag{2.42}$$

where G is given by (2.40) and $G^{(i)}$, $i = 1,2,\ldots$ do not depend on ϵ.

(c) $$J\{U_{\text{opt}}\} = \frac{1}{\epsilon}[\frac{1}{2}x_0^T V_L \tilde{P}^0 V_L^T x_0 + O(\epsilon)] \tag{2.43}$$

i.e, the objective value is predominantly determined by the solution to the "slow" problem.

If Λ_1 is stable and, moreover, if either
(2) $C_0 = V_L^T$, i.e., the objective depends on the "slow" variables only, or
(3) $B_0 = 0$, i.e., the original system is weakly controllable, then the preceding results are true, but in case (2),

$$G = V_L \tilde{P}^0 \tilde{D} \tag{2.44}$$

and in the case (3)

$$G = V_L \tilde{P}^0 V_L^T B_1 \tag{2.45}$$

so that the first stage of the algorithm is not necessary to find the ϵ-optimal control.

[†] This condition is automatically met if (2.21) is obtained by the basic transformation.

Proof of the theorem is cumbersome and is based on the following auxiliary result.

Lemma 2.2. The system (2.3) with the control (2.40) is asymptotically stable. Its eigenvalues are determined by the relations

$$\begin{aligned} &\lambda_i(\epsilon) = \epsilon\tilde{\lambda}_i + o(\epsilon) \ , \ i = 1,\ldots,k \\ &\lambda_i(\epsilon) = \lambda_i^1 + O(\epsilon^{1/p_i}) \ , \ p_i > 0 \ , \ i = k+1,\ldots,m \end{aligned} \tag{2.46}$$

where $\tilde{\lambda}_i$ are the eigenvalues of the matrix $\tilde{H} = \tilde{M} - \tilde{D}\tilde{K}$ and λ_i^1 are the eigenvalues of the matrix $H_1 = \Lambda_1 - D_1K_1$.

Proof of the lemma. It is more convenient to use the transformed structure, i.e., to consider the system (2.21) with the control (2.29). The closed system is described by the equations

$$\begin{aligned} &\dot{\xi}_1 = (\Lambda_1 - D_1K_1)\xi_1 - D_1(\tilde{K} - L)\xi_2 + O(\epsilon) \\ &\dot{\xi}_2 = \epsilon[(\Lambda_{21} - D_{21}K_1)\xi_1 + (\Lambda_{22} - D_{21}\tilde{K} + D_{21}L)\xi_2] \end{aligned} \tag{2.47}$$

Taking into account (2.36), one can show that

$$\Lambda_{22} - D_{21}\tilde{K} + D_{21}L = \tilde{H} - (\Lambda_{21} - D_{21}K_1)H_1^{-1}D_1K_2 \tag{2.48}$$

Let us isolate the "fast" modes by introducing the variables

$$\varsigma_1 = \xi_1 - \rho\xi_2 \ , \ \varsigma_2 = \xi_2 \tag{2.49}$$

where $\rho \triangleq H_1^{-1}D_1K_2$. This nonsingular transformation yields

$$\begin{aligned} &\dot{\varsigma}_1 = H_1\varsigma_1 + O(\epsilon) \\ &\dot{\varsigma}_2 = \epsilon[(\Lambda_{21} - D_{21}K_1)\varsigma_1 + \tilde{H}\varsigma_2] \end{aligned} \tag{2.50}$$

The matrix of the system

$$\left\{\begin{matrix} H_1 + O(\epsilon) & O(\epsilon) \\ \epsilon[\Lambda_{21} - D_{21}K_1] & \epsilon\tilde{H} \end{matrix}\right\} \triangleq \alpha(\epsilon)$$

may be written in the form

$$\alpha(\epsilon) = \alpha_0 + \epsilon\alpha_1 \tag{2.51}$$

where

$$\alpha_0 = \left\{\begin{matrix} H_1 & 0 \\ 0 & 0 \end{matrix}\right\}$$

The eigenvalues of α_0 are the eigenvalues of H_1 along with the k-multiple semi-simple zero eigenvalue. Now the result can be obtained as a consequence of the known results (see, e.g., [36]) or can be established directly from the representation (2.51).

Proof of the theorem. Result (a) is a consequence of Lemma 2.2 since the conditions ensure that the matrixes H_1, $\tilde{H}$ are stable. Note also that (a) holds if u^0 is replaced by any linear feedback with a gain differing from K^0 by $O(\epsilon)$ as follows from the proof of Lemma 2.2. It is evident from (2.50) that the processes are determined only by the slow component ς_2 after a time of the order $c \ln 1/\epsilon$. For $t \geq c \ln 1/\epsilon$, where c is a constant depending on the stability rate of H_1, we have

$$x(t) = (W_R H_1^{-1} D_1 K_2 + V_R)\varsigma_2^0 + O(\epsilon) \ , \ V_L^T x(t) = \varsigma_2^0 + O(\epsilon) \tag{2.52}$$

where ς_2^0 is the solution to the "slow" subsystem

$$\dot{\varsigma}_2^0 = \epsilon \tilde{H} \varsigma_2^0 \ , \ \varsigma_2^0(0) = V_L^T x_0 \tag{2.53}$$

Let us pass to the proof of other results.

We take into account here that the standard scheme of control problem solution is directly applicable if $\epsilon > 0$. Let us put $\Lambda_2 = 0$ in (2.12) and write (2.12) in the form

$$\begin{aligned}
\Phi_{11}(P(\epsilon),\epsilon) \triangleq\ & P_{11}(\Lambda_1 + \epsilon\Lambda_{11}) + \epsilon P_{12}\Lambda_{21} + (\Lambda_1 + \epsilon\Lambda_{11})^T P_{11} + \\
& + \epsilon\Lambda_{21}^T P_{21} - [P_{11}(D_1 + \epsilon D_{11}) + \\
& + \epsilon P_{12} D_{21}] R^{-1} [(D_1^T + \epsilon D_{11}^T) P_{11} + \epsilon D_{21}^T P_{21}] + \bar{Q}_{11}(\epsilon) \\
=\ & 0 \\
\Phi_{12}(P(\epsilon),\epsilon) \triangleq\ & \epsilon P_{12}\Lambda_{22} + [\Lambda_1 + \epsilon\Lambda_{11}]^T P_{12} + \epsilon\Lambda_{21}^T P_{22} + \epsilon P_{11}\Lambda_{12} - \\
& - [P_{11}(D_1 + \epsilon D_{11}) + \epsilon P_{12} D_{21}] R^{-1} [(D_1^T + \epsilon D_{11}^T) P_{12} + \\
& + \epsilon D_{21}^T P_{22}] + \bar{Q}_{12}(\epsilon) \\
=\ & 0 \\
\Phi_{21}(P(\epsilon),\epsilon) =\ & \Phi_{12}^T(P(\epsilon),\epsilon) = 0 \ , \ P_{12}(\epsilon) = P_{21}^T(\epsilon) \\
\Phi_{22}(P(\epsilon),\epsilon) \triangleq\ & \epsilon P_{22}\Lambda_{22} + \epsilon\Lambda_{22}^T P_{22} + \epsilon(\Lambda_{12}^T) P_{12} + \epsilon P_{12}^T \Lambda_{12} - \\
& - [P_{12}^T(D_1 + \epsilon D_{11}) + \epsilon P_{22} D_{21}] R^{-1} [(D_1^T + \epsilon D_{11}^T) P_{12} + \\
& + \epsilon D_{21}^T P_{22}] + \bar{Q}_{22}(\epsilon) \\
=\ & 0
\end{aligned} \tag{2.54}$$

We can rewrite the equations by supposing that $\epsilon P_{22} = \tilde{P}_{22}$

$$\begin{aligned}
P_{11}\Lambda_1 + \Lambda_1^T P_{11} - P_{11} D_1 R^{-1} D_1^T P_{11} + \bar{Q}_{11}(0) &= O(\epsilon) \\
\Lambda_1^T P_{12} + \Lambda_{21}^T \tilde{P}_{22} - P_{11} D_1 R^{-1} D_1^T P_{12} - P_{11} D_1 R^{-1} D_{21}^T \tilde{P}_{22} + \bar{Q}_{12}(0) &= O(\epsilon) \\
\tilde{P}_{11}\Lambda_{22} + \Lambda_{22}^T \tilde{P}_{22} - (P_{21} D_1 + \tilde{P}_{22} D_{21}) R^{-1} (D_1^T P_{12} + D_{21}^T \tilde{P}_{22}) + \bar{Q}_{22}(0) &= O(\epsilon)
\end{aligned} \tag{2.54'}$$

The functions $P_{11}(\epsilon)$, $P_{22}(\epsilon)$, $P_{12}(\epsilon) = P_{21}^T(\epsilon)$ are determined implicitly by (2.54′). It is not hard to see that the reduced system has the solution

$$P_{11} = P_{11}^0 \quad , \quad P_{12} = P_{12}^0 \quad , \quad P_{22} = \tilde{P}^0$$

where P_{11}^0, $\tilde{P}^0$ are the solutions to (2.26) and (2.27) and P_{12}^0 is given by (2.40). Let us write down the Jacobian using the rules of matrix differentiation and taking into account

$$\begin{aligned} H_1 &= \Lambda_1 - D_1 R^{-1} D_1^T P_{11}^0 \\ K_2 &= R^{-1}[D_1^T P_{12}^0 + D_{21}^T \tilde{P}^0] \end{aligned}$$

which follow from (2.36) and (2.40).

We have that

$$J = \begin{vmatrix} E_l \otimes H_1^T + H_1^T \otimes E_l & 0 & 0 & 0 \\ -E_l \otimes (D_1^T K_2)^T & H_1^T \otimes E_k & 0 & (\Lambda_{21} - D_{21}K_1)^T \otimes E_k \\ -(D_1 K_2)^T & 0 & E_k \otimes H_1^T & E_k \otimes (\Lambda_{21} - D_{21}K_1)^T \\ 0 & -(D_1 K_2)^T \otimes E_k & -E_k \otimes (D_1 K_2)^T & E_k \otimes (\Lambda_{22} - D_{21}K_2)^T + \\ & & & +(\Lambda_{22} - D_{21}K_{12})^T \otimes E_k \end{vmatrix}$$

If we post-multiply the second column by $(H_1^T \otimes E_k)^{-1}[(\Lambda_{21} - D_{21}K_1)^T \otimes E_k]$ and the third column by $(E_k \otimes H_1^T)^{-1}[E_k \otimes (\Lambda_{21} - D_{21}K_1)^T]$, and after that subtract these columns from the fourth one, the Jacobian is transformed to block-diagonal form. The fourth diagonal block is equal to

$$\begin{aligned} & E_k \otimes [(\Lambda_{22} - D_{21}K_2)^T + (D_1 K_2)^T (H_1^T)^{-1} (\Lambda_{21} - D_{21}K_1)^T] + \\ & + [(\Lambda_{22} - D_{21}K_2)^T + (D_1 K_2)^T (H_1^T)^{-1} (\Lambda_{21} - D_{21}K_1)^T \otimes E_k = \\ & = E_k \otimes \tilde{H}^T + \tilde{H}^T \otimes E_k \end{aligned}$$

where it is taken into account that, by definition (2.36), we obtain

$$\tilde{H} = \Lambda_{22} - \tilde{D}K_2 = \Lambda_{22} - [D_{21} - (\Lambda_{21} - D_{21}K_1)H_1^{-1}D_1]K_2 \tag{2.55}$$

Thus, we have

$$J = |E_l \otimes H_1^T + H_1^T \otimes E_l|\,|H_1^T \otimes E_k|\,|E_k \otimes H_1^T|\,|E_k \otimes \tilde{H}^T + \tilde{H}^T \otimes E_k| \neq 0$$

by means of the stability of H_1, $\tilde{H}$ and the known properties of the Kronecker product.

This allows us to state the existence of the expansions

$$\begin{aligned} P_{11}(\epsilon) &= P_{11}^0 + \epsilon P_{11}^1 + \cdots \\ P_{21}^T(\epsilon) &= P_{12}(\epsilon) = P_{12}^0 + (\epsilon) P_{12}^1 + \cdots \\ \epsilon P_{22}(\epsilon) &= \tilde{P}_{22}(\epsilon) = \tilde{P}^0 + \epsilon \tilde{P}^1 + \cdots \end{aligned} \tag{2.56}$$

Result (b) of the theorem emerges immediately since

$$u_{\text{opt}} = -R^{-1}[(D_1 + \epsilon D_{11})^T, \epsilon D_{21}^T] \begin{Bmatrix} P_{11}(\epsilon) & P_{12}(\epsilon) \\ P_{21}(\epsilon) & P_{22}(\epsilon) \end{Bmatrix} \begin{Bmatrix} W_L^T \\ V_L^T \end{Bmatrix} x$$

$$= -R^{-1}\{(D_1 + \epsilon D_{11})^T[P_{11}(\epsilon) W_L^T + P_{12}(\epsilon) V_L^T] + + \epsilon D_{21}^T [P_{21}(\epsilon) W_L^T + P_{22}(\epsilon) V_L^T]\} x \tag{2.57}$$

or

$$u_{\text{opt}} = -R^{-1}\{G^T + \epsilon[D_{11}^T(P_{11}(0) W_L^T + P_{12}(0) V_L^T) + D_{21}^T P_{21} W_L^T] + \cdots\} x \tag{2.58}$$

Result (c) is also the direct consequence of (2.56)

$$J\{u_{\text{opt}}\} = \frac{1}{2} x_0^T \{W_L, V_L\} \begin{Bmatrix} P_{11}(\epsilon) & P_{12}(\epsilon) \\ P_{21}(\epsilon) & P_{22}(\epsilon) \end{Bmatrix} \begin{Bmatrix} W_L^T \\ V_L^T \end{Bmatrix} x_0$$

$$= \frac{1}{\epsilon}\frac{1}{2} x_0^T [V_L \tilde{P}^0 V_L^T + \epsilon(W_L P_{11}^0 W_L^T + W_L P_{12}^0 V_L^T + V_L \tilde{P}^1 V_L^T + + V_L P_{21}^0 W_L^T) + \cdots] x_0 \tag{2.59}$$

The results of the second part of the theorem can be proved similarly. In case (2) we obtain

$$\bar{Q}_{11}(0) = 0 \;, \; \bar{Q}_{12}(0) = \bar{Q}_{21}^T(0) = 0$$

and (2.26) has the trivial solution $P_{11}^0 = 0$, which yields (2.44). In case (3) we have $D_0 = 0$ so that (2.26) has a unique solution by the stability of Λ_1, the Jacobian retaining its form with Λ_1 in place of H_1. The final result (2.58) with $D_0 = 0$ gives (2.45). QED.

Let us construct higher-order approximations. To this end it is natural to use (2.57) and the finite segments of the power series (2.56).

Theorem 2.3.

Let the conditions of Theorem 2.2 be fulfilled. Let

$$u_m = -R^{-1} G_m^T x \tag{2.60}$$

where

$$G_m = [W_L P_{11}^{(m)} + V_L P_{21}^{(m)}] W_L^T (B_0 + \epsilon B_1) + \epsilon W_L P_{12}^{(m)} V_L^T B_1 + \\ + V_L \tilde{P}^{(m)} V_L^T B_1 \tag{2.61}$$

and use $P_{11}^{(m)}$, $P_{12}^{(m)}$, $\tilde{P}^{(m)}$, differing from $P_{11}(\epsilon)$, $P_{12}(\epsilon)$, $\tilde{P}(\epsilon)$ by $O(\epsilon^{m+1})$ as a control law in (2.2).

Then there exist ϵ', q such that, over $0 < \epsilon \leq \epsilon'$, the estimate

$$J(u_m) - J(u_{\text{opt}}) \leq q\epsilon^{2m+1} \quad , \quad m = 0,1,\ldots \tag{2.62}$$

is true for the objective value (2.1).

Proof is similar to the proof of Theorem 1.3. However, one needs to take into account that, by Lemma 2.2 the stability rate of the system with the feedback (2.60) is in the first order of ϵ and, hence, that the order of the estimate given by Lemma 1.3 is changed.

Consider further the *dual degenerate case* where the observability condition is not met for $\epsilon = 0$, i.e.,

$$\text{rank } A_y(0) = \text{rank } \{C^T(0) \ , \ A_0^T C^T(0), \ldots, (A_0^T)^{n-1} C^T(0)\} \\ = n - k \quad , \quad k > 0 \tag{2.63}$$

By the transformation described in Lemma 2.1, one can represent (2.2) in the form

$$\dot{\xi}_1 = \Lambda_1 \xi_1 + D_1 u + \epsilon[\Lambda_{11}\xi_1 + \Lambda_{12}\xi_2 + D_{11}u] \tag{2.64}$$

$$\dot{\xi}_2 = \Lambda_2 \xi_2 + D_2 u + \epsilon[\Lambda_{21}\xi_1 + \Lambda_{22}\xi_2 + D_{21}u] \tag{2.65}$$

$$y = \sigma_1 \xi_1 + O(\epsilon) \tag{2.66}$$

Hence, the objective has the form (2.10) but, by (2.65), the conditions

$$\bar{Q}_{22} = 0 \quad , \quad \bar{Q}_{12}(0) = \bar{Q}_{21}(0) = 0 \tag{2.67}$$

are true, i.e., the objective value of the reduced problem is invariant with respect to the behavior of the subsystem (2.65).

If Λ_2 is stable, it is not hard to prove [135] that the control u^0, found as a solution to the aggregated problem

$$\min \ \{\frac{1}{2}\int_0^\infty [\xi_1^T \bar{Q}_{11}\xi_1 + u^T R u] dt \,/\, \dot{\xi}_1 = \Lambda_1\xi_1 + D_1 u \ , \\ \xi_1(0) = \xi_{10} \triangleq W_L^T x_0\} \tag{2.68}$$

is an ϵ-optimal solution to the original problem. Note also that $u^0 = u$ where u is given by (2.25) and (2.26).

The case

$$\Lambda_2 = 0 \tag{2.69}$$

is not so trivial. By using a control u^0 independent of the state variables of the weakly observable subsystem, it is impossible to ensure the asymptotic stability of the system on the whole. On the other hand, it is clear that there exists a weak control, having an "energy" in the order of $O(\epsilon^2)$ and ensuring the stability of the subsystem (2.66) and the system on the whole. Thus, one may suspect that u^1 is a suboptimal control.

The rigorous investigation of this problem may be simplified by means of the duality theorem (Theorem 1.4). In fact, if there is dual degenracy in the original problem, then there is a primal degeneracy in the dual problem. Since the case of primal degeneracy was investigated above, one can use these results effectively to prove the following theorem (the complete proof is contained in [134]).

Theorem 2.4.

Let the pairs $\{A(\epsilon), B(\epsilon)\}$, $\{A^T(\epsilon), C^T(\epsilon)\}$ be completely controllable over $0 < \epsilon \le \epsilon'$ and let the condition (2.63) be fulfilled for ϵ. Then

(1) $J\{u_{\text{opt}}\} = \frac{1}{2} x_0^T P_{\text{opt}}(\epsilon) x_0$ where

$$P_{\text{opt}}(\epsilon) = W_L P_{11}^0 W_L^T + \epsilon P^{(1)} + \cdots \tag{2.70}$$

and where P_{11}^0 is the unique solution to the system (2.26).

(2)

$$\begin{aligned} u_{\text{opt}} &= -K(\epsilon)x \\ K(\epsilon) &= R^{-1}B^T(\epsilon)P_{\text{opt}}(\epsilon) = R^{-1}B_0^T W_L P_{11}^0 W_L^T + \cdots \end{aligned} \tag{2.71}$$

The results obtained above may be used as a key scheme for suboptimal decomposition and aggregation of some classes of practical problems.

5.3. Examples of suboptimal regulator synthesis

5.3.1. Linear model for continuous technological control problems

Let us consider a technological complex, which consist of L aggregates operating continuously and connected with k inventories appropriated for the storage of raw materials, half-finished products, and final goods — one for each kind of resource circulating in the system. The mode of operation of each aggregate depends on its load (consumption of resources being used) and technological controls (supply of fuel, electric power, air, etc.). The product quantity (final or half-finished) is determined by the load as well as the mode of operation. Resources, stored in the inventories, change due to the difference between their consumption during the aggregates' operation and their supply (including external inputs to the complex).

This qualitative description includes a broad class of production system (for more details see [64]). However, for a formalized and quantitative description, the following simplifying assumption will be made:

(1) The nominal state of the complex is characterized by a steadiness of every aggregate operation, a steadiness and a balance between the consumption and supply of every substance, and, as a result, a steadiness of stock levels in the inventories.

(2) The deviation from the nominal state, caused by initial inaccuracies, is small enough to allow the use of linearized models:

$$\dot{s} = y - \sum_j u_j' \ , \ s(0) = s_0 \tag{3.1}$$

$$\epsilon \dot{z}_j = A_j z_j + B_j' u_j' + B_j'' u_j'' \ , \ z_j(0) = z_{j0} \tag{3.2}$$

$$y = \sum_j (G_j z_j + F_j u_j') \tag{3.3}$$

In these relations the vectors $s(t) = (s_j(t))$ denotes, by its components $s_i(t)$, the deviations of all stock levels from the nominal ones. The vector $z_j(t)$ characterizes the deviation of a operational parameters for the j-th aggregate from their nominal values. The vector $y(t) = (y_i(t))$ consists of components $y_i(t)$, which are equal to the deviations of outflow intensity of the i-th resource, summed over all aggregates. The vector u_j' determines the controlled changes of consumption of storaged resources in the j-th aggregate; the vector $u_j''(t)$ determines the changes of technological controls. The matrices $A_j, B_j', B_j'', G_j, F_j$ are determined by the properties of the aggregates and the configuration of the material flows. It is assumed in the sequel that these matrices are constant.

Equation (3.1) describes the changes in stock size, equation (3.2) the dynamics of internal processes in the aggregates; equation (3.3) the relation between the total outflow and the operational parameters as well as the inflows to the aggregates. Using formula (3.3) in (3.1), we obtain a system of equations in the Cauchy form:

$$\begin{aligned} \dot{s} &= \sum_j G_j z_j + \sum_j (F_j - E_j) u_j' \ , \ s(0) = s_0 \\ \epsilon \dot{z}_j &= A_j z_j + B_j' u_j' + B_j'' u_j'' \ , \ z_j(0) = z_{j0} \end{aligned} \tag{3.4}$$

where E_j is the identity matrix of dimension $m_j \times m_j$ where $\dim u_j' = m_j$.

(3) The dynamics of internal processes in the aggregates are much faster than the inventory dynamics (the rate of change of stock levels). This is characterized in (3.4) by the small parameter $\epsilon > 0$ (the analysis carries through for the "slow" time scale).

These assumptions are quite restrictive, but they are natural if initial model deviates are small and the whole system is stable. Note, however, that the latter is not true for a complex that is not controllable. Therefore, the model considered is only reasonable in the case where the control laws assure stability.

Let us make some additional remarks relating to essential properties of the control. First, the control should ensure that deviations from the nominal stock levels are small, since otherwise they may cause an overflow or drain of inventories, which violates the normal operations. This could be avoided by introducing large "damage" reserves, high security levels, and large inventories, but this would entail additional expenditure; therefore, the reducition of stock fluctuations is economical (for more details see also [64]). On the other hand, decreasing the deviations in the operational parameters is also effective, since it allows us to place the nominal state near the constraint boundaries, which are determined by technological requirements (in particular, safety conditions); this is usually more effective ([64], see also Section 2.8). Finally, it should be noted that the control should be directly or indirectly constrained.

The previous considerations motivate, to some extent, the use of a quadratic — with respect to the state and control variables — functional as a performance index:

$$J = \frac{1}{2}\int_0^\infty [s^T Q_0 s + \sum_{i=1}^{L} z_j^T Q_j z_j + \sum_{j=1}^{L} u_j^T R_j u_j] dt \quad , \quad Q_j > 0 \tag{3.5}$$

where $u_j \triangleq (u_j', u_j'')$, $Q_j > 0$, $R_j > 0$, though a functional of this type can only partially meet the technological and economical requirements of the control structure designed for the technological complex; note also that the choice of elements of the matrices Q_j, R_j will be rather arbitrary.

Let us now describe formally how to apply the scheme of suboptimal control structure design presented in the previous section. Note that the system (3.4) has the same form as (2.22) and differs only in notation. It is convenient to show the connections between corresponding symbols

$$\begin{aligned}
&z \triangleq (z_j) \to \xi_i \quad && A \triangleq \mathrm{diag}\,\{A_j\} \to \Lambda_1 \ , \quad G \triangleq \mathrm{diag}\,\{G_j\} \to \Lambda_{21} \\
&s \to \xi_2 && B \triangleq \mathrm{diag}\,\{B_j\} \to D_1 \\
&u \triangleq (u_j) && \Delta F \triangleq \mathrm{diag}\,\{(F_j - E_j, 0\} \to D_{21}
\end{aligned} \tag{3.6}$$

It can be seen that in this case $\Lambda_{22} = 0$ [in the right-hand side of equation (3.4) the "slow" variables s do not appear].

According to the general algorithm, the design of the suboptimal control can be presented as a two-stage procedure. In the first step, the fast part of the process is optimized. We obtain L independent problems of minimization of the functionals

$$\frac{1}{2}\int_0^\infty (z_j Q_j z_j + u_j^T R_j u_j)\mathrm{d}t \tag{3.7}$$

under the constraints

$$\dot{z}_j = A_j z_j + B_j u_j \ , \ z_j(0) = z_{j0}$$

Assuming the controllability of all pairs $\{A_j, B_j\}$ we can design the "fast" control component as follows

$$u_j^f = -K_j^f z_j \tag{3.8}$$

where

$$K_j^f = R_j^{-1} B_j^T P_j^0 \tag{3.9}$$

and P_j^0 are solutions to the independent quadratic equations

$$P_j A_j + A_j^T P_j - P_j B_j R_j^{-1} B_j^T P_j + Q_j = 0 \tag{3.10}$$

In the second step, the aggregated problem of the choice of the "slow" control component is solved, and this accomplishes the coordinating task. This problem has the following form

$$\min \frac{1}{2}\int_0^\infty (s^T \tilde{Q} s + u^T R u)\mathrm{d}t \ , \ R \triangleq \mathrm{diag}\,\{R_j\} \tag{3.11}$$

under the conditions

$$\dot{s} = \tilde{M} s + \tilde{D} u \ , \ s(0) = s_0 \tag{3.12}$$

where matrices $\tilde{M}$, $\tilde{D}$, $\tilde{Q}$ are defined as in equation (2.26) with the proper change of symbols shown in (3.6).

In doing this, we exploit the autonomous character of fast process control and the other specific features of system structure. In fact, since the functional (3.5) is separable with respect to fast and slow variables, then

$$\bar{Q}_{12} = 0$$

and, consequent to (2.33), we must put $L = 0$ in (2.36) and $\tilde{Q} = Q_0$.

Let us recall also that $\Lambda_{22} = 0$. Therefore $\tilde{M} = 0$. The matrix $\tilde{D}$ has a diagonal structure

$$\tilde{D} \triangleq \text{diag}\,\{\tilde{D}_j\} \;\;,\;\; \tilde{D}_j \triangleq \Delta F_j - (G_j - \Delta F_j K_j^b)(A_j - B_j K_j^b)^{-1} B_j \tag{3.13}$$

Thus, the problem of choice of the "slow" control component is reduced to the problem of the optimal stabilization of stock levels

$$\min \{\frac{1}{2}\int_0^\infty (s^T Q_0 s + \sum_j u_j^T R_j u_j)dt \,/\, \dot{s} = \sum_j \tilde{D}_j u_j \;;\; s(0) = s_0\} \tag{3.14}$$

The optimal "slow" control component depends only on the stock level deviations

$$u_j^s = -K_j^s s \tag{3.15}$$

where

$$K_j^s = R_j^{-1} \tilde{D}_j^T \tilde{P}^0$$

and $\tilde{P}^0$ is the solution to the quadratic matrix equation

$$\tilde{P}\tilde{D}R^{-1}\tilde{D}^T\tilde{P} = Q_0 \tag{3.16}$$

Notice that in our case the standard condition of the controllability of $(\tilde{M}, \tilde{D})$ reduces to the requirement: rank $\tilde{D} = k$. Since the symmetric matrices R^{-1}, Q_0 can be written in the form

$$R^{-1} = \rho\rho \;\;,\;\; Q_0 = q_0 q_0 \tag{3.17}$$

then

$$\tilde{P}^0 = q_0(\tilde{D}\rho)^{-1} \tag{3.18}$$

if the right-hand side expression is a symmetric matrix.

Finally, we obtain that

$$u^{\text{opt}} = \{u_j^{\text{opt}}\} \ , \ u_j^{\text{opt}} = u_j^f + u_j^s \tag{3.19}$$

Hence, the procedure of designing the suboptimal regulator for the whole system transforms into the independent synthesis of regulators for the fast internal processes in the aggregates and a fairly simple procedure for the control laws for the slow changes of the stocks. At the same time the structure of the control system is hierarchical: the control of stock levels plays a coordinating part (as indicated in the qualitative approach in [64]). It can easily be seen that the suboptimal coordination problem (3.12) can be constructed without any complicated formulas — it results directly from the hypothesis that the connection between inflows and outflows is quasistatic (which is assured by using the regulators for fast processes), i.e.,

$$y \simeq \sum_j [G(A_j - B_j k_j^f)^{-1} u_j + F_j u_j^1] \tag{3.20}$$

The task of "slow" coordinating control may be formulated, in traditional terms, as an adjustment of set-points of the regulators stabilizing the individual aggregates. It should be emphasized that only the coordinating action requires "global" information; after the coordinating parameters are suplied, each regulator operates separately. This implies that the realization of the described suboptimal control system is quite simple, since it requires only some rational system for the adjustment of the standard regulator set-points. These local regulators exist in all traditional systems and can be used in the proposed control structure.

5.3.2. Ecological system control

The mathematical theory of ecological systems control is not yet well developed. There exist only some probable hypotheses, which are poorly motivated by experimental material. Nevertheless, the social necessity for preserving the environment stimulates the rise of new hypothetic constructions. One of them is described in the sequel. It is based on well-known general ideas, although the formulated control problem is rather specific since some particular attributes are emphasized.

Let $x_i(t)$, $i = 1,\ldots,n$ be the sizes of populations existing in some bounded area at time t, $t \in [0,\infty]$. Let $z(t)$ be a vector, the components of which characterize the state of the environment with regard to the supply of various food products as well as to the level of "pollution" factors that determine the state of the whole macropopulation. Assume that it is possible to influence the system by regulating the food product supply and by cleaning the environment; this effect is represented by the function $v(t)$. Then the dynamic equations for the ecological system may be written as follows:

$$\dot{x}_i = x_i w_i(x, z) \ , \ \dot{z} = F(x, v) \ , \ x \triangleq (x_i) \ , \ i = 1,\ldots,n \tag{3.21}$$

Function $w_i(x, z)$ will be called the accommodation of the i-th micropopulation. Note that in the simplest models of uncompetitive existence the direct influence of $x(t)$ on the accommodation is often neglected. We assume in our hypothesis that the accommodations may be expressed as functions of the sizes of competitive groups. If we divide all the population types into the groups Ω_l, $l = 1,\ldots,L$ and denote

$$x^l = \sum_{i\in\Omega_l} x_i \tag{3.22}$$

then

$$w_i(x, z) = w_i^0(x^l , l = 1,\ldots,L ; z) + \epsilon w_i'(x) \tag{3.23}$$

where $\epsilon > 0$ is a small parameter.

Similarly, we may suppose that the environmental changes are also determined mainly by the sizes of these groups, i.e.,

$$F(x, v) = F^0(x^l , l = 1,\ldots,L ; v) + \epsilon F^1(x) \tag{3.24}$$

Let us assume that for some values $\hat{x}$, $\hat{z}$, $\hat{v}$ equilibrium is attained

$$w_i(\hat{x}, \hat{z}) = 0 \ , \ F(\hat{x}, \hat{v}) = 0 \tag{3.25}$$

and that we are to find a correcting increment $u(t)$ to the control function

$$u(t) = v(t) - \hat{v}$$

such as to stabilize the system in the neighborhood of the equilibrium point.

Let us introduce new variables

$$y_i(t) \triangleq \ln x_i(t) \ , \ y = (y_i , i = 1,\ldots,n)$$

Then

$$\dot{y}_i = \tilde{w}_i(y, z) \ , \ \dot{z} = \tilde{F}(y, v) \tag{3.26}$$

The equilibrium equations (3.25) may be modified in the same way.

The equations (3.26) will now be linearized in the neighborhood of the equilibrium point. If we denote

$$\eta_i = y_i - \hat{y}_i \ , \ \hat{y}_i \triangleq \ln \hat{x}_i \ , \ \varsigma = z - \hat{z}$$

then, regarding (3.22)–(3.24), we obtain

$$\begin{aligned} \dot{\eta}_i &= \sum_l A_{il}^0 \eta^l + \epsilon \sum_j A_{ij}' \eta_j + F_i \varsigma \ , \ i = 1,\ldots,n \\ \dot{\varsigma} &= \sum_l G_l^0 \eta^l + \epsilon \sum_j G_j' \eta_j + Bu \\ \eta^l &= \sum_{i \in \Omega_l} \beta_i \eta_i \ , \ \beta_i \triangleq \hat{x}_i (\prod_{j \in \Omega_l} \hat{x}_j)^{-1} \end{aligned} \tag{3.27}$$

where the matrices A_{il}^0, A_{ij}', F_i, G_l^0, G_j' are composed of coefficients originating from the linearization of functions $\tilde{w}^0$, $\tilde{w}^1$, $\tilde{F}^0$, $\tilde{F}^1$.

We shall now define the performance index. To motivate the use of a standard quadratic functional, we recall the same arguments as in the previous example. However as in the previous case, this form of objective function is not completely adequate. Let us stress that it does not seem rational to use a detailed description of the population microstructure to effect the stabilization by controlling the environment. Therefore, the performance index has the following form

$$J = \frac{1}{2} \int_0^\infty [\sum_l q_l (\eta^l)^2 + \epsilon \eta^T Q^1 \eta + \varsigma^T S \varsigma + u^T R u] dt \tag{3.28}$$

where

$$\eta = (\eta_j) \ , \ q_l > 0 \ , \ S > 0 \ , \ R > 0 \ , \ Q^1 > 0$$

It is easy to see that problem (3.27)–(3.28) relates to the class of weakly controllable problems described in the previous section; the structure of such problems admits an approximate aggregation.

In fact, if we sum equations (3.27) over the groups $i \in \Omega_l$ with weighting coefficients β_i then we obtain

$$\begin{aligned} \dot{\eta}^l &= \sum_l \bar{A}_l^0 \eta^l + F_i \varsigma + \epsilon \sum_j \bar{A}_{ij}^1 \eta_j \\ \dot{\varsigma} &= \sum_l G_l^0 \eta^l + Bu + \epsilon \sum_j G_j^1 \eta_j \end{aligned} \tag{3.29}$$

where the meaning of the new symbols is obvious. The reduced system ($\epsilon = 0$) (3.28) and (3.29) includes only aggregated variables.

As stated in the previous section, the main part of the problem solution constitutes finding the optimal control for the reduced problem:

$$\min \frac{1}{2}\int_0^\infty [\sum_l q_l(\eta^l)^2 + \varsigma^T S\varsigma + u^T Ru]dt \tag{3.30}$$

such that

$$\dot{\eta}^l = \sum_l \bar{A}_l^0 \eta^l + \bar{F}_i \varsigma \ , \ \dot{\varsigma} = \sum_l G_l^0 \eta^l + Bu \tag{3.31}$$

The reduced solution cannot ensure the stabilization of the "detailed" macropopulation structure. Though a common number of individuals in each group l, $l = 1,\ldots,L$ converges asymptotically to the stable level

$$\hat{x}^l = \sum_{i\in\Omega_l} \hat{x}_i$$

the partition of population within the group is nevertheless determined by the initial conditions and small perturbations. The methods developed in the previous section permit the design of a correcting control, which ensures internal stabilization also.

5.4. Control of oscillatory systems

The degenerate case corresponding to the proximity of a system to the set of unstabilizable systems was considered in Section 5.2. The situation in which the weakly controllable subsystem has a zero eigenvalue (for $\epsilon = 0$) was investigated in detail and its practical importance established. The other critical situation that arises, where the eigenvalues are purely imaginary, is also of importance since it characterizes the control of oscillatory systems.

Let us begin with a description of the control object as a linear mechanical oscillatory system that is nearly conservative.

Let $q = (q_l \ , l = 1,\ldots,m)$ be the matrix columns of generalized displacements from the origin as a stable state of equilibrium. The dynamic equations can be written in the form

$$\ddot{q} + Lq = \epsilon F \tag{4.1}$$

where L is a symmetric ($n \times n$) matrix having n positive semi-simple eigenvalues and where ϵF is a matrix column of nonconservative generalized forces, with $\epsilon > 0$

a scalar parameter characterizing their smallness with respect to the conservative ones.

Let us suppose that

$$F = \Gamma \dot{q} + Gu \quad , \quad \Gamma, G = \text{constant} \tag{4.2}$$

where u is the vector of control forces.

The hypothesis of the smallness of control influence can be motivated by one of the following arguments:

(1) If the elastic system is sufficiently rigid, then the application of external forces results in small static displacements.
(2) The smallness is conditioned by a limitation on energy losses in the control.
(3) The oscillatory character may disappear if one allows large control forces.

One can rewrite (4.1) and (4.2) in Cauchy form, letting

$$x = (x_k \quad , \quad k = 1,\ldots,2n)$$

$$x_{2l-1} = q_l \quad ; \quad x_{2l} = \dot{q}_l \quad , \quad l = 1,\ldots,n$$

Then

$$\dot{x} = (A_0 + \epsilon A_1)x + \epsilon D u \tag{4.3}$$

where the matrices A_0, A_1, D are evidently determined by L, Γ, G.

Note that A_0 has $2n$ semi-simple purely imaginary eigenvalues $\pm i\omega_l$, $l = 1,\ldots,n$, where ω_l are the fundamental frequencies of the reduced conservative system.

If q_l are normal coordinates, then A_0 has a block-diagonal structure

$$A_0 = \text{diag}\,\{A_0^l\} = \text{diag}\left\{\begin{Vmatrix} 0 & 1 \\ \omega_l^2 & 0 \end{Vmatrix}\right\} \tag{4.4}$$

By definition, the matrices A_1, D have zero odd rows. Let the performance index be written in the form

$$J\{u\} = \frac{1}{2}\int_0^\infty [x^T Q x + u^T R u]\mathrm{d}t \quad , \quad Q,R > 0 \tag{4.5}$$

If one consider $v \triangleq \epsilon u$ as a control, then

$$J\{u\} \triangleq \bar{J}\{v\} = \frac{1}{2}\int_0^\infty [x^T Qx + \frac{1}{\epsilon^2} v^T Rv] dt \tag{4.6}$$

The smallness of ϵ corresponds to the greater importance of energetic losses in the control as noted above.

The optimal control is equal to

$$u_{\text{opt}} = -K_{\text{opt}}(\epsilon)x \tag{4.7}$$

where

$$K_{\text{opt}}(\epsilon) = \epsilon R^{-1} D^t P_{\text{opt}}(\epsilon) \tag{4.8}$$

$P_{\text{opt}}(\epsilon)$ being the positive definite solution to the equation

$$(A_0 + \epsilon A_1)^T P + P(A_0 + \epsilon A_1) - \epsilon^2 PDR^{-1}D^T P + Q = 0 \tag{4.9}$$

It is supposed here that the controllability condition is met for $\epsilon > 0$, i.e.,

$$\text{rank } \|\epsilon D, \epsilon(A_0 + \epsilon A_1)D, \ldots, \epsilon(A_0 + \epsilon A_1)^{2n-1}D\| = 2n \tag{4.10}$$

Evidently, this is not true for $\epsilon = 0$.

Let us look for $P_{\text{opt}}(\epsilon)$ for $\epsilon > 0$ in the form

$$P_{\text{opt}}(\epsilon) = \frac{1}{\epsilon}[P^{(0)} + \epsilon P^{(1)} + \cdots + \epsilon^i P^{(i)} + \cdots] \tag{4.11}$$

Substituting (4.11) in (4.9) and equating coefficients in powers of ϵ, one can obtain

$$\begin{aligned} A_0^T P^{(0)} + P^{(0)} A_0 &= 0 \\ A_0^T P^{(1)} + P^{(1)} A_0 &= -[A_1^T P^{(0)} + P^{(0)} A_1] + P^{(0)} DR^{-1} D^T P^{(0)} - Q \\ &\vdots \\ A_0^T P^{(i)} + P^{(i)} A_0 &= F^{(i)}[P^{(0)}, \ldots, P^{(i-1)}] \\ &\vdots \end{aligned} \tag{4.12}$$

The relations (4.12) are an infinite system of linear matrix equations that determine the coefficients $P^{(i)}$ recursively. However, it is not hard to see that these equations are degenerate, which determines the specific character of their solution. To begin with, let us consider the simplest example of a system with one degree of freedom. For that case

$$\begin{aligned} A_0 &= \begin{Vmatrix} 0 & 1 \\ -\omega^2 & 0 \end{Vmatrix} \;;\; D = \begin{Vmatrix} 0 \\ g \end{Vmatrix} \;;\; A_1 = \begin{Vmatrix} 0 & 0 \\ 0 & \gamma \end{Vmatrix} \;; \\ Q &= \begin{Vmatrix} q_{11} & q_{12} \\ q_{12} & q_{22} \end{Vmatrix} \;;\; R = 1 \end{aligned} \tag{4.13}$$

and the equations (4.12) have the form

$$\begin{Vmatrix} 0 & -\omega^2 \\ 1 & 0 \end{Vmatrix} \begin{Vmatrix} p_{11}^{(i)} & p_{12}^{(i)} \\ p_{12}^{(i)} & p_{22}^{(i)} \end{Vmatrix} + \begin{Vmatrix} p_{11}^{(i)} & p_{12}^{(i)} \\ p_{12}^{(i)} & p_{22}^{(i)} \end{Vmatrix} \begin{Vmatrix} 0 & 1 \\ -\omega^2 & 0 \end{Vmatrix} = F^{(i)}(P^{(0)},\ldots,P^{(i-1)})$$

where $F^{(0)} = \|0\|$.

This is equivalent to the scalar equations

$$-2\omega^2 p_{12}^{(i)} = f_{11}^{(i)} \;;\; -\omega^2 p_{22}^{(i)} + p_{11}^{(i)} = f_{12}^{(i)} \;;\; 2p_{12}^{(i)} = f_{22}^{(i)} \;,\; i = 0,1,\ldots$$

from which it follows that

$$p_{12}^{(0)} = 0 \;;\; p_{11}^{(0)} = \omega^2 p_{22}^{(0)} \triangleq c_0\omega^2 \tag{4.14}$$

Thus,

$$p^{(0)} = c_0 \begin{Vmatrix} \omega^2 & 0 \\ 0 & 1 \end{Vmatrix} \tag{4.15}$$

where the constant c_0 can be determined only from an existence condition for a solution to the equation for the first-order approximation.

This condition is

$$f_{11}^{(1)} = -\omega^2 f_{22}^{(1)}$$

Taking into account the structure of $F^{(1)}$, one can obtain

$$q_{11} = \omega^2[-2\gamma c_0 + c_0^2 g - q_{22}]$$

The positive solution to this quadratic equation

$$c_0 = \frac{\gamma}{g} + \sqrt{(\frac{\gamma}{g})^2 + \frac{1}{g\omega^2}(q_{11} + q_{22}\omega^2)}$$

determines $P^{(0)}$ and

$$K_{\text{opt}}(0) = \|0g\|\, c_0 \begin{Vmatrix} \omega^2 & 0 \\ 0 & 1 \end{Vmatrix}$$

Hence, the zero-order approximation to the optimal control law has the form

$$u^0 = -gc_0x_2 = -gc_0\dot{q}$$

i.e., the suboptimal control uses only velocity measurements so that the main part of the control energy must be spent on damping, a dissipation of the energy of the oscillations.

Let us find the solution to (4.9) for $n > 1$. Let A_0 be written in the form (4.4), i.e., let the normal coordinates be used. Then the equations (4.12) may be written as follows

$$\text{diag}\,\{A_l^T\}\{P_{lk}^{(i)}\} + \{P_{lk}^{(i)}\}\,\text{diag}\,\{A_l\} = \{F_{lk}^{(i)}\} \tag{4.16}$$

where $P_{lk}^{(i)}$ are (2×2)-blocks of $P^{(i)}$ and $F_{lk}^{(i)}$ are (2×2)-blocks of $F^{(i)}$, with $F^{(0)}$ equal to zero. By block-multiplying, one can obtain

$$A_l^TP_{ll}^{(i)} + P_{ll}^{(i)}A_l = F_{ll}^{(i)} \quad , \; l = 1,\ldots,n \tag{4.17}$$

$$A_l^TP_{lk}^{(i)} + P_{lk}^{(i)}A_k = F_{lk}^{(i)} \quad , \; l,k = 1,\ldots,n \;\; ; \; l \neq k \tag{4.18}$$

The equations (4.17) can be solved by the scheme given above so that

$$P_{ll}^i = c_l^i \begin{Vmatrix} \omega_l^2 & 0 \\ 0 & 1 \end{Vmatrix} + \begin{Vmatrix} f_{2l-1,2l}^{(i)} & ; & \frac{1}{2}f_{2l,2l}^{(i)} \\ \frac{1}{2}f_{2l,2l}^{(i)} & ; & 0 \end{Vmatrix}$$

If all fundamental frequencies are distinct, then the equations (4.18) have the unique solution $P_{lk}^{(i)}$, $l \neq k$, writing these down explicitly in the elements of $F_{lk}^{(i)}$

$$P^{(i)} = \sum_{l=1}^{n} c_l^i\,\text{diag}\left\{\delta_{lk}\begin{Vmatrix} \omega_k^2 & 0 \\ 0 & 1 \end{Vmatrix}\right\} + \prod F^{(i)} \tag{4.19}$$

where δ_{lk} is the Kronecker delta and $\prod$ is the linear operator over $F^{(i)}$ operating in accordance with the described procedure.

The constants c_l^i are determined by the conditions

$$f_{2l-1,2l-1}^{(i+1)} + \omega_l^2 f_{2l,2l}^{(i+1)} = 0 \tag{4.20}$$

which can be written in the form

$$e_{2l-1}^T F^{(i+1)} e_{2l-1} + \omega_l^2 e_{2l}^T F^{(i+1)} e_{2l} = 0$$

or, by the eigenvectors of A_0,

$$\bar{u}_l^T F^{(i+1)} u_l = 0 \quad , \quad i = 0,1,\dots \quad , \quad u_l \triangleq e_{2l-1} + i\omega_l e_{2l}$$

where

$$F^{(1)} = -[A_1^T P^{(0)} + P^{(0)} A_1] + P^{(0)} S P^{(0)} - Q \quad , \quad S \triangleq DR^{-1}D^T \tag{4.21}$$

and

$$F^{(i+1)} = -[A_1^T P^{(i)} + P^{(i)} A_1] + \sum_{q=0}^{i} P^{(q)} S P^{(i-q)} \tag{4.22}$$

Equations (4.20) and (4.21) yield the quadratic equations for

$$s_{2l,2l}(c_l^0)^2 - 2\gamma_{ll} c_l^0 - \frac{1}{\omega_l^2} \alpha_l^0 = 0$$

where

$$\alpha_l^0 = q_{2l-1,2l-1} + \omega_l^2 q_{2l,2l}$$

and $s_{2l,2l}$ are the corresponding elements of the matrix $s \triangleq DR^{-1}D^T$. Thus, we obtain a natural generalization of the simplest result

$$c_l^0 = \frac{\gamma_{l,l}}{s_{2l,2l}} + \sqrt{\left(\frac{\gamma_{l,l}}{s_{2l,2l}}\right)^2 + \frac{1}{s_{2l,2l}\omega_l^2}(q_{2l-1,2l-1} + q_{2l,2l}\omega_l^2)} \tag{4.23}$$

Since the matrix $P^{(0)}$ is diagonal

$$P^{(0)} = \sum_{l=1}^{n} c_l^0 \,\mathrm{diag}\left\{\delta_{lk}\begin{Vmatrix}\omega_k^2 & 0\\ 0 & 1\end{Vmatrix}\right\}$$

then $K_{\mathrm{opt}}(0) = R^{-1}D^TP^{(0)}$ has zero odd columns. Hence, the suboptimal control is based on the even components of the vector, i.e., on the generalized velocities:

$$u = -R^{-1}G^Tc^0\dot{q} \tag{4.24}$$

where $c^0 = \mathrm{diag}\,\{c_l^0\}$.

The system in which this suboptimal control is used is described by the equations

$$\ddot{q} + \epsilon[GR^{-1}G^Tc^0 - \Gamma]\dot{q} + Lq = 0 \tag{4.25}$$

To find the coefficients $P^{(i)}$, $i = 1,2,\ldots$, which determine the higher approximations, it is sufficient to solve linear equations provided by (4.20) and (4.22) so that

$$\bar{u}_l[A_1^TP^{(i)} + P^{(i)}A_1]u_l = 2\gamma_{ll}\omega_l^2c_l^i = \bar{u}_l^T\sum_{q=0}^{i} P^qsP^{i-q}u_l \tag{4.26}$$

Thus, the application of the perturbation method gives an elementary procedure for the construction of the control law. However, this procedure is formal, and the efficacy of such a law is not clear. A justification of the method is given by the following theorem.

Theorem 4.1.

Let the eigenvalues of the matrix A_0 be purely imaginary and simple. Let the pair $\{A_0, D\}$ be completely controllable. Then there exists ϵ' such that over $0 < \epsilon \leq \epsilon'$

(1) The matrix function $\epsilon P(\epsilon)$ is analytic

$$\epsilon P(\epsilon) = P^{(0)} + \epsilon P^{(1)} + \cdots + \epsilon^iP^{(i)} + \cdots \triangleq P_i(\epsilon) + O(\epsilon^{i+1}) \tag{4.27}$$

(2) The control

$$u = u^{(i)} \triangleq -R^{-1}D^T P_i(\epsilon)x \tag{4.28}$$

ensures the objective value J_i differing from the optimal value J^* by $O(\epsilon^{2i+1})$.

The proof of the theorem is given in [143]. We note only that there is a close connection between the algorithm under consideration and the theory of critical cases for singularly perturbed differential equations [142].

Example. Let $\Gamma = 0$, let u be scalar and let the performance index be given in the form

$$\int_0^\infty (\sum_l \rho_l^2 q_l^2 + u^2)dt \tag{4.29}$$

then

$$c_l^0 = \frac{\rho}{\omega_l |g_l|}$$

and

$$u_0 = -\sum_l \frac{\rho_l}{\omega_l}\dot{q}_l \tag{4.30}$$

so that the equation of the closed system can be written as follows

$$\ddot{q}_l + \epsilon \sum_m \frac{\rho_m}{\omega_m}\dot{q}_m + \omega_l^2 q_l = 0$$

Note that the system

$$\ddot{q}_l + \epsilon \frac{\rho_l}{\omega_l}\dot{q}_l + \omega_l^2 q_l = 0$$

with the feedback introduced by the rate of change of each independent mode has the same eigenvalues to within $o(\epsilon)$ [5]. The system is robust and stable for any $\epsilon > 0$. If the coefficients ρ_l characterizing the importance of the displacements along the different coordinate axes coincide, then the suboptimal gains are inversely proportional to the fundamental frequencies. This implies the natural and simple practical rule.

Let us consider further the general control problem for systems containing both a weakly controllable oscillatory part and a strongly controllable one. Recall that the system (3.8) can be interpreted in this way if the matrix Λ_2 has only pure imaginary eigenvalues.

The basic matrix quadratic equation corresponding to (3.8) has the form (3.12). Let us seek a solution to (3.12) letting

$$P_{22}(\epsilon) = \frac{1}{\epsilon}\tilde{P}_{22}(\epsilon) \tag{4.31}$$

and supposing that P_{11}, P_{12}, P_{22} are analytic functions of ϵ.

If this is true, then a zero-order approximation to the optimal control is given by the formula

$$u_0(x) = -K_0 x = -R^{-1}D_1^T P_{11}^0 \xi_1 - R^{-1}(D_1^T P_{12}^0 + D_2^T \tilde{P}_{22}^0)\xi_2 \tag{4.32}$$

where P_{11}^0, P_{12}^0, $\tilde{P}_{22}^0$ are the main terms of the ϵ-power expansions of the corresponding functions.

Substituting the expansions in (3.12) and equating coefficients in ϵ, one can obtain matrix equations determining the coefficients of those expansions. Write down the first two equations

$$P_{11}^0\Lambda_1 + \Lambda_1^T P_{11}^0 - P_{11}^0 s_{11} P_{11}^0 + \bar{Q}_{11}(0) = 0 \tag{4.33a}$$

$$P_{12}^0\Lambda_2 + H_1^T P_{12}^0 - P_{11}^0 s_{12}\tilde{P}_{22}^0 + \bar{Q}_{12}(0) = 0 \tag{4.33b}$$

$$\tilde{P}_{22}^0\Lambda_2 + \Lambda_2^T\tilde{P}_{22}^0 = 0 \tag{4.33c}$$

$$P_{11}^1 H_1 + H_1^T P_{11}^1 = \varphi_{11}(P_{11}^0, P_{12}^0, \tilde{P}_{22}^0) - \bar{Q}_{11}'(0) \tag{4.34a}$$

$$P_{12}^1\Lambda_1 + H_1^T P_{12}^1 = P_{11}^1 s_{11} P_{12}^0 + \varphi_{12}(P_{11}^0, P_{12}^0, \tilde{P}_{22}^0) - \bar{Q}_{12}'(0) \tag{4.34b}$$

$$\begin{aligned} P_{22}^1\Lambda_2 + \Lambda_2^T P_{22}^1 = {} & -\tilde{P}_{22}^0\Lambda_{22} - \Lambda_{22}^T\tilde{P}_{22}^0 + \tilde{P}_{22}^0 s_{22}\tilde{P}_{22}^0 + P_{21}^0 s_{11} P_{12}^0 - \\ & - \bar{Q}_{22}(0) \end{aligned} \tag{4.34c}$$

where

$$H_1 = \Lambda_1 - s_{11}P_{11}^0 \;;\; s_{11} = D_1 R^{-1} D_1^T$$

$$s_{12} = D_1 R^{-1} D_{21}^T \;;\; s_{22} = D_2 R^{-1} D_{21}^T$$

Note that, as in other degenerate situations, the matrix P_{11}^0 determining a zero-order approximation to the control of the strongly controlled subsystem can be found independently by equation (4.33a).

The matrix P_{12}^0 can be determined as a solution to the linear equation (4.33b), since H_1 is stable. However, the right-hand side in (4.33b) is a linear function of $\tilde{P}_{22}^0$ and, hence,

$$P_{12}^0 = \tilde{\prod} \tilde{P}_{22}^0 \tag{4.35}$$

where $\tilde{\prod}$ is a linear operator.

In turn, $\tilde{P}_{22}^0$ cannot be determined by the zero-order approximation equation (4.33c). As we have seen above, under the isolated consideration of the weakly controllable oscillatory system, $\tilde{P}_{22}^0$ has a form

$$\tilde{P}_{22}^0 = \sum_{l=1}^{r} c_l^0 \operatorname{diag}\left\{\delta_{lk} \begin{Vmatrix} \omega_k^2 & 0 \\ 0 & 1 \end{Vmatrix}\right\}, \quad r = \frac{1}{2}(n-k) \tag{4.36}$$

where ω_k are the fundamental frequencies of the reduced oscillatory subsystem, and where c_l^0 are constants that can be found if one uses the existence condition for the following order of approximation. Substituting (4.35), we obtain

$$\tilde{P}_{22}^1 \Lambda_2 + \Lambda_2^T \tilde{P}_{22}^1 = \tilde{F}^1 \tag{4.37}$$

where $\tilde{F}^1$ is a quadratic matrix-function of c_l^0, $l = 1,\ldots,n$. As in (4.20) and (4.21) it is not hard to show that the conditions

$$\bar{u}_l^T \tilde{F}^1 u_l = 0 \tag{4.38}$$

where u_l is an eigenvector of Λ_2 corresponding to the eigenvalue $i\omega_l$ give r scalar quadratic equations allowing one to find c_l^0 independently.

It turns out that this algebraic procedure can be described by the following algorithm. To construct a suboptimal control law

$$u_0 = -K_1^0 \xi_1 - K_2^0 \xi_2 \tag{4.39}$$

approximately minimizing the performance index

$$J\{u\} = \frac{1}{2}\int_0^\infty [\xi_1^T Q_{11} \xi_1 + 2\xi_1^T Q_{12} \xi_2 + \xi_2^T Q_{22} \xi_{22} + u^T R u] dt \tag{4.40}$$

such that

$$\begin{aligned} \dot{\xi}_1 &= \Lambda_1 \xi_1 + D_1 u + \epsilon(\Lambda_{11}\xi_1 + \Lambda_{12}\xi_2 + D_{11}u) \\ \dot{\xi}_2 &= \Lambda_2 \xi_2 \qquad\qquad + \epsilon(\Lambda_{21}\xi_1 + \Lambda_{22}\xi_2 + D_{21}u) \end{aligned} \tag{4.41}$$

one needs to

(1) Find K_1^0 by the formula

$$K_1^0 = R^{-1} D_1^T P_{11}^0 \tag{4.42}$$

where P_{11}^0 is the positive definite solution to the equation

$$P_{11}\Lambda_1 + \Lambda_1^T P_{11} - P_{11} D_1 R^{-1} D_1^T P_{11} + Q_{11} = 0 \tag{4.43}$$

(2) Calculate the eigenvalues $\pm i\omega_l$, $l = 1,\dots,r$; $r = 1/2(n-k)$ and corresponding left and right eigenvectors of the matrix, i.e., the solutions to the system†

$$\begin{aligned} \Lambda_2^T v_l &= i\omega_l v_l \ ; \ \Lambda_2^T \bar{v}_l = -i\omega_l \bar{v}_l \\ \Lambda_2 y_l &= i\omega_l y_l \ ; \ \Lambda_2 \bar{y}_l = -i\omega_l \bar{y}_l \end{aligned} \tag{4.44}$$

subject to the normalizing conditions

$$\bar{v}_l^T v_l = 1 \ ; \ \bar{v}_l^T \bar{y}_l = 1 \ ; \ v_l^T y_l = 1 \tag{4.45}$$

(3) Calculate the values φ_l, $\bar{\varphi}_l$ of the first subsystem frequency response for the fundamental frequencies of the second subsystem

$$\varphi_l = (\Lambda_1 - D_1 K_1^0 - i\omega_l E_k)^{-1} \ ; \ \bar{\varphi}_l = (\Lambda_1 - D_1 K_1^0 - i\omega_l E_k)^{-1} \tag{4.46}$$

(4) Transform the properties of the oscillatory subsystem, taking into account all weak connections with the first subsystem. To this end, calculate the matrices

$$\tilde{\Lambda}_{22} = \Lambda_{22} - D_{21} L \ ; \ \tilde{Q}_{22} = Q_{22} - L^T R L \ ; \ \tilde{D}_{22} = D_{21} + G D_1 \tag{4.47}$$

† Here and later, $\bar{x}$ is the vector conjugate to x

where

$$L = R^{-1}D_1^T\tilde{Q}_{12}$$

$$\tilde{Q}_{12} = -\sum_{l=1}^{r}\bar{\varphi}_l^TQ_{12}y_lv_l^T - \sum_{l=1}^{r}\varphi_l^TQ_{12}\bar{y}_l\bar{v}_l^T$$

$$G = -\sum_{l=1}^{r}y_lv_l^T(\Lambda_{12} - D_{21}K_1^0)\varphi_l - \sum_{l=1}^{K}\bar{y}_l\bar{v}_l^T(\Lambda_{12} - D_{21}k_1^0)\bar{\varphi}_l$$

(5) Find a suboptimal control for the transformed weakly controlled subsystem, i.e., find

$$\tilde{u}_2 = -\tilde{K}_2\xi_2 \tag{4.48}$$

approximately minimizing the transformed performance index

$$\frac{1}{2}\int_0^\infty[\xi_2^T\tilde{Q}_{22}\xi_2 + u^TRu]\mathrm{d}t \tag{4.49}$$

such that

$$\dot{\xi}_2 = \tilde{\Lambda}_{22}\xi_2 + \epsilon\tilde{D}_{21}u \ ; \ \xi_2(0) = \xi_{21} \tag{4.50}$$

This can be realized by means of the following explicit formulae

$$\tilde{K}_2 = R^{-1}\tilde{D}_{21}^T\tilde{P}_{22}^0 \tag{4.51}$$

$$\tilde{P}_{22}^0 = \sum_{l=1}^{r}c_l^0(v_l\bar{v}_l^T + \bar{v}_lv_l^T) \tag{4.52}$$

where c_l^0 are the positive solutions to the scalar quadratic equations

$$\begin{gathered} c_l^2 - a_lc_l - b_l = 0 \\ a_l \triangleq \frac{v_l^T\tilde{\Lambda}_{22}y_l + \bar{V}_l^T\tilde{\Lambda}_{22}\bar{y}_l}{\bar{v}_l^T\tilde{D}_{21}R^{-1}\tilde{D}_{21}v_l} \ ; \ b_l \triangleq \frac{\bar{y}_l^T\tilde{Q}_{22}y_l}{\bar{v}_l^T\tilde{D}_{21}R^{-1}\tilde{D}_{21}v_l} \end{gathered} \tag{4.53}$$

(6) Find K_2^0 as follows

$$K_2^0 = \tilde{K}_2^0 + R^{-1}D_1^T\tilde{Q}_{12}^T \tag{4.54}$$

Theorem 4.2.

Let the eigenvalues of Λ_2 be simple and purely imaginary. Let the conditions

$$v_l^T \tilde{D}_{22} \neq 0 \quad , \quad l = 1,\ldots,r \tag{4.55}$$

be fulfilled. Then for any small $\epsilon > 0$

(a) The system (4.41) is completely controllable.

(b) The optimal feedback gain $K(\epsilon)$ may be written in the form

$$K(\epsilon) = K^0 + \epsilon K^1 + \cdots + \epsilon^m K^m + O(\epsilon^{m+1}) = K_m(\epsilon) + O(\epsilon^{m+1}) \tag{4.56}$$

where $K^0 = (K_1^0 \,, K_2^0)$ is calculated by the algorithm described above.

(c) $$\frac{J\{u_0\} - J\{u_{\text{opt}}\}}{J\{u_0\}} = O(\epsilon^2) \tag{4.57}$$

(d) Using the gain $K_m(\epsilon)$ given by the expansion (4.56), one can obtain a suboptimal control u_m such that

$$J\{u_m\} - J\{u_{\text{opt}}\} = O(\epsilon^{2m+1})$$

In particular,

$$\begin{aligned} J\{u_{\text{opt}}\} &= J_0 + O(\epsilon) \\ &= \frac{1}{2\epsilon}\xi_{20}^T \tilde{P}_{22}^0 \xi_{20} + \frac{1}{2}\xi_{10}^T P_{11}^0 \xi_{10} + \xi_{10}^T P_{12}^0 \xi_{20} + O(\epsilon) \end{aligned} \tag{4.58}$$

where

$$P_{12}^0 = \tilde{Q}_{12} + G^T \tilde{P}_{22}^0$$

We cannot give a complete proof of the theorem here, but the physical meaning of the result will be clear from the preceding considerations. There is a close connection between this result and the theory of averaging for differential equations with fast oscillating coefficients. In fact, by the transformation,

$$\xi_2 = e^{\Lambda_2 t}\tilde{\xi}_2$$

one can see that $\tilde{\xi}_2$ is a slow-changing vector and, after a time interval $O(\ln 1/\epsilon)$, the relation

$$\xi_1 = \delta \xi_2$$

is established wherein δ is a constant matrix determined by the frequency response of the controlled subsystem around the fundamental frequencies of the oscillatory subsystem. By using this relation one can pass to the optimal control problem for objects described by differential equations with oscillatory coefficients in "standard" form [129]. This approach utilizes some known results [130] but we have preferred to present a more direct algebraic scheme here.

Note that the equations (4.41) can be interpreted as a mathematical model of a real system including a controllable basic part that is weakly connected with weakly controllable and flexible dynamic subsystems. The suboptimal control is constructed for the basic object and then for each oscillatory mode independently but accounting for the feedback of the basic part.

A class of similar but significantly different problems, which can be interpreted as that of linear regulator design for a very inflexible or "quasi-rigid" body, is, perhaps, more important for practical purposes. It can be shown that the corresponding model is generally representable in the form

$$\begin{aligned} \dot{\xi}_1 &= \sqrt{\epsilon}\,[\Lambda_1 \xi_1 + D_1 u + O(\epsilon)] \ , \\ \dot{\xi}_2 &= \Lambda_2 \xi_2 + \epsilon D_2 u + O(\epsilon) \end{aligned} \tag{4.59}$$

with Λ_2 a matrix with pure imaginary eigenvalues. The difference from (4.41) is expressed by the multiplier $\sqrt{\epsilon}$. The variables ξ_1 describe a "rigid" motion, and the variables ξ_2 describe the mutual oscillations of components. It is not hard to show that a suboptimal control for such a system can be found more simply than that for (4.41). It is sufficient here to solve the control problems for each subsystem independently, and this can be done separately for each mode of the oscillatory subsystem.

We shall outline the simplest example of such a system, permitting us to explain the structure of the equations (4.59) and, moreover, to give a practical estimate of the computational efficiency of the perturbation method.

Example. Consider a simple mechanical model in which two ideal pendulae of equal length are connected by a spring. Control forces F_1, F_2 are applied to each oscillating mass center. Let x_1, x_2 be the displacements from stable equilibrium. Then, using standard notations, one can write the equations

$$\begin{aligned} \frac{d^2 x_1}{d\tau^2} + \frac{g}{l} x_1 + \frac{c}{m_1}(x_1 - x_2) &= \frac{1}{m_1} F_1 \\ \frac{d^2 x_2}{d\tau^2} + \frac{g}{l} x_2 + \frac{c}{m_2}(x_2 - x_1) &= \frac{1}{m} F_2 \end{aligned}$$

or, introducing the normal coordinates,

$$\frac{d^2 q_1}{d\tau^2} + \omega_1^2 q_1 = \frac{1}{m_1 + m_2}(F_1 + F_2)$$
$$\frac{d^2 q_2}{d\tau^2} + \omega_2^2 q_2 = \frac{1}{m_1}F_1 - \frac{1}{m_2}F_2$$

where

$$q_1 = \frac{m_1 x_1 + m_2 x_2}{m_1 + m_2} \;;\; q_2 = x_1 - x_2$$
$$\omega_1^2 = \frac{g}{l} \;;\; \omega_2^2 = \frac{g}{l} + c(\frac{1}{m_1} + \frac{1}{m_2})$$

The first equation describes the motion of the mass center and the second characterizes the mutual oscillations.

Let the performance index be of the form

$$J\{u\} = \int_0^\infty [x_1^2 + x_2^2 + r(F_1^2 + F_2^2)]dt$$

It is clear from a physical point of view that if the rigidity of the spring is large, i.e., if ω_2^2 is large, then a suboptimal control can be found from a simplified model in which the two material points are replaced by one of their combined mass. It is, of course, necessary to add some damping to stabilize high-frequency oscillations. This heuristic approach, as a method to obtain a suboptimal control, coincides with the scheme described above.

One can write the equations in the standard form (4.59) by letting

$$\omega_2^2 = \frac{1}{\epsilon}\omega_1^2 \;;\; t = \frac{\omega_1}{\sqrt{\epsilon}}\tau \;;\; \xi_1 = (\xi_{11}, \xi_{12}) \;;\; \xi_2 = (\xi_{21}, \xi_{22})$$
$$\xi_{11} = q_1 \;;\; \xi_{12} = \frac{1}{\sqrt{\epsilon}}\frac{dq_1}{dt} \;;\; \xi_{21} = q_2 \;;\; \xi_{22} = \frac{dq_2}{dt}$$
$$u_1 = \frac{1}{m_1\omega_1^2}F_1 \;;\; u_2 = \frac{1}{m_2\omega_1^2}F_2 \;;\; \mu = \frac{m_2}{m_1}$$

Then

$$\frac{d\xi_1}{dt} = \sqrt{\epsilon}[\Lambda_1\xi_1 + D_1 u]$$

$$\frac{d\xi_2}{dt} = \Lambda_2\xi_2 + \epsilon D_2 u$$

where

$$\Lambda_1 = \Lambda_2 = \begin{Vmatrix} 0 & 1 \\ -1 & 0 \end{Vmatrix} \; ; \; D_1 = \begin{Vmatrix} 0 & 0 \\ (1+\mu)^{-1} & \mu(1+\mu)^{-1} \end{Vmatrix} \; ; \; D_1 = \begin{Vmatrix} 0 & 0 \\ -1 & -1 \end{Vmatrix}$$

Analysis shows that the decomposition scheme gives an objective value equal to the optimal one to within 5% if the ratio of fundamental frequencies is more than 1.3.

It is of interest to note that a standard code for the solution to the basic matrix quadratic equation by the Newton–Raphson method is ineffectual if the ratio is more than 5 since ill-conditioned matrices arise at each iteration.

On the other hand, if the frequencies are close, i.e., if the rigidity of the spring is small, one can look for a suboptimal control law by considering each pendulum independently, as follows from the zero-order approximation of the theory of weakly coupled systems (see Section 5.1). It was found, for our example, that the two simple approaches shown above give an admissible precision as spring rigidity varies from 0 to ∞! One may suppose that the perturbation method will give good results for many real problems, even if the parameter value is not small.

5.5. LQG problems

Let us recall some of linear regulator design theory, taking into account random perturbations in the model and random state measurement noise. The optimization problem can be formulated in the following standard form.

Find a control law in the form

$$u(t) = F\{y(\tau), \tau \leq t\} \tag{5.1}$$

where F is a linear operator minimizing the performance index (here and later in this section, E denotes the operator of mathematical expectation)

$$J\{u\} = \lim_{\substack{t_1\to\infty \\ t_0\to-\infty}} \frac{1}{t_1 - t_0} E\int_{t_0}^{t_1} (x^T Q x + u^T R u) dt \tag{5.2}$$

such that

$$y(t) = Cx + \eta(t) \tag{5.3}$$

$$\dot{x} = Ax + Bu + \varsigma(t) \tag{5.4}$$

If the random processes $\eta(t)$, $\varsigma(t)$ are Gaussian "white" noise, then (5.1)–(5.4) is called a stationary linear-quadratic-Gaussian (LQG) optimal control problem.

Let

$$\begin{aligned} E\{\eta(t)\eta^T(t+\tau)\} &= Y\delta(\tau) \ ; \\ E\{\varsigma(t)\varsigma^T(t+\tau)\} &= Z\delta(\tau) \ ; \\ E\{\eta(t)\varsigma^T(t+\tau)\} &= 0 \end{aligned} \tag{5.5}$$

where the constant matrices Y, Z are given.

It is well known (see, e.g., [12]) that

$$u_{\text{opt}}(t) = -K_{\text{opt}}\hat{x}(t) \tag{5.6}$$

where $\hat{x}(t)$ is the optimal estimate of $x(t)$ satisfying the Kalman filter equation

$$\dot{\hat{x}} = A\hat{x} + Bu + L_{\text{opt}}[y - C\hat{x}] \tag{5.7}$$

The matrices K_{opt}, L_{opt} of optimal gains can be found by the formula

$$K_{\text{opt}} = R^{-1}B^TP_{\text{opt}} \ ; \ L_{\text{opt}}^T = Y^{-1}cN_{\text{opt}} \tag{5.8}$$

where P_{opt},N_{opt} are the positive definite solutions to the matrix quadratic equations

$$PA + A^TP - PBR^{-1}B^TP + Q = 0 \tag{5.9}$$

$$NA^T + AN - NC^TY^{-1}CN + Z = 0 \tag{5.10}$$

If $Q > 0$; $R > 0$; $Z > 0$; $Y > 0$ and the matrix pairs $(A\,,B)$, $(A^T\,,C^T)$ are both completely controllable, then the desired solutions exist.

It is evident that all the preceding results of the perturbation theory are applicable in the case of "complete observations", i.e., if $\eta(t) = 0$, and if C is a nonsingular matrix. The same is true of the "pure" filtering problem of optimal estimation of system state from the given observations $y(t)$. However, some new details arise in the complete LQG problem since the separation theorem is valid for optimal solutions but, in general, does not hold for suboptimal ones.

Let us consider the most interesting situation of primal degeneracy.

Let

$$A = A(\epsilon) = A_0 + \epsilon A_1 \ ; \ B = B(\epsilon) = B_0 + \epsilon B_1 \ ; \ C = C(\epsilon) = C_0 + \epsilon C_1$$

and let

$$\text{rank}\,\{B_0, A_0B_0, \ldots, A_0^{n-1}B_0\} = k < n \tag{5.11}$$

If the conditions of Lemma 2.1 are met, there exists a nonsingular transformation such that

$$\dot{\xi}_1 = \Lambda_1\xi_1 + D_1u + \epsilon[\Lambda_{11}\xi_1 + \Lambda_{12}\xi_2 + D_{11}u] + \varsigma_1 \tag{5.12}$$

$$\dot{\xi}_2 = \Lambda_2\xi_2 + \epsilon[\Lambda_{12}\xi_1 + \Lambda_{22}\xi_2 + D_{21}u] + \varsigma_2 \tag{5.13}$$

$$y = \sigma_1\xi_1 + \sigma_2\xi_2 + \eta \tag{5.14}$$

where

$$\xi = (\xi_1, \xi_2) \;\; ; \;\; \dim \xi_1 = k \;\; ; \;\; \dim \xi_2 = n - k$$

$$\varsigma_1 = W_L^T\varsigma \;\; ; \;\; \varsigma_2 = V_L^T\varsigma \;\; ; \;\; \sigma_1 = cW_R \;\; ; \;\; \sigma_2 = cV_R$$

with the construction of the constant matrices W_R, V_R, W_L, V_L as shown in the proof of the lemma.

The objective (5.2) takes the form

$$H\{u\} = \lim_{\substack{t_1\to\infty \\ t_0\to-\infty}} \frac{1}{t_1 - t_0} E\int_{t_0}^{t_1} (\xi^T\bar{Q}\xi + u^TRu)\mathrm{d}t \tag{5.15}$$

where

$$\bar{Q} = \begin{Bmatrix} \bar{Q}_{11} & \bar{Q}_{12} \\ \bar{Q}_{21} & \bar{Q}_{22} \end{Bmatrix}, \;\; \bar{Q}_{11} = W_R^TQW_R \;\; ;$$

$$\bar{Q}_{22} = V_R^TQV_R \;\; ; \;\; \bar{Q}_{12} = \bar{Q}_{21}^T = W_R^TQV_R$$

We consider the most important case, $\Lambda_2 = 0$, which is equivalent, as we know, to the case of singular perturbations. Let us denote

$$P = P(\epsilon) = \begin{Bmatrix} P_{11}(\epsilon) & P_{12}(\epsilon) \\ P_{21}(\epsilon) & P_{22}(\epsilon) \end{Bmatrix} \;\; ; \;\; N = N(\epsilon) = \begin{Bmatrix} n_{11}(\epsilon) & n_{12}(\epsilon) \\ n_{21}(\epsilon) & n_{22}(\epsilon) \end{Bmatrix}$$

Equation (2.54) yields the expansion (2.56) for $P_{11}(\epsilon)$. The formula (2.58) gives

$$K_{opt}(\epsilon) = R^{-1}G(\epsilon) \tag{5.16}$$

where

$$G(\epsilon) = G + \epsilon G_1 + \cdots$$

and where the principal term is given by (2.40).

Equation (5.10) can be similarly analyzed to obtain expansions for $N_{opt}(\epsilon)$. The structure of the expansions depends on properties of the random factors.

Let the noise intensity Y be large, i.e., let

$$Y = \frac{1}{\epsilon^2} Y_0 \tag{5.17}$$

(some other hypotheses are discussed below). Substituting (5.17) in (5.10) we obtain

$$NA^T + AN - N(\epsilon C^T) Y_0^{-1} (\epsilon C) N + Z = 0$$

This equation can be investigated directly. It is evident, however, that the same equation appears if one considers the following dual LQ problem

$$\begin{aligned} &\min \int_0^\infty (x^T Z x + u^T Y_0 u) dt \\ &\dot{x} = -A^T x + \epsilon C^T u \end{aligned} \tag{5.18}$$

and that the properties of the optimal control to this weakly controllable system were investigated in Section 5.2. By Theorem 2.2 [the result (3)] we obtain

$$L^T(\epsilon) = \epsilon[Y_0^{-1} C \tilde{n}_{11} V_R^T + O(\epsilon)] \tag{5.19}$$

$\tilde{n}_{11} > 0$ being the solution to the matrix equation

$$n_{11}\Lambda_{22}^T + \Lambda_{22} n_{11} - n_{11} W_R^T C^T Y_0^{-1} C W_R^T n_{11} + Z_{11} = 0 \; ; \; Z_{11} \triangleq V_L^T Z V_L \tag{5.20}$$

Thus, the asymptotic structure of the optimal control law can be written in the form

$$u_{\text{opt}}(\epsilon, t) = -R^{-1}\{D_1^T P_{11} W_L^T + D_1^T P_{12} V_L^T + D_{21}^T \tilde{P} V_L^T + O(\epsilon)\}\hat{x}(\epsilon, t) \tag{5.21}$$

where $\hat{x}(\epsilon, t)$ is the solution to the stochastic equation

$$\dot{\hat{x}} = A(\epsilon)\hat{x} + Bu_{\text{opt}}(\epsilon, t) + \epsilon[V_R \tilde{n}_{11} V_R^T C^T Y_0^{-1} + O(\epsilon)](y(t) - C\hat{x}) \tag{5.22}$$

which is determined by the measurement $y(t)$, the initial values $\hat{x}(0)$ being arbitrary.

Let

$$\hat{x} = W_R \hat{\xi}_1 + V_R \hat{\xi}_2 \;\; ; \;\; \xi_1 = W_L^T \hat{x} \;\; ; \;\; \hat{\xi}_2 = V_L^T \hat{x} \tag{5.23}$$

Then, by the properties of the transforming matrices, one can obtain

$$\begin{aligned} \dot{\hat{\xi}}_1 &= \Lambda_1 \hat{\xi}_1 + D_1 u_{\text{opt}}(0, t) + O(\epsilon)[y(t) - C\hat{x}] \;, \\ \dot{\hat{\xi}}_2 &= \epsilon[\Lambda_{12} \hat{\xi}_1 + \Lambda_{22} \hat{\xi}_2 + D_{21} u_{\text{opt}}(0, t) + \epsilon(L_0 + O(\epsilon))[y(t) - C\hat{x}] \;, \\ L_0 &\triangleq \tilde{n}_{11} V_R^T C^T Y_0^{-1} \end{aligned} \tag{5.24}$$

and

$$u_{\text{opt}}(\epsilon, t) = -R^{-1}[D_1^T P_{11} + O(\epsilon)]\hat{\xi}_1 - R^{-1}[D_1^T P_{12} + D_{21}^T \tilde{P} + O(\epsilon)]\hat{\xi}_2 \tag{5.25}$$

Supposing Λ_1 to be stable and following the known heuristic, one can accept a suboptimal control in the form

$$u_0 = -K_1 \hat{\xi}_1 - K_2 \hat{\xi}_2 \;,\; K_1 = R^{-1} D_1^T P_{11} \;;\; K_2 = R^{-1}[D_1^T P_{12} + D_{12}^T \tilde{P}] \tag{5.26}$$

where the estimates $\hat{\xi}_1$, $\hat{\xi}_2$ are found as the outputs of a reduced filter

$$\Lambda_1 \hat{\xi}_1 + D_1 u_0 = 0 \tag{5.27}$$

$$\dot{\hat{\xi}}_2 = \epsilon[\Lambda_{12} \hat{\xi}_1 + \Lambda_{22} \hat{\xi}_2 + D_{21} u_0 + L_0(y - CW_R \hat{\xi}_1 - CV_R \hat{\xi}_2)] \tag{5.28}$$

Excluding $\hat{\xi}_1$ yields a suboptimal filter giving the estimate $\hat{\xi}_2$ and a suboptimal control which is based on this estimate only

$$u_0 = -\hat{K}_2\hat{\xi}_2 \tag{5.29}$$

$$\dot{\hat{\xi}}_2 = \epsilon[\hat{\Lambda}_{12}\hat{\xi}_2 + L_0 y] \tag{5.30}$$

where

$$\hat{K}_2 = K_1 H^{-1} D_1 K_2 + K_2 \ ,$$

$$\hat{\Lambda}_{12} = \Lambda_{12} H^{-1} D_1 K_2 - D_{21}\hat{K}_2 - L_0 C W_R H^{-1} D_1 K_2 - L_0 C V_R + \Lambda_{22}$$

$$H_1 = \Lambda_1 - D_1 K_1$$

The practical advantages of the control law defined by (5.29) and (5.30) are evident. It is essential not only that the optimization procedure is simplified here, but that the same is true of the device that realizes the resulting control. The efficiency of the control (5.29) and (5.30) is characterized by the following result.

Theorem 5.1.

Let the conditions of Theorem 2.2 be fulfilled, let the matrix Λ_1 be stable, $Y_0 > 0$, $Z_{11} > 0$, let the pair (Λ_{22}^T, C) be completely controllable.

Then there exists ϵ' such that, for $\epsilon \in [0, \epsilon']$, the estimate

$$\frac{J^*(\epsilon) - J_0(\epsilon)}{J_0(\epsilon)} = O(\epsilon) \tag{5.31}$$

holds, where $J^*(\epsilon)$ is the optimal objective value, and where $J_0(\epsilon)$ is the objective value achieved by the application of the control (5.29) and (5.30) to the system (5.3) and (5.4).

Proof of this result and of those following is based on a known relation (see, e.g., [121]), which we state as a lemma.

Lemma 5.1. Let A be stable, let $Q > 0$, ς be the "white" noise having the intensity matrix $Z > 0$. Then

$$\lim_{\substack{t_1 \to \infty \\ t_0 \to -\infty}} \frac{1}{t_1 - t_0} E \int_{t_0}^{t_1} x^T Q x dt = trGZG^T P \tag{5.32}$$

if $x(t)$ solves the stochastic equation

$$\dot{x} = Ax + G\varsigma \tag{5.33}$$

under any fixed initial value $x(0)$, and where P solves the linear matrix equation

$$A^TP + PA + Q = 0 \tag{5.34}$$

The proof of the theorem is cumbersome, and we shall present a sketch only.

1. The equations of the system with the optimal feedback and with the suboptimal one can be written in the standard form of (5.33). Substituting (5.3) and (5.6) in (5.4) and (5.7) and denoting $\tilde{x} = (x, \hat{x})$, $\tilde{\eta} = (\varsigma, \eta)$ one can obtain a vector equation in the form

$$\dot{\tilde{x}} = \tilde{A}(\epsilon)\tilde{x} + \tilde{G}(\epsilon)\tilde{\eta} \tag{5.35}$$

instead of (5.4) and (5.7). Similarly, substituting $u = u_0$, given by (5.29) in (5.4) and (5.3) in (5.30) we have

$$\dot{\bar{x}} = \bar{A}(\epsilon)\bar{x} + \bar{G}(\epsilon)\tilde{\eta} \tag{5.36}$$

where $\bar{x} = (x, \hat{\xi}_2)$. The structure of the matrices $\tilde{A}(\epsilon)$, $\tilde{G}(\epsilon)$; $\bar{A}(\epsilon)$, $\bar{G}(\epsilon)$ is evident.

2. It can be proved that there exists a ϵ'' such that $\bar{A}(\epsilon)$ is stable for $\epsilon \in [0, \epsilon'']$.

3. It is not hard to see that both $\tilde{A}(0)$ and $\bar{A}(0)$ have a zero $2k$-multiple semi-simple eigenvalue and further to find an explicit expression for the right and left eigenvectors $\tilde{V}_R$, $\tilde{V}_L$; $\bar{V}_R$, $\bar{V}_L$ associated with that eigenvalue.

4. To find the main terms of the ϵ-expansions of solutions to equations of (5.34) type

$$\tilde{A}^T(\epsilon)\tilde{P} + \tilde{P}\tilde{A}(\epsilon) + \tilde{Q} = 0 \tag{5.37}$$

$$\bar{A}^T(\epsilon)\bar{P} + \bar{P}\bar{A}(\epsilon) + \bar{Q} = 0 \tag{5.38}$$

the following lemma can be used.

Lemma 5.2. Let zero be a k-multiple semi-simple eigenvalue of the matrix A_0. Let all other eigenvalues of A_0 have negative real parts. Let the matrix A_1 be such that the $(k \times k)$ matrix

$$M = V_L^T A_1 V_R \tag{5.39}$$

is stable, where V_L, V_R are matrices of left and right eigenvectors of A_0 associated with zero eigenvalues, and

$$V_L^T V_R = E_k \tag{5.40}$$

Then the solution $P(\epsilon)$ to the equation

$$[A_0 + \epsilon A_1]^T P + P[A_0 + \epsilon A_1] + Q = 0 \tag{5.41}$$

may be written in the form

$$P(\epsilon) = \frac{1}{\epsilon}[V_L d V_L^T + O(\epsilon)] \tag{5.42}$$

where d solves the equations

$$M^T d + dM + V_R^T Q V_R = 0 \tag{5.43}$$

Proof of the lemma is similar to the proof of Lemma 5.1, Section 3.5, where an almost identical problem was considered with the matrix equation written in the "common" form instead of the "Liapunov" structure inherent in (5.41).

5. By Lemma 5.2 one can show that

$$\tilde{P}(\epsilon) = \frac{1}{\epsilon}\{\tilde{V}_L d \tilde{V}_L^T + O(\epsilon)\} \tag{5.44}$$

$$\bar{P}(\epsilon) = \frac{1}{\epsilon}\{\bar{V}_L d \bar{V}_L^T + O(\epsilon)\} \tag{5.45}$$

with one and the same matrix d in (5.44) and (5.45). This matrix is determined by an equation of (5.43) type with coefficients which can be given explicitly.

6. By virtue of Lemma 5.1, it is sufficient to state the relation

$$tr\{\tilde{G}\tilde{Z}\tilde{G}^T \tilde{V}_L d \tilde{V}_L^T\} = tr\{\bar{G}\tilde{Z}\bar{G}^T \bar{V}_L d \bar{V}_L^T\} + O(\epsilon) \tag{5.46}$$

where $\tilde{Z} = \text{diag}\ \{Z, Y\}$, $\tilde{G} \triangleq \tilde{G}(0)$; $\bar{G} \triangleq \bar{G}(0)$ in order to prove the theorem.

Let us turn to the original statement of the problem. The hypothesis (5.17) seems unnatural. Therefore, an analysis of other possible situations is also of interest, all the more so as different variants have been described by other authors [122, 123].

For case of comparison, it is convenient to transform (5.12) and (5.13) by introducing the "slow" time scale $\tau = \epsilon t$. Then, for $\Lambda_2 = 0$, we obtain

$$\begin{aligned} \epsilon\frac{d\xi_1}{d\tau} &= \Lambda_1\xi_1 + D_1 u + \epsilon[\Lambda_{11}\xi_1 + \Lambda_{12}\xi_2 + D_{11}u] + \tilde{\varsigma}_1(\tau)\ , \\ \frac{d\xi_2}{d\tau} &= \Lambda_{21}\xi_1 + \Lambda_{22}\xi_2 + D_{21}u + \tilde{\varsigma}_2(\tau)\ , \\ y(\tau) &= \sigma_1\xi_1(\tau) + \sigma_2\xi_2(\tau) + \tilde{\eta}(\tau) \end{aligned} \tag{5.47}$$

where $\tilde{\xi}_1$, $\tilde{\xi}_2$, $\tilde{\eta}$ are "white" noises with the intensity matrices

$$\tilde{Z}_1 = \epsilon Z_1 \ , \ \tilde{Z}_2 = \frac{1}{\epsilon} Z_2 \ ; \ \tilde{Y} = \epsilon Y = \frac{1}{\epsilon} Y^0 \tag{5.48}$$

(the transformation of time leads to a specific transformation of the intensities — see, e.g., [124]).

Thus, the original relations are transformed to the singularly perturbed equations, in which the "fast" subsystem is weakly excited compared with the "slow" one, and where the measurement noise level has the same order as the excitation of the "slow" subsystem so that the "signal-noise" ratio is independent of ϵ. At first glance, it is of more interest to analyze the LQG problem for (5.47), supposing that the intensities $\tilde{Z}_1$, $\tilde{Z}_2$, $\tilde{Y}$ do not depend on ϵ as was done in [122].

As usual it is not hard to show that the optimal control has the form

$$u = -K(\epsilon)(\hat{\xi}_1, \hat{\xi}_2) = -K_1(\epsilon)\hat{\xi}_1 - K_2(\epsilon)\hat{\xi}_2 \tag{5.49}$$

where

$$\begin{aligned} \epsilon\frac{\mathrm{d}\hat{\xi}_1}{\mathrm{d}\tau} &= \Lambda_1\hat{\xi}_1 + D_1 u + \epsilon[\Lambda_{11}\hat{\xi}_1 + \Lambda_{12}\hat{\xi}_2 + D_{11}u] + \\ &\quad + L_1(\epsilon)[y - \sigma_1\hat{\xi}_1 + \sigma_2\hat{\xi}_2] \end{aligned} \tag{5.50}$$

$$\frac{\mathrm{d}\hat{\xi}_2}{\mathrm{d}\tau} = \Lambda_{21}\hat{\xi}_1 + \Lambda_{22}\hat{\xi}_2 + D_{21}u + L_2(\epsilon)[y - \sigma_1\hat{\xi}_1 - \sigma_2\hat{\xi}_2] \tag{5.51}$$

the asymptotic behavior of $K(\epsilon)$ is the same as that in (5.16), but the main terms of the filter gain $L(\epsilon)$ are different:

$$L(\epsilon) = (\frac{1}{\epsilon}[n_{11}\sigma_1^T + O(\epsilon)]\,\tilde{Y}^{-1} \ ; \ [n_{12}\sigma_1^T + n_{22}\sigma_2^T + O(\epsilon)]\,\tilde{Y}^{-1}) \tag{5.52}$$

where n_{11}, n_{12}, n_{22} are ϵ-independent matrices, which can be found from (5.10) by scheme described in Section 5.2. It is essential that

$$L_1(\epsilon) = O(\frac{1}{\epsilon}) = \frac{1}{\epsilon}[L_1 + O(\epsilon)]$$

Therefore, the corresponding term plays a dominant role in the right-hand side of (5.50). Hence, the estimate $\hat{\xi}_1$ dominates over $\hat{\xi}_2$ so that the second term in the control law (5.49) is negligible. Though these arguments are not precise, the following rigorous result can be shown.

Theorem 5.2.

Let all the conditions ensuring solvability of the matrix equation and determining the main terms of the ϵ-expansions of $K(\epsilon)$, $L(\epsilon)$ be fulfilled.

Then there exists $\epsilon_1 > 0$ such that, for $\epsilon \in [0, \epsilon_1]$, the control

$$\begin{aligned} \bar{u} &= -K_1\bar{\xi}_1 - K_z\bar{\xi}_2 \\ \epsilon\frac{d\bar{\xi}_1}{d\tau} &= \Lambda_1\bar{\xi}_1 + \epsilon F_z\bar{\xi}_2 - D_1K_1\bar{\xi}_1 + L_1(y - C_1\bar{\xi}_1) \\ \frac{d\bar{\xi}_2}{d\tau} &= M_z\bar{\xi}_2 + L_z(y - C_z\bar{\xi}_2) \end{aligned} \tag{5.53}$$

used in system (5.47) gives an objective value $\bar{J}(\epsilon)$ satisfying the estimate

$$\frac{J^*(\epsilon) - \bar{J}(\epsilon)}{\bar{J}(\epsilon)} = O(\epsilon) \tag{5.54}$$

if the matrices K_z, L_z, M_z, C_z are arbitrary, but ensure stability of system on the whole when ϵ is small.

The optimal trajectory and the suboptimal one are equal to within $O(\epsilon)$

$$\lim_{\substack{t_1\to\infty \\ t_0\to-\infty}} \frac{1}{t_1 - t_0} E\int_{t_0}^{t_1}\{[\xi_{1,\mathrm{opt}} - \xi_{1,\mathrm{subopt}}]^2 - [\hat{\xi}_1 - \bar{\xi}_1]^2\}dt = O(\epsilon) \tag{5.55}$$

Remark. It can be shown that a special choice of the matrices K_z, L_z, M_z, C_z gives a heuristic control law constructed in [122]. However, that control cannot ensure a more effective result than any arbitrary law that satisfies the stability condition mentioned in the theorem.

Two other cases are of some interest.

Corollary 1. Denote

$$\begin{aligned} \bar{M} &= \Lambda_{22} + \tilde{D}R^{-1}D_1^TP_{11}(\Gamma_1^{-1})^T n_{11}\sigma_1^T\tilde{Y}^{-1}\sigma_2 \\ \bar{D} &= \tilde{D}[E - R^{-1}D_1^TP_{11}(\Gamma_1^{-1})^TD_1] \\ \bar{\sigma} &= \sigma_2 - \sigma_1H_1^{-1}D_1R^{-1}D_1^TP_{11}(\Gamma_1^{-1})^Tn_{11}\sigma_1^T\tilde{Y}^{-1}\sigma_2 \end{aligned} \tag{5.56}$$

where

$$\Gamma_1 = \Lambda_1^T - n_{11}\sigma_1^T\tilde{Y}^{-1}\sigma_1 \ ; \ H_1 = \Lambda_1 - D_1K_1 \tag{5.57}$$

n_{11} is the positive definite solution to

$$n_{11}\Lambda_1^T + \Lambda_1 n_{11} - n_{11}\sigma_1^T \tilde{Y}^{-1}\sigma_1 n_{11} + \tilde{Z}_1 = 0 \qquad (5.58)$$

and where the matrices $\tilde{D}$, P_{11} are as defined in Section 5.2.

Let the pairs $\{\bar{M}, \bar{D}\}$, $\{\bar{M}^T, \bar{\sigma}\}$ be completely controllable. Let

$$M_0 \triangleq \bar{M} - \bar{D}K_0 + \bar{L}_0[\sigma_1 H_1^{-1} D_1 R^{-1} P_{11}(T_1^{-1})^T D_1 - \sigma_1 H_1^{-1} D_1]\bar{K}_0$$

where $\bar{L}_0$, $\bar{K}_0$ are arbitrary but ensure the stability of the matrices $\bar{M} - \bar{D}K_0$; $\bar{M} - \bar{L}_0\bar{\sigma}$.

Then there exists a decomposed suboptimal filter

$$\begin{aligned} \epsilon\frac{d\bar{\xi}_1}{d\tau} &= H_1\bar{\xi}_1 + L_1(y - \sigma_1\bar{\xi}_1) \\ \frac{d\bar{\xi}_2}{d\tau} &= M_0\bar{\xi}_1 + L_0(y - \bar{\sigma}\bar{\xi}_2) \end{aligned} \qquad (5.59)$$

ensuring the efficiency shown in Theorem 5.2.

Corollary 2. Let $\bar{M}$ be stable. Then the same precision is ensured by the aggregated control law

$$\begin{aligned} \bar{u} &= -K_1\bar{\xi}_1 \\ \epsilon\frac{d\bar{\xi}_1}{d\tau} &= H_1\bar{\xi}_1 + L_1(y - \sigma_1\bar{\xi}_1) \end{aligned} \qquad (5.60)$$

such that suboptimality is achieved by means of a feedback based on estimates of the "fast" variables only.

Recall, however, that this result may be interesting only from a pure theoretical point of view since any mathematical model of the "fast" subsystem is, as a rule, insufficiently accurate. This is a consequence of the unsuccessful characterization of the random factors. In fact, if the intensity of the "white" noise $\tilde{\zeta}_1$ in (5.47) is not small, then this means a great difference exists between the random component (a "diffusions" term) and the deterministic one (a "drift") determining the speed of the "fast" variables. Therefore, the control must suppress random fluctuations of "fast" variables such that their variances are of the order of $1/\epsilon$.

To ensure the like order of "diffusion" and the "drift" terms, we require that

$$\tilde{Z}_1 = \epsilon Z_{10} \qquad (5.61)$$

As a consequence, the solution $N(\epsilon)$ to the equation (5.10) will be regular so that

$$L_1(\epsilon) = O(\epsilon) \;\; ; \;\; L_2(\epsilon) = L_2 + O(\epsilon) \;\; ; \;\; L_2 = n_{22}\sigma_2^T\tilde{Y}^{-1} + O(\epsilon) \tag{5.62}$$

where n_{22} is a positive definite solution to the "slow" block of that equation, i.e.,

$$n_{22}\Lambda_{22}^T + \Lambda_{22}n_{22} - n_{22}\sigma_2^T\tilde{Y}^{-1}\sigma_2 n_{22} + \tilde{Z}_2 = 0 \tag{5.63}$$

which exists if the pair $(\Lambda_{22}^T, \sigma_2^T)$ is completely controllable.

The asymptotic structure of the optimal control leads to a method of constructing a suboptimal one. It is natural to replace the optimal filter by its reduced description and to neglect small terms in the optimal feedback gain $K(\epsilon)$.

After that, the control will have the following simple form

$$\bar{u} = -\bar{K}_2\hat{\xi}_2^0 \tag{5.64}$$

$$\begin{aligned} \frac{d\hat{\xi}_2^0}{d\tau} = & [\Lambda_{22} - \tilde{D}R^{-1}(\tilde{D}^T\tilde{P} - L)]\hat{\xi}_2^0 + \\ & + L_2\{y - [\sigma_1 H_1^{-1}D_1R^{-1}(D_1^TP_{12} + D_1^T\tilde{P}) + \sigma_2]\hat{\xi}_2^0\} \end{aligned} \tag{5.65}$$

where the matrixes P_{11}; P_{12}, $\tilde{P}$, H_1, $\tilde{D}$, L are the same as in Section 5.2.

Theorem 5.3.

Let all of the conditions ensuring the existence of the positive definite matrices P_{11}, $\tilde{P}$, n_{22} be fulfilled. Let Λ_1 be stable. Then

(1) $J\{u_{\text{opt}}\} = J + O(\epsilon)$ (5.66)

where

$$J = tr\left\{\begin{Vmatrix} \tilde{Z}_2 \; ; & 0 \\ 0 \; ; & n_{22}\sigma_2^T\tilde{Y}^{-1}\sigma_2 n_{22} \end{Vmatrix}\hat{d}\right\} + tr\{Z_{10}q_1\} \tag{5.67}$$

and where $\hat{d}(2k \times 2k)$, $q_1(n-k \times n-k)$ solve the linear matrix equations

$$\hat{P}^Td + dP = \hat{R} \;\; ; \;\; \Lambda_1^Tq_{11} + q_{11}\Lambda_1 + Q_{11} = 0 \tag{5.68}$$

with the coefficients

$$\hat{P} \triangleq \left\| \begin{matrix} \Lambda_{22} & ; & -\tilde{D}r^{-1}(\tilde{D}^T\tilde{P} - L) \\ n_{22}\sigma_2^T\tilde{Y}^{-1}\sigma_2 & ; & \Lambda_{22} - \tilde{D}R^{-1}(\tilde{D}^T\tilde{P} - L) - n_{22}\sigma_2^T\tilde{Y}^{-1}\sigma_2 \end{matrix} \right\|$$

$$\hat{R} \triangleq \left\| \begin{matrix} Q_{22} & ; & Q_{21}H_1^{-1}D_1R^{-1}(D_1^TP_{12}+D_1^T\tilde{P}) \\ (P_{12}^TD_1+\tilde{P}D_{21})R^{-1}D_1^TH_1^TQ_{12} & ; & (P_{12}^TD_1+\tilde{P}D_{21})R^{-1}(D_1^TP_{12}+D_{21}^T\tilde{P}) \end{matrix} \right\|$$

(2) $J\{\bar{u}\} = J + O(\epsilon) = J\{u_{\text{opt}}\} + O(\epsilon)$ (5.69)

(3) $\lim\limits_{\substack{t_1\to\infty \\ t_0\to-\infty}} \dfrac{1}{t_1 - t_0} E\int_{t_0}^{t_1} [(\xi_1 - \xi_1^0)^2 + (\xi_2 - \xi_2^0)^2 + (\hat{\xi}_2 - \hat{\xi}_2^0)^2]\mathrm{d}t = O(\epsilon)$ (5.70)

Proof can be shown by the same scheme as that of Theorem 5.1. Theorems 5.1 and 5.3 give a reasonable basis for suboptimal analytical design of controls for "large-scale" dynamic systems, corresponding with the heuristic that the neglect of small time constants in mathematical models is permissible. A similar result appears in [123], the difference in formulation being overcome by a transformation of variables. However, a proof given in [123] is not complete since the notion of order $O(\epsilon)$ was used with respect to Gaussian random values, which may exceed arbitrary large levels.

Example. To illustrate the scheme let us consider a numerical example. Let the description of the "slow" time scale be as in (5.47)

$$\epsilon\dot{\xi}_1 = -3\xi_1 + u + \tilde{\zeta}_1 \ ;$$
$$\dot{\xi}_2 = 5u + \tilde{\zeta}_2 \ ;$$
$$y = \xi_1 + \xi_2 + \tilde{\eta}$$

with the performance index

$$J\{u\} = \lim_{\substack{t_1\to\infty \\ t_0\to-\infty}} E\int_{t_0}^{t_1} (16\xi_1^2 + 9\xi_2^2 + u^2)\mathrm{d}t$$

In accordance with hypothesis (5.61), put

$$\tilde{Z}_1 = 6\epsilon \ ; \ Z_2 = 1 \ ; \ \tilde{Y} = 1$$

The precise solution has the form

$$u_{\mathrm{opt}} = -\left\|\frac{1}{\epsilon}, 5\right\| \cdot \left\|\begin{matrix} P_{11}(\epsilon) & ; & P_{12}(\epsilon) \\ P_{12}(\epsilon) & ; & P_{22}(\epsilon) \end{matrix}\right\| \cdot \left\|\begin{matrix} \hat{\xi}_1 \\ \hat{\xi}_2 \end{matrix}\right\| = -K_1(\epsilon)\hat{\xi}_1 - K_1(\epsilon)\hat{\xi}_2$$

where

$$\dot{\hat{\xi}}_1 = -\frac{3}{\epsilon}\hat{\xi}_1 + \frac{1}{\epsilon}u_{\mathrm{opt}} + L_1(\epsilon)(y - \hat{\xi}_1 - \hat{\xi}_2)$$
$$\dot{\hat{\xi}}_2 = 5u_{\mathrm{opt}} + L_2(\epsilon)(y - \hat{\xi}_1 - \hat{\xi}_2)$$

and

$$K_1(\epsilon) = \frac{1}{\epsilon}P_{11}(\epsilon) + 5P_{12}(\epsilon) \ ; \ K_2(\epsilon) = \frac{1}{\epsilon}P_{12}(\epsilon) + 5P_{22}(\epsilon) \ ;$$
$$L_1(\epsilon) = n_{11}(\epsilon) + n_{12}(\epsilon) \ ; \ L_2(\epsilon) = n_{12}(\epsilon) + n_{22}(\epsilon)$$

The entries of matrices $P(\epsilon)$, $N(\epsilon)$ can be written explicitly

$$P_{11} = \epsilon\bar{P}_{11} \ ; \ P_{22} = \frac{1}{5}(3 + \frac{\bar{P}_{11}}{1+5\epsilon}) \ ;$$
$$P_{12} = \frac{\epsilon\bar{P}_{11}}{1+5\epsilon} \ ; \ \bar{P}_{11} \triangleq 3(1+5\epsilon)^2\left[-1 + \sqrt{1 + \frac{16}{9(1+5\epsilon)^2}}\right]$$
$$n_{11} = -\frac{3+\epsilon}{\epsilon}n_{12} \ ; \ n_{12} = 1 - \sqrt{\frac{3+3\epsilon}{1+3\epsilon}} \ ; \ n_{22} = \sqrt{\frac{3+3\epsilon}{1+3\epsilon}}$$

The optimal objective value is equal to

$$J^*(\epsilon) = P_{11}(n_{11} + n_{12})^2 - \frac{2P_{11}}{1+5\epsilon}(n_{11} + n_{12}) + P_{22} + 16n_{11} + 9n_{22}$$

One can show that

$$J^*(\epsilon) \geq 26 - 14.37\epsilon - 58.03\epsilon^2 \ ; \ 0 \leq \epsilon \leq 1$$

Let us construct a suboptimal control by the algorithm described above. We obtain

$$u_0 = -1.8\hat{\xi}_2$$

$$\frac{d\hat{\xi}_2}{dt} = -9\hat{\xi}_2 + y - 0.4\hat{\xi}_2$$

By Lemma 5.1 the corresponding objective value is given by the formula

$$J\{u_0\} = \frac{6}{\epsilon}x_{11} + x_{22} + x_{33}$$

where $\{x_{ij}\} \triangleq X$ solves the equation

$$\begin{Vmatrix} -\frac{3}{\epsilon} & 0 & 1 \\ 0 & 0 & 1 \\ -\frac{1.8}{\epsilon} & -9 & -9.4 \end{Vmatrix} X + X \begin{Vmatrix} -\frac{3}{\epsilon} & 0 & \frac{-1.8}{\epsilon} \\ 0 & 0 & -9 \\ 1 & 1 & -9.4 \end{Vmatrix} = \begin{Vmatrix} -16 & 0 & 0 \\ 0 & -9 & 0 \\ 0 & 0 & -3.24 \end{Vmatrix}$$

The following estimate can be used

$$J\{u_{\text{opt}}\} \leq 26 - 0.64\epsilon + 67.55\epsilon^2 \quad ; \quad 0 \leq \epsilon \leq 1$$

from which we obtain an estimate of the relative error

$$\delta \triangleq \frac{|J^*(\epsilon) - J\{u_0\}|}{J^*(0)} \leq 0.53\epsilon + 4.83\epsilon^2$$

so that $\epsilon = 0.1$ yields $\delta \leq 0.1$. This estimate is very rough. A direct calculation gives $\delta = 0.004$ for $\epsilon = 0.1$, and $\delta = 0.013$ for $\epsilon = 0.5$.

CHAPTER 6

Nonlinear Optimal Control Problems

In the preceding presentation, the peculiarity of dynamic systems subject to optimization was not fully exhibited. In fact, perturbation theory was used only for investigating the algebraic equations that explicitly determine optimal or suboptimal control laws.

In this chapter, the perturbation method is developed for nonlinear systems considered over a finite time interval, and where one needs to take into account constraints imposed on the control functions at any moment, belonging to the interval. As a rule, there are no explicit solutions to such problems. A formal analysis can be carried out by comparing the solutions to the perturbed differential equations and the reduced ones under various feasible control functions.

Two situations are investigated in detail. The first part of the chapter (Sections 6.1–6.4) is devoted to problems that are near to simpler ones (in particular, to linear problems). As in Chapter 1, the perturbation method proves to be efficient for systems that are weakly different from directly aggregatable or decomposable ones. In the second part (Section 6.5), singularly perturbed systems are investigated. This section is closely connected with Section 5.2. Let us review the principal results.

Section 6.1 is devoted to optimal control problems with objective functionals of the state variables that are unconstrained at the right end of the time interval (so-called terminal problems). It is supposed here that the Pontriagin maximum principle is applicable and, moreover that the maximizing control depends continuously on the state and conjugated variables. Estimates of differences between the optimal control and the reduced one are given as well as estimates of the loss in the objective value arising if the reduced control is applied to the perturbed system. A scheme is suggested which allows approximation of the optimum to within any given precision by means of solving the reduced problem and by the recursive integration of systems of first-order differential equations. Under the same assumptions, some problems are investigated, among which are weakly controllable systems (with a unique reduced solution) and quasilinear systems with a quadratic objective function.

Terminal problems are also considered in Section 6.2, but with the fiat that the reduced problems may have nonunique solutions. A way to isolate an asymptotically optimal perturbed solution from the reduced optimum set is shown. To this end, an auxiliary problem is considered, which is close in structure to that introduced for investigating mathematical programs in Chapter 1. However, one needs a complete description of the reduced solutions set to use the suggested scheme, and this is difficult from a practical point of view. Therefore, another procedure is constructed in Section 6.3. This procedure is based on difference approximations of the perturbed problem to the reduced one. If grid points are approached, the approximate solution tends to the optimal control and trajectory.

The terminal problem for weakly controllable systems is investigated in Section 6.4. The reduced solution gives no additional information about the optimal control, i.e., the nonuniqueness occurs. However, the auxiliary problem, having here the form of a maximum principle, easily admits a suboptimal control. This control ensures an objective value differing from the optimal one by $O(\epsilon)$. Conditions are stated that ensure the convergence of the suboptimal control to the optimal one for $\epsilon \to 0$. A scheme of successive approximations is also shown.

Singularly perturbed terminal problems are considered in Section 6.5. A natural decomposition procedure is constructed here on the basis of the classical theory of perturbations of finite-dimensional operators. Estimates are given for the efficiency of suboptimal controls, which are obtained by an independent consideration of "slow" and "fast" subsystems.

6.1. The maximum principle and smooth solutions

Assume that the perturbed problem is of the form

$$\begin{aligned} J\{\bar{u}(t,\epsilon)\} \triangleq \min \{ & G(x(T)) \,/\, \dot{x} = f(x,u,t,\epsilon) \,; \\ & x(0) = x_0 \,,\, u(t) \in U \,,\, t \in [0,T]\} \end{aligned} \tag{1.1}$$

where $x(t)$ is an n-vector of state variables, $u(t)$ is an m-vector of controls, and $U \subset E^m$. The constraints on the functions in (1.1) will be specified later in the exposition of formal results.

Suppose that in the class of admissible functions $u(t)$ there exists a bounded control $\bar{u}(t,\epsilon)$ and that the corresponding trajectory $\bar{x}(t,\epsilon)$, $t \in [0,T]$, $0 \leq \epsilon \leq \epsilon_0$ attains the solution to (1.1). The main goal of the investigation is, as usual, to establish the connections between $\bar{u}(t,\epsilon)$, $\bar{x}(t,\epsilon)$, with ϵ a small positive scalar, and the solution to the reduced problem $\bar{u}(t,0)$, $\bar{x}(t,0)$ and to then evaluate the difference between the associated performance indices. It is also of interest to find an estimate of the performance index defined on the trajectory of the perturbed system when the reduced optimal control $\bar{u}(t,0)$ is used.

Let us suppose that the maximum principle allows us to find a solution. Then the usual procedure can be described as follows: introduce the Hamiltonian

$$H(x, \lambda, u, t, \epsilon) \triangleq \lambda^T f(x, u, t, \epsilon) \tag{1.2}$$

and write the primal and conjugate systems in standard form, taking into account the boundary conditions

$$\begin{aligned} \dot{x} &= \nabla_\lambda H = f \quad , \quad x(0) = x_0 \\ \dot{\lambda} &= -\nabla_x H \triangleq g \;\; , \;\; \lambda(T) = -\nabla_x G[x(T)] \end{aligned} \tag{1.3}$$

Define the function

$$\varphi(x, \lambda, t, \epsilon) \triangleq \arg\max \{H(x, \lambda, u, t, \epsilon) \,/\, u \in U\} \tag{1.4}$$

and substitute it into the right hand sides of equations (1.3). [Throughout this section we only consider the situation in which the solution to (1.3) is unique.] The solution $\bar{x}(t, \epsilon)$, $\bar{\lambda}(t, \epsilon)$ to the boundary value problem

$$\begin{aligned} \dot{x} &= f[x, \varphi(x, \lambda, t, \epsilon), t, \epsilon] \;\; , \;\; x(0) = x_0 \\ \dot{\lambda} &= g[x, \varphi(x, \lambda, t, \epsilon,), t, \epsilon] \;\; , \;\; \lambda(T) = -\Delta_x G(x(T)) \end{aligned} \tag{1.5}$$

allows us to construct the optimal control

$$\bar{u}(t, \epsilon) = \varphi[\bar{x}(t, \epsilon) \; , \; \bar{\lambda}(t, \epsilon) \; , \; t \; , \; \epsilon] \tag{1.6}$$

So, the maximum principle is reduced by solving the initial problem to that of:

(1) Finding the maximum in u for the Hamiltonian (1.4), with λ, x, t, ϵ as the variables.
(2) Solving the boundary value problem (1.5).

Both stages are supposed to allow a simpler realization for the reduced problem ($\epsilon = 0$).

Let us try to estimate the attainable accuracy when the reduced problem is considered instead of the perturbed problem. For this purpose, let us consider, based on classic results due to H. Poincaré [71, 54], some properties of the boundary value problem (1.5). Rewrite it in the form

$$\dot{z} = F(z, t, \epsilon) \tag{1.7}$$

$$\Psi[z(0), z(T, \epsilon)] = 0 \tag{1.8}$$

where $z \triangleq (x, \lambda)$, $F \triangleq (f, g)$ and (1.8) include the boundary conditions at both ends

$$\Psi[z(0), z(T, \epsilon)] \triangleq (x(0) - x_0, \lambda(T, \epsilon) + \nabla_x G[x(T, \epsilon)]) \tag{1.9}$$

Let us compare the solution to (1.7) and (1.8) with that to the reduced problem

$$\dot{\varsigma} = F(\varsigma, t, 0) \tag{1.10}$$

$$\Psi[\varsigma(0), \varsigma(T)] = 0 \tag{1.11}$$

Note that the initial values of the vector $z(t)$ or $\varsigma(t)$ are not, in general, fixed and that $z(0) \neq \varsigma(0)$. Establishing the correspondence between the initial condition of the perturbed and reduced problems is the main feature distinguishing our analysis from that given for the Poincaré problem.

Theorem 1.1.

Let

(1) The function $F(z, t, \epsilon)$ be twice continuously differentiable in z and differentiable in t, ϵ for

$$(t, z, \epsilon) \in D \triangleq \{(t, z, \epsilon) / t \in [0, T], z \in Z \subset E^{2n}, 0 \leq \epsilon \leq \epsilon_0\}$$

(2) The solution $\varsigma(t, z^0)$ to the reduced problem under the initial conditions

$$\varsigma(0, z^0) = z^0 \ , \ z^0 \in Z' \subset Z$$

be defined for all t and localized in some neighborhood of Z.

(3) The equation

$$\Psi[\varsigma(0), \varsigma(T)] = 0$$

have the solution

$$\varsigma(0) = \varsigma_0^0 \in \text{int } Z'$$

where $\varsigma(0)$, and $\varsigma(T)$ are connected by the Cauchy problem for the reduced system (1.10).

(4) The Jacobian (with respect to the variables z^0) of the system

$$\Psi[z(0), z(T, \epsilon)] = 0$$

be nonzero when $\epsilon = 0$, $z(0) = \varsigma_0^0$ where $z(0)$, $z(T, \epsilon)$ are connected by the Cauchy problem for the perturbed system (1.7).

(5) The function $\Psi(\cdot)$ be twice differentiable in some neighborhood of $(\varsigma_0^0, \varsigma(T, \varsigma_0^0))$.

Then there exists $\epsilon' > 0$ such that, for $0 \leq \epsilon \leq \epsilon'$, the solution to the perturbed problem (1.7) and (1.8) exists in some neighborhood of the reduced trajectory and, moreover,

$$\|z(t, \epsilon) - \varsigma(t)\| \leq c\,\epsilon \ , \quad c = \text{constant} \tag{1.12}$$

Statement of the theorem and its proof are similar to those in [61], where the accuracy of the averaging scheme was under consideration.

Proof. Let $z(t, \epsilon, z^0)$ be the solution to the Cauchy problem

$$\dot{z} = F(z, t, \epsilon) \ , \quad z(0) = z^0 \subset \text{int } Z' \tag{1.13}$$

Let us consider

$$\Psi[z^0, z(T, \epsilon, z^0)]$$

as a vector-function of z^0, ϵ. Let us denote by $\Delta(z^0, \epsilon)$ the determinant of the matrix $\nabla_{z^0}\Psi$ and prove the asymptotic formula

$$\Delta(z^0, \epsilon) = \Delta_0(z^0) + c\epsilon \tag{1.14}$$

By the definition,

$$\nabla_{z^0}\Psi = \Psi_1[z^0, z(T, \epsilon, z^0)] + \Psi_2[z^0, z(T, \epsilon, z^0)]\nabla_{z^0}z(T, \epsilon, z^0) \tag{1.15}$$

where Ψ_1 is the matrix of derivatives with respect to the first group of the variables, and Ψ_2 similarly with respect to the second.

The matrix $\nabla_{z^0}z(T, \epsilon, z^0)$ can be found by integrating the system

$$\frac{\mathrm{d}\,W(t, \epsilon)}{\mathrm{d}t} = \nabla_z F(z, t, \epsilon)\, W(t, \epsilon) \ , \quad W(0, \epsilon) = E \tag{1.16}$$

for

$$W(T,\epsilon) = \nabla_{z^0} z(T,\epsilon,z^0) \tag{1.17}$$

Using in (1.13) and (1.16) the Poincaré theorem that gives the dependence of the Cauchy problem solution on its parameter, we obtain

$$z(T,\epsilon,z^0) = \varsigma(T,z^0) + O(\epsilon) \tag{1.18}$$

$$W(T,\epsilon) = W(T,0) + O(\epsilon) \tag{1.19}$$

where $\varsigma(T,z^0)$ is found by integrating the reduced system with the same initial conditions as in the perturbed system. Since, by the assumptions, the functions Ψ_1, Ψ_2 are differentiable, we have that

$$\nabla_{z^0}\Psi = \Psi_1[z^0,\varsigma(T,z^0)] + \Psi_2[z^0,\varsigma(T,z^0)]\,W(T,0) + O(\epsilon)$$

which proves (1.14).

By applying the implicit function theorem, it is not difficult to verify that the equation

$$\Psi[z^0, z(T,\epsilon,z^0)] = 0$$

defines, for $0 \le \epsilon \le \epsilon'$, the unique function $z^0(\epsilon)$, which satisfies the relation

$$z^0(0) = \varsigma_0^0$$

with ς_0^0 as the root of the equation

$$\Psi[\varsigma(0),\varsigma(T,\varsigma(0))] = 0 \tag{1.20}$$

Moreover,

$$\|z^0(\epsilon) - \varsigma_0^0\| \le c\epsilon \tag{1.21}$$

Let us proceed to prove the basic result. As was shown above, the solution $z(t,\epsilon)$ to the boundary value problem satisfies the condition

$$z(0,\epsilon) = z^0(\epsilon)$$

So, taking into account (1.21) and the differentiability of the perturbed problem solution with respect to the initial values, we obtain

$$\|z(t,\epsilon) - z(t,\epsilon,\varsigma_0^0)\| \leq c\epsilon \quad , \quad c = \text{constant} \tag{1.22}$$

On the other hand, by the Poincaré theorem,

$$\|z(t,\epsilon,\varsigma_0^0) - \varsigma(t)\| \leq c\epsilon \tag{1.23}$$

since ς_0^0 is the initial value of the trajectory of the reduced problem. The relations (1.22) and (1.23) lead immediately to the desired inequality (1.12). QED.

Let us try to apply the result just obtained to the investigation of the optimization problem.

In order that the conditions of Theorem 1.1 be fulfilled for the problem (1.5), the following conditions must hold:

(A) The function $f(x,u,t,\epsilon)$ must be twice differentiable in x, u and differentiable in t, ϵ for (x,u,t,ϵ) within corresponding domains.

(B) The function $\varphi(x,\lambda,t,\epsilon) \triangleq \varphi(z,t,\epsilon)$ must be twice differentiable in z and t, ϵ.

The fulfillment of these conditions will be assumed throughout this section. The latter condition is especially restrictive since the smooth dependence on parameters is inherent in mathematical programming solutions in only a few situations (see Chapter 1). In particular, this occurs if the Hamiltonian is strongly convex on U. However, as follows from the proof of the theorem, this smoothness must occur in some finite neighborhood of the trajectory of the reduced system, and not necessarily everywhere.

Theorem 1.2.

For the system (1.5) presented in the form of the boundary value problem (1.7) and (1.8), let all the conditions of Theorem 1.1 be satisfied and let the solution $z(t,\epsilon)$ be unique. Then there exists $\epsilon' > 0$ such that, for $0 \leq \epsilon \leq \epsilon'$, the following estimates of the difference between the solutions of the perturbed and reduced problem (1.1) are valid:

$$\|\bar{u}(t,\epsilon) - \bar{u}(t,0)\| \leq c\epsilon \tag{1.24}$$

$$\|\bar{x}(t,\epsilon) - \bar{x}(t,0)\| \leq c\epsilon \quad , \quad c = \text{constant} \tag{1.25}$$

Proof. Let $\bar{x}(t,\epsilon)$ be the optimal trajectory of (1.1). Then it satisfies the necessary conditions in the form of the maximum principle, i.e., the components of $\bar{x}(t,\epsilon)$ form the first group of components of the solution vector $z(t,\epsilon) = (x(t,\epsilon), \lambda(t,\epsilon))$ to the boundary value problem such that the later is unique. Moreover, in this case

$$\bar{u}(t,\epsilon) = \varphi[z(t,\epsilon), t, \epsilon] \tag{1.26}$$

Consider the solution $\varsigma(t,\varsigma^0)$ to the system (1.10) with initial condition

$$\varsigma^0 \triangleq z(0,\epsilon)$$

By the Poincaré theorem

$$\varsigma(t,\varsigma^0) = z(t,\epsilon) + O(\epsilon)$$

and, consequently,

$$\Psi[\varsigma^0, \varsigma(T,\varsigma^0)] = O(\epsilon)$$

from which

$$\|\varsigma^0 - \varsigma(0,\varsigma_0^0)\| \leq c\epsilon \ , \ \varsigma(0,\varsigma_0^0) \triangleq \varsigma_0^0$$

or

$$\|z(0,\epsilon) - \varsigma_0^0\| \leq c\epsilon$$

Thus, the whole trajectory $z(t,\epsilon)$ is localized in an ϵ-pipe including $\varsigma(t,\varsigma_0^0)$:

$$\|z(t,\epsilon) - \varsigma(t,\varsigma_0^0)\| \leq c\epsilon \tag{1.27}$$

where $\varsigma(t,\varsigma_0^0)$ is the vector of trajectories $\bar{x}(t,0)$, $\lambda(t,0)$ of the boundary value problem uniquely determining the solution to the reduced problem. This, together with the postulated smoothness, [see condition (B)] leads to the desired result.

Remark. The fulfillment of the conditions of Theorem 1.1 for the system (1.5) allows us to conclude that the optimal control exists in the class of functions with bounded derivative.

Let us assume that this control is uniformly bounded in ϵ

$$\|\bar{u}(t,\epsilon)\| \leq M = \text{constant}$$

Then, if the solution to the reduced problem is unique, the relations (1.24) and (1.25) remain valid as before. In fact, they follow from Theorem 1.1 if one establishes the limits

$$\bar{u}(t,\epsilon) \xrightarrow[\epsilon\to 0]{c} \bar{u}(t,0) \ , \ \bar{x}(t,\epsilon) \xrightarrow[\epsilon\to 0]{c} \bar{x}(t,0) \ , \ \bar{\lambda}(t,\epsilon) \xrightarrow[\epsilon\to 0]{c} \bar{\lambda}(t,0) \qquad (1.27')$$

As is easy to show (as will be done in Section 3.7 for an essentially more general situation),

$$\lim_{\epsilon\to 0} |J\{\bar{u}(t,\epsilon)\} - J_0\{\bar{u}(t,0)\}| = 0 \qquad (1.27'')$$

Suppose now that (1.27′) does not hold. By the Arzelà theorem [38], there exists a subsequence such that

$$\bar{u}(t,\epsilon) \xrightarrow[\epsilon\to 0]{c} \bar{u}(t)$$

with

$$\bar{u}(t) \neq \bar{u}(t,0)$$

That is, $\bar{u}(t)$ is an optimal control to the reduced problem that is not equal to $\bar{u}(t,0)$. This contradicts its uniqueness.

By means of the method of successive approximations, one can improve the efficiency of the reduced control.

Theorem 1.3.

Let the successive approximation scheme be applied to the system

$$\begin{aligned} \dot{x} &= f(x,u,t,\epsilon) \quad &, \quad x(0) &= x_0 \\ \dot{\lambda} &= g(x,\lambda,u,t,\epsilon) \quad &, \quad \lambda(T) &= -\nabla_x G[x(T)] \\ u &= \varphi(x,\lambda,t,\epsilon) \quad &, \quad t &\in [0,T] \end{aligned} \qquad (1.28)$$

i.e., construct sequences $\{x^l\}$, $\{\lambda^l\}$, $\{u^l\}$ satisfying the following equations:

$$\begin{aligned} \dot{x}^0 &= f(x^0, u^0, t, 0) \quad , \quad x^0(0) = x_0 \ ; \\ \dot{\lambda}^0 &= g(x^0, \lambda^0, u^0, t, 0) \ , \ \lambda^0(T) = -\nabla_x G[x^0(T)] \ ; \\ u^0 &= \varphi(x^0, \lambda^0, t, 0) \quad , \quad t \in [0, T] \end{aligned} \tag{1.29}$$

$$\begin{aligned} \dot{x}^l =& \ f(x^l, u^l, t, 0) + f(x^{l-1}, u^{l-1}, t, \epsilon) - \\ & - f(x^{l-1}, u^{l-1}, t, 0) \ , \ x^l(o) = x_0 \ ; \\ \dot{\lambda}^l =& \ g(x^l, \lambda^l, u^l, 0) + g(x^{l-1}, \lambda^{l-1}, u^{l-1}, t, \epsilon) - \\ & - g(x^{l-1}, \lambda^{l-1}, u^{l-1}, t, 0) \ , \ \lambda^l(T) = -\nabla_x G[x^l(T)] \ ; \\ u^l(t) =& \ \varphi(x^l(t), \lambda^l(t), t, \epsilon) \ , \ l = 1,2,\ldots \end{aligned} \tag{1.30}$$

Let the conditions of Theorem 1.1 and conditions (A) and (B) (partly implying them) be satisfied. Then there exists a positive $\hat{\epsilon}$ such that, for $0 \leq \epsilon \leq \hat{\epsilon}$, the following estimates are valid

$$\|\bar{x}(t, \acute{\epsilon}) - x^l(t)\| = O(\epsilon^{l+1}) \tag{1.31}$$

$$\|\bar{u}(t, \epsilon) - u^l(t)\| = O(\epsilon^{l+1}) \tag{1.32}$$

Let $\bar{x}^l(t)$ be the solution to the Cauchy problem

$$\dot{x} = f[x, u^l(t), t, \epsilon] \ , \ x(0) = x_0 \tag{1.33}$$

Then

$$\|\bar{x}(t, \epsilon) - \bar{x}^l(t)\| = O(\epsilon^{l+1}) \tag{1.34}$$

Proof. Note, first of all, that the problem of zero-order approximation coincides with the reduced boundary value problem (1.10) and (1.11), and, by Theorem 1.2, that the evaluations (1.31) and (1.32) hold for $l = 0$. The construction of high-order approximations requires the solution of boundary value problems that are close in structure to the reduced one.

By induction, one can easily verify that all successive approximations are within an ϵ-pipe of the reduced solution

$$\begin{aligned} |x^l(t) - x^0(t)| &= O(\epsilon) \\ |\lambda^l(t) - \lambda^0(t)| &= O(\epsilon) \\ |u^l(t) - u^0(t)| &= O(\epsilon) \end{aligned} \tag{1.35}$$

Let us further compare the successive approximation trajectories. By definition, we have that

$$x^k(t) = x_0 + \int_0^t [f(x^k, u^k, t, 0) + \epsilon\beta(x^{k-1}, u^{k-1}, t, \epsilon)]dt$$

where, according to (1.35), the notation

$$\epsilon\beta[\cdot] = f[x^{k-1}, u^{k-1}, t, \epsilon] - f[x^{k-1}, u^{k-1}, t, 0] \ , \ \overline{\lim_{\epsilon \to 0}} \|\beta\| < \text{constant}$$

is used. This implies that

$$\begin{aligned} \|x^{k+1}(t) - x^k(t)\| \le & \int_0^t \|f(x^{k+1}, u^{k+1}, \tau, 0) - f(x^k, u^k, \tau, 0)\|d\tau + \\ & + \epsilon\int_0^t \|\beta(x^k, u^k, \tau, \epsilon) - \beta(x^{k-1}, u^{k-1}, \tau, \epsilon)\|d\tau \\ \le & \ c_1\int_0^t \|x^{k+1} - x^k\|d\tau + \epsilon c_2\int_0^t \|x^k - x^{k-1}\|d\tau + \\ & + \epsilon c_3\int_0^t \|u^k - u^{k-1}\|d\tau + c_0\int_0^t \|\lambda^{k+1} - \lambda^k\|d\tau \ , \\ & t \in [0, T] \ , \ c_i = \text{constant} \end{aligned} \tag{1.36}$$

Similarly, we obtain from the second group of equations (1.30)

$$\begin{aligned} \|\lambda^{k+1}(t) - \lambda^k(t)\| \le & \ c_4\int_t^T \|\lambda^{k+1} - \lambda^k\|d\tau + c_5\int_t^T \|x^{k+1} - x^k\|d\tau + \\ & + \epsilon c_6\int_t^T \|x^k - x^{k-1}\|d\tau + \epsilon c_7\int_t^T \|\lambda^k - \lambda^{k-1}\|d\tau + \\ & + \epsilon c_8\int_t^T \|u^k - u^{k-1}\|d\tau + c_9\|x^{k+1}(T) - x^k(T)\| \ , \\ & t \in [0, T] \ , \ c_i = \text{constant} \end{aligned} \tag{1.37}$$

Since the function $\varphi(\cdot)$ is differentiable in x, λ, we have

$$\begin{aligned} \|u^k(t) - u^{k-1}(t)\| \le & \ c_{10}\|x^k(t) - x^{k-1}(t)\| + c_{11}\|\lambda^k(t) - \lambda^{k-1}(t)\| \ , \\ & t \in [0, T] \ , \ c_i = \text{constant} \end{aligned} \tag{1.38}$$

Substituting (1.38) into (1.36) and (1.37) and applying the Gronwall–Bellman lemma, we obtain the estimates

$$\begin{aligned} \|\Delta x^k(t)\| \leq\ & \epsilon k_1\int_0^t \|\Delta x^{k-1}(\tau)\|\mathrm{d}\tau + \epsilon k_2\int_0^t \|\lambda^{k-1}(\tau)\|\mathrm{d}\tau + k_0\int_0^t \|\Delta\lambda^k\|\mathrm{d}\tau \\ \|\Delta\lambda^k(t)\| \leq\ & k_3\int_t^T \|\Delta x^k(\tau)\|\mathrm{d}\tau + k_4\|\Delta x^k(T)\| + \epsilon k_5\int_t^T \|\Delta\lambda^{k-1}(\tau)\|\mathrm{d}\tau + \\ & + \epsilon k_6\int_t^T \|\Delta x^{k-1}(\tau)\|\mathrm{d}\tau \ , \ k_i = \text{constant} \end{aligned} \tag{1.39}$$

where

$$\Delta x^k \triangleq x^{k+1} - x^k \ , \ \Delta\lambda^k \triangleq \lambda^{k+1} - \lambda^k$$

Denote

$$\Delta_x^k \triangleq \max_t \|\Delta x^k(t)\| \ , \ \Delta_\lambda^k \triangleq \max_t \|\Delta\lambda^k(t)\|$$

Then, strengthening (1.39), we obtain

$$\begin{aligned} &\Delta_x^k \leq \epsilon\bar{k}_1\Delta_x^{k-1} + \epsilon\bar{k}_2\Delta_\lambda^{k-1} + \bar{k}_0\Delta_\lambda^k \\ &\Delta_\lambda^k \leq \bar{k}_4\Delta_x^k + \epsilon\bar{k}_5\Delta_\lambda^{k-1} + \epsilon\bar{k}_6\Delta_x^{k-1} \ , \ \bar{k}_i = \text{constant} \end{aligned}$$

from which, finally,

$$\begin{aligned} &\max_t \|z^{k+1}(t) - z^k(t)\| \leq \epsilon\tilde{k} \max_t\|z^k(t) - z^{k-1}(t)\| \\ &z^k(t) \triangleq (x^k(t) \ , \lambda^k(t)) \ , \ \tilde{k} = \text{constant} \end{aligned} \tag{1.40}$$

Using the Weierstrass theorem for uniform convergence allows us to conclude that there exists a positive $\hat{\epsilon}$ such that, for $0 \leq \epsilon \leq \hat{\epsilon}$, the series

$$z^0(t) + \sum_{k=0}^{\infty} \Delta z^k(t)$$

converges uniformly with respect to t or, what is just the same, the sequence $\{z^k(t)\}$ converges uniformly, from which the uniform convergence of the sequence $\{u^k(t)\}$ follows automatically. The limits of the indicated sequences satisfy system (1.28) and, by definition, constitute the solution to the optimization problem

$$\bar{z}(t,\epsilon) = (\bar{x}(t,\epsilon), \bar{\lambda}(t,\epsilon)) \ , \ \bar{u}(t,\epsilon)$$

Since

$$\bar{z}(t,\epsilon) - z^l(\epsilon) = \sum_{k=l}^{\infty} \Delta z^k(t)$$

the main results follows from (1.31).

To finish the proof, let us note that the estimate (1.35) is obtained as a consequence of the Gronwall–Bellman lemma, since

$$\begin{aligned} \bar{x}(t,\epsilon) - \bar{x}^l(t) &= \int_0^t [f(\bar{x}, \bar{u}, t, \epsilon) - f(\bar{x}^l, u^l, t, \epsilon)] dt \Rightarrow \\ \Rightarrow \|\bar{x}(t,\epsilon) - \bar{x}^l\| &\le c \int_0^t \|\bar{u} - u^l\| dt = O(\epsilon^{l+1}) \end{aligned} \tag{1.41}$$

It is particularly worthwhile to attract attention to the case when the initial problem is of the form

$$\min \{G(x(T)) + \int_0^T \bar{f}(x, u, t, \epsilon) dt\} \tag{1.41$'$}$$

subject to

$$\dot{x} = f_0(x,t) + \epsilon f_1(x,u,t,\epsilon) \ , \ x(0) = x_0 \ , \ u(t) \in U \tag{1.42}$$

This description emphasizes the fact that, on the one hand, the system is weakly controlled and, on the other, the cost of the controls influences the performance index in the main term (see Section 6.4 for the comparison). Evidently, problem (1.41′) and (1.42) can be transformed to standard from; however, in this form its special features will not be so clearly expressed. At the same time, the problem (1.41′) and (1.42) allows us to simplify the construction of the reduced solution and to improve the accuracy of the successive approximations.

Theorem 1.4.

Let conditions (A) and (B) and the conditions of Theorem 1.1 be satisfied for the problem (1.41′) and (1.42). (It is assumed obvious here that the problem is reduced to the standard form if the dimension of the state space is increased by unity.) Let the solution to the reduced problem

$$\bar{u}(t,0) = \arg\max\{-\bar{f}[\bar{x}(t,0), u, t, 0] \,/\, u \in U\} \tag{1.43}$$

be unique, where $\bar{x}(t,0)$ is the solution to the Cauchy problem

$$\dot{x} = f_0(x,t) \ , \ x(0) = x_0 \tag{1.44}$$

Use the following scheme of successive approximations

$$\begin{aligned} \dot{x}^l &= f_0(x^l, t) + \epsilon f_1(x^{l-1}, u^{l-1}, t, \epsilon) \ , \ x^l(0) = x_0 \\ \dot{\lambda}^l &= g_0(x^l, \lambda^l, t) + \epsilon g_1(x^{l-1}, \lambda^{l-1}, u^{l-1}, t, \epsilon) \\ \lambda^l(T) &= -\nabla_x G[x^l(T)] \\ u^l(t) &= \varphi[x^l(t), \lambda^l(t), t, \epsilon] \triangleq \\ &\triangleq \arg\max\{-\bar{f}(x^l, u, t, \epsilon) + \epsilon(\lambda^l(t))^T f_1(x^l, u, t, \epsilon) \,/\, u \in U\} \end{aligned} \tag{1.45}$$

with $x^0(t) \triangleq \bar{x}(t,0)$, $u^0(t) \triangleq \bar{u}(t,0)$ and with $\lambda^0(t)$ as the solution to the problem

$$\dot{\lambda}^0(t) = g_0[x^0(t), \lambda^0(t)] \ , \ \lambda^0(T) = -\nabla_x G(x^0(T)) \tag{1.46}$$

conjugate to (1.44).

Then, for the sequences $\{x^l(t)\}$, $\{\lambda^l(t)\}$, the results of Theorem 1.3 are valid. Unlike Theorem 1.3, the sequences $\{x^l(t)\}$, $\{\lambda^l(t)\}$ are obtained as the solutions to the Cauchy problems and not the boundary value ones. The proof is similar to that of Theorem 1.3.

Let us consider another particular case of practical importance, in which a quasilinear system

$$\dot{x} = Ax + \epsilon f(x,t) + Bu \ , \ x(0) = x_0 \ , \ t \in [0, T] \tag{1.47}$$

with the performance index

$$\frac{1}{2} x^T(T) H x(t) + \frac{1}{2} \int_0^T [x^T Q x + u^T R u] dt \tag{1.48}$$

is optimized where the matrices A, B, Q, R depend, in general, on t. The well-known technique for solving the reduced problem permits the generation of different schemes of successive approximation. Let us give one of the known results without proof.

Theorem 1.5 [39].

Let the matrices $A(t)$, $B(t)$, $Q(t)$, $R(t)$ be measurable and bounded for $t \in [0, T]$ and, moreover, let

$$Q(t) \geq 0 \ , \ R(t) > 0 \ , \ H > 0$$

uniformly with respect to $t \in [0, T]$. Let the function $f(x, t)$ be measurable in both arguments. Let it satisfy Lipschitz conditions and let its norm $\|f(x, t)\|$ have no more than quadratic growth in x for $t \in [0, T]$. Let the reduced control be constructed as the feedback law

$$u^0(t) = -R^{-1}(t)B^T(t)P(t)x(t)$$

where $P(t) \geq 0$ is the solution to the matrix equation

$$-\dot{P} = PA + A^TP - PBR^{-1}B^TP + Q \ , \ P(T) = H \tag{1.49}$$

and where the successive approximations $u^l(t)$ are constructed as the results of minimization of the objective (1.48) calculated for each $l = 1,2,\ldots,$ on the trajectories of the linear system

$$\dot{x} = Ax + Bu + \epsilon f[\tilde{x}^{l-1}(t), t] \ , \ x(0) = x_0 \ , \ t \in [0, T] \tag{1.50}$$

where $\tilde{x}^{l-1}$ is the solution to the preceding optimization problem. Then there exists $\hat{\epsilon} > 0$ such that, for $0 \leq \epsilon \leq \hat{\epsilon}$, the estimate

$$|\bar{x}(t, \epsilon) - \tilde{x}^l(t)| = O(\epsilon^{l+1}) \tag{1.51}$$

holds, where $\bar{x}(t, \epsilon)$ is the unique solution to the problem (1.47) and (1.48).

If the control $u^l(t)$ is applied to the system (1.47) and (1.48), the deflection from the optimal trajectory and the optimal value of the performance index can be estimated to within $O(\epsilon^{l+1})$.

The proof given in [39] uses similar ideas to those described. Note only that, in contradiction to the instructions of [39], approximations of more than the first order cannot be done in the form of a feedback since even $\tilde{x}^0(t)$ in (1.50) depends on fixed initial conditions.

A suboptimal synthesis is constructed in [109] for the polynominal in x, $f(x, t)$. This is done by means of the expansion of the solution to the dynamic programming equations that are equivalent to problem (1.47) and (1.48). We shall also demonstrate a result concerning the weakly controlled system

$$\dot{x} = f(x, t) + \epsilon Bu \quad , \quad x(0) = x_0 \tag{1.52}$$

with the quadratic performance index (1.48).

Theorem 1.6.

Let there exist a unique solution $x^0(t)$ to the reduced problem

$$\dot{x} = f(x, t) \quad , \quad x(0) = x_0$$

where $f(x, t)$ is continuously differentiable with respect to t, x. Then there exists $\hat{\epsilon} > 0$ such that, for $0 \le \epsilon \le \hat{\epsilon}$, the solutions $\bar{x}^l(t)$, $\bar{u}^l(t)$ to the successive problems

$$\begin{aligned} \dot{\bar{x}}^l &= f(\bar{x}^l, t) + \epsilon^2 BR^{-1}B^T\bar{\lambda}^{l-1}(t) \, , \, \bar{x}^l(0) = x_0 \; ; \\ \dot{\bar{\lambda}}^{l-1} &= g[\bar{x}^{l-1}(t), \bar{\lambda}^{l-1}(t), t] + Q\bar{x}^{l-1}(t) \, , \, \bar{\lambda}^{l-1}(T) = -H\bar{x}^{l-1}(T) \; ; \\ \bar{u}^l(t) &= -\epsilon R^{-1}B^T\bar{\lambda}^{l-1}(t) \, , \, \bar{u}^0(t) = 0 \, , \, g(x, \lambda, t) \triangleq -(\nabla_x f(x, t))^T\lambda \end{aligned} \tag{1.53}$$

satisfy the conditions

$$\|\bar{x}(t, \epsilon) - x^l(t)\| = O(\epsilon^{2l+2}) \quad , \quad \|\bar{u}(t, \epsilon) - u^l(t)\| = O(\epsilon^{2l+2}) \tag{1.54}$$

where $\bar{x}(t, \epsilon)$, $\bar{u}(t, \epsilon)$ constitute the unique solution to the optimization problem (1.48) and (1.52).

The proof is given in [40] and is analogous to that of Theorem 1.3, but the existence of the perturbed solution and its uniqueness are proved according to the general scheme of the monograph [71].

Note in conclusion that the results described can be applied to an approximate decomposition. This is sufficiently clear in the case of weakly connected systems with a separable (in evaluation of the subsystems quality) objective.

The most popular range of problems is related to the state stabilization of a solid body (airplane, missile, satellite, gyroplatform). It is known that, if angles of rotation (referred to the standard axes) are small, the equations of motion can be written in a linear form and are decomposed into independent subsystems. However, when the angles are not small, nonlinear interconnections must be taken into account. Application of the perturbation method allows one to use the independent linear control as the reduced one to construct corrections that make the solution consistent with weakly nonlinear interconnections. Calculations (see, in particular, [109]) show that the first approximation gives good results.

6.2. The general terminal problem

Consider the problem consisting of minimization of the performance index

$$\inf G(x_\epsilon(T)) \triangleq G_\epsilon^* \tag{2.1}$$

where the n-vector function $x_\epsilon(t)$, $t \in [0, T]$ is a trajectory of the system described by the following differential equation

$$\dot{x}_\epsilon = f(x_\epsilon, u, t) + \epsilon\gamma(x_\epsilon, u, t) \quad , \quad x(0) = x_0 \tag{2.2}$$

provided that

$$u \in U \tag{2.3}$$

and where U is a compact set in R^m. The corresponding reduced problem is of the form

$$\inf G(x(T)) \triangleq G^* \tag{2.4}$$

$$\dot{x} = f(x, u, t) \tag{2.5}$$

$$u \in U \tag{2.6}$$

Assume that the solution to (2.4)–(2.6) is not unique. Let us denote by $\{u^\beta\}$ the family of functions for which each is an optimal control of the problem (2.4)–(2.6) and, conversely that any control function that is optimal for (2.4)–(2.6) is an element of $\{u^\beta\}$. With any control $u_\beta \in \{u^\beta\}$ we associate the trajectories $x_\beta(t)$, $\lambda_\beta(t)$ of the primal and conjugate systems.

$$\dot{x}_\beta(t) = f(x_\beta, u_\beta, t), \; x(0) = x_0 \tag{2.7}$$

$$\dot{\lambda}_\beta(t) = -f_x^T(x_\beta, u_\beta, t)\lambda_\beta(t), \; \lambda_\beta(T) = -G_x(x_\beta(T)) \tag{2.8}$$

(Here and in what follows we use the abbreviated notations G_x, f_x instead of $\nabla_x G$, $\nabla_x f$.)

By analogy with the discrete time problem (see Section 2.7), we can formulate the auxiliary problem as follows. Find the measurable function $u^*(t)$ maximizing the objective

$$\Phi^* \triangleq \max_\beta \int_0^T \lambda_\beta^T(\tau)\gamma(x_\beta(\tau), u_\beta(\tau), \tau)\mathrm{d}\tau \tag{2.9}$$

on the set of functions $\{u^\beta\}$.

As in Section 2.7, the auxiliary problem takes the simplest form when the nonuniqueness is caused by an aggregation of controls, i.e., when

$$f(x, u, t) \triangleq \tilde{f}(x, \psi(u, t), t) \quad , \quad \psi(\cdot) \in \tilde{U} \subset R^k \quad , \quad k < m$$

Denoting by $\psi^*(t)$ the optimal control to the aggregated problem

$$\begin{aligned} &\dot{x} = \tilde{f}(x, \psi, t) \ , \ x(0) = x_0 \\ &\psi \in \tilde{U} \ , \ \inf G(x(T)) \end{aligned}$$

and by $\tilde{x}(t)$, $\tilde{\lambda}(t)$ the associated trajectories of the primal and conjugate systems

$$\begin{aligned} &\dot{\tilde{x}} = \tilde{f}(\tilde{x}, \psi^*(t), t) \ , \ \tilde{x}(0) = x_0 \\ &\dot{\tilde{\lambda}} = -f_x^T(\tilde{x}, \psi^*(t), t)\tilde{\lambda} \ , \ \lambda(T) = -G_x(\tilde{x}(T)) \end{aligned}$$

one can reduce the problem (2.9) to that of maximization of the objective

$$\max_u\{(\tilde{\lambda}(t))^T\gamma(\tilde{x}(t), u, t) \,/\, u \in U^*(t)\}$$

where the sets

$$U^*(t) \triangleq \{u \,/\, u \in U \ , \ \psi(u, t) = \psi^*(t)\}$$

are supposed to be nonvoid for all t.

An especially simple form, which is close to the maximum principle, is exhibited by the auxiliary problem for weakly controlled systems, which will be investigated in detail in Section 6.4. As an illustration, one can also consider the next example that is, in a certain sense, the inverse with regard to the case of weak control.

Consider the problem of finding the vector of controls u that minimizes the objective

$$G(x(T)) = -I^T x(T) \ , \ I^T \triangleq (1,1,\ldots,1) \tag{2.10}$$

subject to

$$\dot{x} = f(x, u, t) \triangleq u + \epsilon\gamma(x, u, t) \tag{2.11}$$

$$U = \{u \,/\, I^T u \leq 1 \ , \ u \geq 0\} \tag{2.12}$$

Clearly the optimal reduced controls must be in the set

$$U^* = \{u \,/\, I^T u = 1 \ , \ u \geq 0\} \tag{2.13}$$

and also

$$\bar{\lambda}(t) = I$$

The auxiliary problem is written here as a standard optimal control problem

$$\max \{\int_0^T I^T \gamma(x, u, t) \mathrm{d}t \,/\, \dot{x} = u \,,\, x(0) = x_0 \,,\, u \in U^*\} \tag{2.14}$$

and a suboptimal control synthesis can easily be found by determining

$$u^*(t) = \arg\max \{I^T \gamma(x, u, t) \,/\, u \in U^*\} \tag{2.15}$$

Let us proceed to the formal investigation of the relationship between the auxiliary and the original perturbed problem. Assume that the solution $u^*(t)$ to the auxiliary problem exists and is of bounded full variation (the full variation of a vector-function being the sum of the full variations of its components):

$$\overset{T}{\underset{0}{V}}(u^*) \leq k \;,\; k = \text{constant} \tag{2.16}$$

It is then natural to look for a solution to the perturbed problem in a class of functions also having the bounded full variation

$$\overset{T}{\underset{0}{V}}(u) \leq \hat{k} \;,\; \hat{k} = \text{constant} \;,\; \hat{k} \geq k \tag{2.17}$$

By the Helly convergence theorem for sequences of functions of bounded variation [55], this solution $u_\epsilon^*(t)$ always exists in the indicated class. The following result can be used for its approximate construction.

Theorem 2.1.

Let the functions $f(\cdot)$, $f_x(\cdot)$, $\gamma(\cdot)$, $G(\cdot)$ be continuous and continuously differentiable with respect to x, u. Then

(1) The optimal value G_ϵ^* of the perturbed problem is approximated by optimal values of the reduced problem G and the auxiliary one Φ^*:

$$G_\epsilon^* = G^* - \epsilon \Phi^* + o(\epsilon)$$

(2) $u^*(t)$ is a suboptimal control for the perturbed problem, i.e.,

$$G^*_\epsilon = \tilde{G}_\epsilon + o(\epsilon)$$

where

$$\tilde{G}_\epsilon \triangleq G[x_\epsilon(T)]$$

and where $x_\epsilon(t)$, $t \in [0, T]$ is the solution to the system (2.2) with $u = u^*(t)$.

(3) If the solution to the auxiliary problem is unique (that is, if $u^*(t)$ is unique up to the equivalence), then

$$u^*_\epsilon(t) \underset{\epsilon \to 0}{\overset{\text{measure}}{\longrightarrow}} u^*(t) \ , \ x^*_\epsilon(t) \underset{\epsilon \to 0}{\overset{c}{\longrightarrow}} x^*(t)$$

where $x^*_\epsilon(t)$ is a trajectory of the perturbed system (2.2) with $u^*_\epsilon(t)$ a control function, and where $x^*(t)$ is a trajectory of the reduced system (2.5) corresponding to $u^*(t)$.

Proof. Let $x_\epsilon(t)$ be the solution to the Cauchy problem

$$\dot{x}_\epsilon = f(x_\epsilon, u^*, t) + \epsilon\gamma(x_\epsilon, u^*, t, \epsilon) \ , \ x_\epsilon(0) = x_0 \tag{2.18}$$

Let us show first that

$$x_\epsilon(t) = x^*(t) + \epsilon\tilde{x}_1(t, \epsilon) \ , \ \|\tilde{x}_1(t, \epsilon)\| \le M = \text{constant} \tag{2.19}$$

Rewriting the corresponding differential equations in the equivalent integral form, we can obtain

$$x_\epsilon(t) - x^*(t) = \int_0^t [f(x_\epsilon, u^*, \tau) - f(x^*, u^*, \tau)]d\tau + \epsilon\int_0^t \gamma(x_\epsilon, u^*, \tau, \epsilon)d\tau$$

This implies the estimate

$$\|x_\epsilon(t) - x^*(t)\| \le k_1\int_0^t \|x_\epsilon(\tau) - x^*(\tau)\|d\tau + k_2\epsilon \tag{2.20}$$

Application of the Gronwall–Bellman lemma to (2.20) allows us to justify (2.19). Let us substitute (2.19) into (2.18). Then, in view of the definition of $x^*(t)$,

$$\dot{\tilde{x}}_1(t, \epsilon) = f_x(x^*(t), u^*(t), t)\tilde{x}_1(t, \epsilon) + \gamma(x^*(t), u^*(t), t, 0) + O(\epsilon) \ , \ \tilde{x}_1(0, \epsilon) = 0 \tag{2.21}$$

We have, further,

$$G(x_\epsilon(T)) = G(x^*(T)) + \epsilon G_x^T(x^*(T))\tilde{x}_1(T,\epsilon) + o(\epsilon) \tag{2.22}$$

Let us denote by $\lambda^*(t)$ the solution to the Cauchy problem

$$\dot{\lambda}^*(t) = -f_x^T(x^*(t), u^*(t), t)\lambda^*(t) \quad , \quad \lambda^*(T) = -G_x(x^*(T)) \tag{2.23}$$

For the sake of brevity, we shall write

$$\gamma(x, u, t, \epsilon) = \gamma(x, u, t)$$

We may now write

$$\begin{aligned} G_x^T(x^*(T))\tilde{x}_1(T,\epsilon) &= -(\lambda^*(T))^T\tilde{x}_1(T,\epsilon) \\ &= -\int_0^T \frac{\mathrm{d}}{\mathrm{d}\tau}[(\lambda^*(\tau))^T\tilde{x}_1(\tau,\epsilon)]\mathrm{d}\tau \\ &= -\int_0^T (\lambda^*(\tau))^T\gamma(x^*(\tau), u^*(\tau), \tau)\mathrm{d}\tau + O(\epsilon) \end{aligned}$$

Substituting the latter into (2.22), we obtain

$$\tilde{G}_\epsilon \triangleq G(x_\epsilon(T)) = G^* - \epsilon\int_0^T (\lambda^*(\tau))^T\gamma(x^*(\tau), u^*(\tau), \tau)\mathrm{d}\tau + o(\epsilon) \tag{2.24}$$

The similar expansion for G_ϵ^* is obtained in the same way. To this end the solution to the system (2.5) is constructed for $u = u_\epsilon^*(t)$

$$\dot{x}_{u_\epsilon^*} = f(x_{u_\epsilon^*}, u_\epsilon^*, t) \quad , \quad x_{u_\epsilon^*}(0) = x_0$$

Then, analogous to (2.19), the representation

$$x_\epsilon^*(t) = x_{u_\epsilon^*}(t) + \epsilon\check{x}_1(t,\epsilon)$$

is verified, from which

$$G_\epsilon^* \triangleq G(x_\epsilon^*(T)) = G(x_{u_\epsilon^*}(T)) + \epsilon G_x^T(x_{u_\epsilon^*}(T))\check{x}_1(T,\epsilon) + o(\epsilon)$$

It follows that

$$G_\epsilon^* = G(x_{u_\epsilon^*}(T)) - \epsilon\int_0^T \lambda_{u_\epsilon^*}^T(\tau)\gamma(x_{u_\epsilon^*}(\tau), u_\epsilon^*(\tau), \tau)\mathrm{d}\tau + o(\epsilon) \tag{2.25}$$

where $\lambda_{u_\epsilon^*}(\tau)$ is the solution to the Cauchy problem

$$\dot{\lambda}_{u_\epsilon^*}(\tau) = -f_x^T(x_{u_\epsilon^*}, u_\epsilon^*, \tau)\lambda_{u_\epsilon^*}(\tau) \ , \ \lambda_{u_\epsilon^*}(T) = -G_x(x_{u_\epsilon^*}(T))$$

Let us now compare (2.24) and (2.25). By definition, we have

$$G_\epsilon^* \leq \tilde{G}_\epsilon \tag{2.26}$$

$$G^* \leq G(x_{u_\epsilon^*}(T)) \Rightarrow \tag{2.27}$$

$$\int_0^T \lambda_{u_\epsilon^*}^T(\tau)\gamma(x_{u_\epsilon^*}(\tau), u_\epsilon^*(\tau), \tau) \geq \int_0^T (\lambda^*(\tau))^T \gamma(x^*(\tau), u^*(\tau), \tau)\mathrm{d}\tau + \\ + \nu(\epsilon) \ , \ \nu(\epsilon) \underset{\epsilon \to 0}{\to} 0 \tag{2.28}$$

We now show that the following limit exists

$$\lim_{\epsilon \to 0} \int_0^T \lambda_{u_\epsilon^*}^T(\tau)\gamma(x_{u_\epsilon^*}(\tau), u_\epsilon^*(\tau), \tau)\mathrm{d}\tau = \int_0^T (\lambda^*(\tau))^T \gamma(x^*(\tau), u^*(\tau), \tau)\mathrm{d}\tau \tag{2.29}$$

Assume the contrary. Then, by virtue of (2.28), there exists some subsequence $\{\epsilon'\}$ such that

$$\lim_{\epsilon' \to 0} \int_0^T \lambda_{u_\epsilon^*}^T(\tau)\gamma(x_{u_\epsilon^*}(\tau), u_\epsilon^*(\tau), \tau)\mathrm{d}\tau > \int_0^T (\lambda^*(\tau))^T \gamma(x^*(\tau), u^*(\tau), \tau)\mathrm{d}\tau \tag{2.30}$$

From the Helly theorem [55], we have that there exists a subsequence of the sequence $\{\epsilon'\}$ such that there exists the pointwise limit

$$\lim_{\epsilon'' \to 0} u_{\epsilon''}^*(\tau) = \hat{u}(\tau) \ , \ \overset{T}{\underset{0}{V}}(\hat{u}) \leq \hat{k} \tag{2.31}$$

Taking into account the smoothness of the function $f(\cdot)$, one can easily prove the following inequality

$$\|x_{u_\epsilon^*}(t) - x_{\hat{u}}(t)\| \leq k_1 \int_0^t \|x_{u_\epsilon^*} - x_{\hat{u}}\|\mathrm{d}\tau + k_2 \int_0^t \|u_\epsilon^* - \hat{u}\|\mathrm{d}\tau$$

where $x_{\hat{u}}(t)$ is the trajectory of the reduced system (2.5) corresponding to $u(t) = \hat{u}(t)$. Applying the Gronwall–Bellman lemma, one can obtain

$$\|x_{u_\epsilon^*}(t) - x_{\hat{u}}(t)\| \leq k_3 \int_0^T \|u_{\epsilon''}^*(\tau) - \hat{u}(\tau)\|\mathrm{d}\tau$$

from which, by the Lebesgue theorem [55],

$$x_{u^*_{\epsilon''}}(t) \xrightarrow{c} x_{\hat{u}}(t) \quad , \quad \epsilon'' \to 0 \tag{2.32}$$

In exactly the same way, one can establish that

$$\lambda_{u^*_{\epsilon''}}(t) \xrightarrow{c} \lambda_{\hat{u}}(t) \quad , \quad \epsilon'' \to 0 \tag{2.33}$$

with $\lambda_{\hat{u}}(t)$ as the solution to the Cauchy problem

$$\dot{\lambda}_{\hat{u}} = -f_x^T(x_{\hat{u}}, \hat{u}, \tau)\lambda_{\hat{u}} \quad , \quad \lambda_{\hat{u}}(T) = -G_x(x_{\hat{u}}(T))$$

The relations (2.32) and (2.33) imply

$$\lim_{\epsilon''\to 0} \int_0^T \lambda^T_{u^*_\epsilon}(\tau)\gamma(x_{u^*_\epsilon}(\tau), u^*_\epsilon(\tau), \tau)d\tau = \int_0^T \lambda^T_{\hat{u}}(\tau)\gamma(x_{\hat{u}}(\tau), \hat{u}(\tau), \tau)d\tau \Rightarrow \tag{2.34}$$

$$\int_0^T \lambda^T_{\hat{u}}(\tau)\gamma(x_{\hat{u}}(\tau), \hat{u}(\tau), \tau) > \int_0^T (\lambda^*(\tau))^T\gamma(x^*(\tau), u^*(\tau), \tau) \tag{2.35}$$

On the other hand, according to (2.24)–(2.26),

$$\lim_{\epsilon''\to 0} G^*_{\epsilon''} = G(x_{\hat{u}}(T)) \le \overline{\lim_{\epsilon\to 0}}\, \tilde{G}_\epsilon = G^* \Rightarrow G(x_{\hat{u}}(T)) = G^* \tag{2.36}$$

Thus, $\hat{u}(\tau)$ is the optimal control for the reduced problem.

Then, however, (2.35) contradicts the definition of $u^*(\tau)$ as a solution to the auxiliary problem. This contradiction proves (2.29) which, in view of (2.24) and (2.25), implies the validity of the first and the second part of the theorem. The third part will be proved if (given that the auxiliary problem has a unique solution) we show:

$$u^*_\epsilon(t) \xrightarrow[\epsilon\to 0]{L_2} u^*(t) \tag{2.37}$$

Assume the contrary. Then there exists a positive δ and a subsequence $\{\epsilon'\}$ such that

$$\int_0^T \| u^*_{\epsilon'}(\tau) - u^*(\tau)\| d\tau > \delta \quad , \quad \epsilon' \to 0 \tag{2.38}$$

Choose any subsequence $\{\epsilon''\}$ that ensures the existence of the limit (2.31). As above, it is easy to verify that $\hat{u}(\tau)$ is the solution to the auxiliary problem. At the same time, taking the limit in (2.38) along the subsequence $\{\epsilon''\}$ gives

$$\int_0^T \| \hat{u}(\tau) - u^*(\tau)\| d\tau > \delta$$

This contradicts the uniqueness condition. The proof of continuous convergence of the trajectories is analogous to the proof of (2.32). QED.

6.3. Difference approximations

In the preceding section, we investigated a procedure that permits the approximate solution of (2.1)–(2.3) by constructing the set of solutions to the reduced problem (2.4)–(2.6) and then solving the auxiliary problem (2.9). However, in the general case, the determination of the set $\{u^\beta\}$ of optimal reduced controls is rather difficult. Moreover, this set can have a complex structure that does not allow standard variation. In this section, another algorithm for the approximate solution to the problem (2.1)–(2.3) is developed. Instead of considering the reduced and auxiliary problems, we deal with the associated difference approximations. As above, we look for the solution in a class of functions of bounded variation

$$\overset{T}{\underset{0}{V}}(u) \leq \hat{k} \ , \ \hat{k} = \text{constant} \tag{3.1}$$

For the perturbed problem, the discrete approximation of the order n is defined as the following finite-dimensional mathematical programming problem

$$\min G(x_{(n,n)}) \triangleq G^*_{n\epsilon} \tag{3.2}$$

$$\begin{aligned} x_{(n,i)} = \ & x_{(n,i-1)} + f(x_{(n,i-1)}, u_{(n,i-1)}, t_{(n,i-1)})\Delta + \\ & + \epsilon\gamma(x_{(n,i-1)}, u_{(n,i-1)}, t_{(n,i-1)})\Delta \ , \ i = 1,\ldots,n \ , \ x_{(n,0)} \triangleq x_0 \end{aligned} \tag{3.3}$$

$$u_{(n,i)} \in U \ , \ i = 0,1,\ldots,n-1 \tag{3.4}$$

$$\sum_{i=1}^{n} |u_{(n,i)} - u_{(n,i-1)}| \leq \hat{k} \tag{3.5}$$

Here $\Delta = T/n$, $t_{(n,i)} = iT/n$, $i = 1,2,\ldots,n$. Condition (3.5) is a consequence of the constraint (3.1) on a class of admissible solutions in the original problem.

The reduced [with respect to (3.2)–(3.5)] problem is of the form

$$\min G(x_{(n,n)}) \triangleq G^*_n \tag{3.6}$$

$$\begin{aligned} x_{(n,i)} = \ & x_{(n,i-1)} + f(x_{(n,i-1)}, u_{(n,i-1)}, t_{(n,i-1)})\Delta \ , \\ & i = 1,\ldots,n \ , \ x_{(n,0)} \triangleq x_0 \end{aligned} \tag{3.7}$$

$$u_{(n,i)} \in U \ , \ i = 0,1,\ldots,n-1 \tag{3.8}$$

$$\sum_{i=1}^{n} |u_{(n,i)} - u_{(n,i-1)}| \leq \hat{k} \tag{3.9}$$

Obviously, this problem is that of discrete approximation of the order n for (2.4)–(2.6). Let $\lambda_{(n,i)}(x^n, u^n)$, $i = 1,2,\ldots,n$, be the Lagrange multipliers vector corresponding to the feasible solution $x^n \triangleq x_{(n,i)}$, $i = 1,2,\ldots,n$, $u^n = u_{(n,i)}$, $i = 1,2,\ldots,n-1$, to the problem (3.6)–(3.9). By definition, the vector $\lambda_{(n,i)}$ is the solution to the system

$$\begin{aligned} \lambda_{(n,i)}(x^n, u^n) = {} & \lambda_{(n,i+1)}(x^n, u^n) + \\ & + f_x^T(x_{(n,i)}, u_{(n,i)}, t_{(n,i)}) \cdot \lambda_{(n,i+1)}(x^n, u^n)\Delta \ , \\ & i = n-1,\ldots,1 \ , \ \lambda_{(n,n)} = -G_x(x_{n,n}) \end{aligned} \tag{3.10}$$

The approximate auxiliary problem of index M is the following mathematical program

$$\begin{aligned} \max \ & \{ \sum_{i=1}^{n} \lambda_{(n,i)}^T(x^n, u^n)\gamma(x_{(n,i-1)}, u_{(n,i-1)}, t_{(n,i-1)})\Delta \ / \\ & (x^n, u^n) \in Q_n^M \} \end{aligned} \tag{3.11}$$

where Q_n^M is the set of feasible solutions to (3.6)–(3.9) such that corresponding values of the objective function are equal to the optimal one to within $M\Delta$, i.e.,

$$\begin{aligned} Q_n^M = {} & \{ (x^n, u^n) \ / \ G(x_{(n,n)}) \leq G_n^* + M\Delta \ , \\ & x_{(n,i)} = x_{(n,i-1)} + f(x_{(n,i-1)}, u_{(n,i-1)}, t_{(n,i-1)})\Delta \ , \\ & i = 1,\ldots,n \ , \ x_{(n,0)} \triangleq x_0 \ , \\ & \sum_{i+1}^{n} |u_{(n,i)} - u_{(n,i-1)}| \leq \hat{k} \ , \ u_{(n,i)} \in U \} \end{aligned} \tag{3.12}$$

The difference between (3.11) and (3.12) and the auxiliary problem associated with (3.2)–(3.5) is obvious: the former has a wider feasible set.

Let us define the piecewise constant control

$$u_n^M(t) = u_{(n,i)}^M \ , \ t_{(n,i)} < t \leq t_{(n,i+1)} \ , \ i = 0,1,\ldots,n-1 \tag{3.13}$$

where $u_{(n,i)}^M$, $i = 0,1,\ldots,n-1$, $x_{(n,i)}^M$, $i = 1,\ldots,n$ is the solution to (3.11). This control is feasible for the original perturbed problem (2.1)–(2.3). We now show that its application is as efficient as that of the differential auxiliary problem solution [to within $O(\Delta)$].

Theorem 3.1.

Let the functions $f(\cdot)$, $f_x(\cdot)$, $G(\cdot)$, $\gamma(\cdot)$ be continuous and let them have continuous derivatives with respect to x. Let these functions satisfy the Lipschitz conditions with respect to u and t. Then there exists some positive constant M_0 such that, for all $M \geq M_0$,

$$G_\epsilon^* \leq \tilde{G}_{n\epsilon}^M \leq G_\epsilon^* + (M + L)\Delta + o(\epsilon) \tag{3.14}$$

where $G_{n\epsilon}^M$ is the value of the performance index (2.1) on the trajectory (2.2) with $u_n^M(t)$ as a control and where G_ϵ^* is the optimal value of the performance index. The inequality (3.14) is uniform with respect to Δ, and L is a constant which does not depend on Δ, ϵ. (The way to find L, M_0 is indicated below.)

The proof is based on the following two lemmata.

Lemma 3.1. Let all the conditions of the theorem be fulfilled. Let $\{\tilde{u}_{(n,i)}, i = 0,1,\ldots,n-1, \tilde{x}_{(n,i)}, i = 1,\ldots,n\}$ be a feasible solution to the problem (3.6)–(3.9), and let $\{\tilde{\lambda}_{(n,i)}, i = 1,\ldots,n\}$ be the corresponding Lagrange vector. Then there exist constants k_1, k_2 such that

$$\|\tilde{x}_n(t_{(n,i)}) - \tilde{x}_{(n,i)}\| \leq k_1\Delta \tag{3.15}$$

$$\|\tilde{\lambda}_n(t_{(n,i)}) - \tilde{\lambda}_{(n,i)}\| \leq k_2\Delta \tag{3.16}$$

where $\tilde{x}_n(t)$, $\tilde{\lambda}_n(t)$ are the trajectories of the reduced system (2.5) and its conjugate, which are obtained by using the control $\tilde{u}_n(t)$, $\tilde{u}_n(t)$, defined as follows.

$$\tilde{u}_n(t) = \tilde{u}_{(n,i)} \quad , \quad t_{(n,i)} \leq t \leq t_{(n,i+1)} \quad , \quad i = 0,1,\ldots,n-1 \tag{3.17}$$

Moreover, for any function $\psi(x, \lambda, u, t)$ satisfying the Lipschitz condition, one can find a constant k_3 such that

$$\left| \sum_{i=1}^{n} \psi(\tilde{x}_{(n,i-1)}, \tilde{\lambda}_{(n,i)}, \tilde{u}_{(n,i-1)}, t_{(n,i-1)})\Delta - \int_0^T \psi(\tilde{x}_n(\tau), \tilde{\lambda}_n(\tau), \tilde{u}_n(\tau), \tau)d\tau \right| \leq k_3\Delta \tag{3.18}$$

Lemma 3.2. Let the conditions of the theorem be fulfilled. Let $u(t)$ be a feasible control in (2.4)–(2.6) with bounded full variation (3.1), and let $x(t)$, $\lambda(t)$ be the primal and conjugate trajectories of the reduced system (2.5). If $x_{(n,i)}$, $\lambda_{(n,i)}$, $i = 1,\ldots,n$ are the trajectories of the discrete systems (3.7) and (3.10) with

$$u_{(n,i)} \triangleq u(t_{(n,i)}) \quad , \quad i = 0,1,\ldots,n-1 \tag{3.19}$$

then there exist constants k_4, k_5 such that

$$\|x(t_{(n,i)}) - x_{(n,i)}\| \leq k_4\Delta \tag{3.20}$$

$$\|\lambda(t_{(n,i)}) - \lambda_{(n,i)}\| \leq k_5\Delta \tag{3.21}$$

For an arbitrary function $\psi(x, \lambda, u, t)$ satisfying the Lipschitz conditions, there exists a constant k_6 such that

$$\left| \sum_{i=1}^{n} \psi(x_{(n,i-1)}, \lambda_{(n,i)}, u_{(n,i-1)}, t_{(n,i-1)})\Delta - \int_0^T \psi(x(\tau), \lambda(\tau), u(\tau), \tau)d\tau \right| \leq k_6\Delta \tag{3.22}$$

Proof of Lemma 3.1. By definition, we have

$$\begin{aligned}
\|\tilde{x}_n(t_{(n,i)}) - \tilde{x}_{(n,i)}\| \leq{} & \|\tilde{x}_n(t_{(n,i-1)}) - \tilde{x}_{(n,i-1)}\| + \\
& + \int_{t_{(n,i-1)}}^{t_{(n,i)}} \|f(\tilde{x}_n(\tau), \tilde{u}_n(\tau), \tau) - f(\tilde{x}_{(n,i-1)}, \tilde{u}_{(n,i-1)}, t_{(n,i-1)})\|d\tau \\
\leq{} & \|\tilde{x}_n(t_{(n,i-1)}) - \tilde{x}_{(n,i-1)}\| + L_1 \int_{t_{(n,i-1)}}^{t_{(n,i)}} \|\tilde{x}_n(\tau) - \tilde{x}_{(n,i-1)}\|d\tau + \\
& + L_2\Delta^2 \\
\leq{} & \|\tilde{x}_n(t_{(n,i-1)}) - \tilde{x}_{(n,i-1)}\|(1 + L_1\Delta) + (L_2 + L_1L_3)\Delta^2
\end{aligned}$$

where L_1, L_2 are the Lipschitz constants, $L_3 \triangleq \max_{(x,u,t)\in D} \|f(x, u, t,)\|$, with D a sufficiently large closed domain. Thus we can write

$$\|\tilde{x}_n(t_{(n,i)}) - \tilde{x}_{(n,i)}\| \leq y_{(n,i)} \quad , \quad i = 0,1,\ldots,n$$

where $y_{(n,i)}$ is the solution to the linear difference equation

$$y_{(n,i)} = y_{(n,i-1)}(1 + L_1\Delta) + (L_2 + L_1L_3)\Delta^2 \quad , \quad y(n,0) = 0$$

As is easy to verify, its solution is of the form

$$\begin{aligned}
y_{(n,i)} &= \frac{L_1L_3 + L_2}{L_1}(1 + L_1\Delta)^i\Delta - \frac{L_1L_3 + L_2}{L_1}\Delta \\
&\leq \Delta\frac{L_1L_3 + L_2}{L_1}\left[\left(1 + \frac{TL_1}{n}\right)^n - 1\right] \\
&\leq \frac{L_1L_3 + L_2}{L_1}[e^{TL_1} - 1]\Delta \\
&\triangleq k_1\Delta
\end{aligned}$$

The latter leads to the required (3.16). The inequality (3.17) may be similarly proved, and (3.18) is a simple implication of (3.15), (3.16), and (3.17).

Proof of Lemma 3.2. As in the preceding lemma, we can write

$$\begin{aligned}\|x(t_{(n,i)}) - x_{(n,i)}\| \le\ & \|x(t_{(n,i-1)}) - x_{(n,i-1)}\| + \\ & + \int_{t_{(n,i-1)}}^{t_{(n,i)}} \|f(x(\tau), u(\tau), \tau) - f(x_{(n,i-1)}, u_{(n,i-1)}, t_{(n,i-1)})\| d\tau \\ \le\ & \|x(t_{(n,i-1)}) - x_{(n,i-1)}\| + L_1 \int_{t_{(n,i-1)}}^{t_{(n,i)}} \|x(\tau) - x_{(n,i-1)}\| d\tau + \\ & + L_4 \int_{t_{(n,i-1)}}^{t_{(n,i)}} \|u(\tau) - u_{(n,i-1)}\| d\tau + L_2\Delta^2 \\ \le\ & \|x(t_{(n,i-1)}) - x_{(n,i-1)}\|(1 + L_1\Delta) + \\ & + L_4 \int_{t_{(n,i-1)}}^{t_{(n,i)}} \|u(\tau) - u_{(n,i-1)}\| d\tau + (L_2 + L_1L_3)\Delta^2\end{aligned}$$

using, in the last inequality, the fact that the Euclidean norm is not greater than the Chebyshev norm

$$\|u(\tau) - u_{(n,i-1)}\| \le |u(\tau) - u_{(n,i-1)}|$$

Let us denote by s_i the full variation of $u(\tau)$ on the interval $[t_{(n,i-1)}, t_{(n,i)}]$

$$s_i \triangleq \mathop{V}_{t_{(n,i-1)}}^{t_{(n,i)}} (u) \tag{3.23}$$

According to (3.1)

$$\sum_{i=1}^{n} s_i \le \hat{k} \tag{3.24}$$

Taking into account the notations introduced above, we have

$$\begin{aligned}& \int_{t_{(n,i-1)}}^{t_{(n,i)}} |u(\tau) - u_{(n,i-1)}| d\tau \le s_i\Delta \Rightarrow \\ & \|x(t_{(n,i)} - x_{(n,i)}\| \le \|x(t_{(n,i-1)}) - x_{(n,i-1)}\|(1 + L_1\Delta) + \\ & + L_4 s_i\Delta + (L_2 + L_1L_3)\Delta^2\end{aligned} \tag{3.25}$$

Consequently, the following estimates are valid:

$$\|x(t_{(n,i)}) - x_{(n,i)}\| \le y_{(n,i)} \quad , \quad i = 0,1,\ldots,n \tag{3.26}$$

where $y_{(n,i)}$ is the solution to the difference equation

$$y_{(n,i)} = y_{(n,i-1)}(1 + L_1\Delta) + L_4 s_i\Delta + (L_1 + L_1 L_3)\Delta^2 \ , \ y_{(n,0)} \triangleq 0$$

This simple substitution indicates that its solution can be written in the form

$$\begin{aligned} y_{(n,i)} &= \sum_{j=1}^{i}(1 + L_1\Delta)^{i-j}[L_4 s_j\Delta + (L_2 + L_1 L_3)\Delta^2] \ , \ i = 0,1,\dots,n \Rightarrow \\ y_{(n,i)} &\le (1 + L_1\Delta)^n [L_4\Delta \sum_{j=1}^{n} s_j + (L_2 + L_1 L_3)\Delta] \\ &\le e^{L_1 T}[L_4\hat{k} + L_2 + L_1 L_3]\Delta \triangleq k_4\Delta \end{aligned}$$

which, together with (3.26), proves (3.20). The proof of (3.21) is analogous.

Let us prove the validity of (3.22). We have

$$\begin{aligned} \delta &\triangleq |\sum_{i=1}^{n}\psi(x_{(n,i-1)}, \lambda_{(n,i)}, u_{(n,i-1)}, t_{(n,i-1)})\Delta - \int_0^T \psi(x(\tau), \lambda(\tau), u(\tau), \tau)d\tau| \\ &\le \sum_{i=1}^{n}\int_{t_{(n,i-1)}}^{t_{(n,i)}} |\psi(x_{(n,i-1)}, \lambda_{(n,i)}, u_{(n,i-1)}, t_{(n,i-1)}) - \psi(x(\tau), \lambda(\tau), u(\tau), \tau)|d\tau \\ &\le \sum_{i=1}^{n}\int_{t_{(n,i-1)}}^{t_{(n,i)}} [L_6\|x_{(n,i-1)} - x(\tau)\| + L_7\|\lambda_{(n,i)} - \lambda(\tau)\| + \\ &\quad + L_8|u_{(n,i-1)} - u(\tau)| + L_9(\tau - t_{(n,i-1)})]d\tau \end{aligned}$$

where $L_6,\dots,L_9$ are the Lipschitz constants of the function $\psi(\cdot)$. Using further (3.20) and (3.21) and also the fact that, by definition of $x(t)$, $\lambda(t)$,

$$\|x(\tau) - x(t_{(n,i-1)})\| \le L_3\Delta \ , \ \|\lambda(\tau) - \lambda(t_{(n,i)})\| \le L_{10}\Delta$$

we obtain

$$\begin{aligned} \delta &\le \sum_{i=1}^{n}[L_6(k_4 + L_3)\Delta^2 + L_7(k_5 + L_{10})\Delta^2 + L_9\Delta^2 + L_8 s_i\Delta] \\ &\le \Delta[L_6(k_4 + L_3) + L_7(k_5 + k_{10}) + L_9]T + \Delta L_8\hat{k} \\ &\triangleq k_6\Delta \end{aligned}$$

Proof of Theorem 3.1. Let $\{\hat{u}_{(n,i)},\ i = 0,1,\dots,n-1,\ \hat{x}_{(n,i)},\ i = 1,\dots,n\}$ be an arbitrary solution to the problem (3.6)–(3.9) and let $\{\lambda_{(n,i)},\ i = 1,\dots,n\}$ be the corresponding Lagrange vector. Let us construct the piecewise constant control function $\hat{u}(t)$

$$\hat{u}(t) = \hat{u}_{(n,i)} \ , \ t_{(n,i)} \le t \le t_{(n,i+1)} \ , \ i = 0,1,\dots,n-1$$

Consider the trajectories $\hat{x}_n(t)$, $\hat{\lambda}_n(t)$ of the system (2.5) and its conjugate. By Lemma 3.1, there exists a constant $\hat{L}_1$ such that

$$|G(\hat{x}_{(n,n)}) - G(\hat{x}_n(T))| \leq \hat{L}_1\Delta \Rightarrow G(\hat{x}_{(n,n)}) + \hat{L}_1\Delta \geq G(\hat{x}_n(T))$$

Then, since

$$G(\hat{x}_{(n,n)}) \triangleq G_n^* \ , \ \ G(\hat{x}_n(T)) \geq G^*$$

we can write

$$G_n^* + \hat{L}_1\Delta \geq G^* \tag{3.27}$$

Now let $u^*(t)$ be a solution to the auxiliary problem belonging to the class of functions (3.1), and let $x^*(t)$, $\lambda^*(t)$ be the trajectories of the primal and conjugate system (2.5) corresponding to this control.

Let us set

$$u^*_{(n,i)} = u^*(t_{(n,i)}) \ , \ \ i = 0,1,\ldots,n-1$$

and form the vectors $x^*_{(n,i)}$, $\lambda^*_{(n,i)}$, $i = 1,2,\ldots,n$, these being the solutions to (3.7) and (3.10) with controls $\{u^*_{(n,i)}, i = 0,1,\ldots,n-1\}$. In accordance with Lemma 3.2, there exists the constant $\hat{L}_2$ for which

$$|G(x^*_{(n,n)}) - G(x^*(T))| \leq \hat{L}_2\Delta \Rightarrow G(x^*(T)) + \hat{L}_2\Delta \geq G(x^*_{(n,n)}) \tag{3.28}$$

At the same time,

$$G(x^*(T)) \triangleq G^*$$

Hence, in view of (3.27), we can write

$$G_n^* + (\hat{L}_1 + \hat{L}_2)\Delta \geq G(x^*_{(n,n)}) \tag{3.29}$$

Let us set

$$M_0 \triangleq \hat{L}_1 + \hat{L}_2 \tag{3.30}$$

Then, for $M \geq M_0$, the vectors $\{u^*_{(n,i)}, i = 0,1,\ldots,n-1, x^*_{(n,i)}, i = 1,\ldots,n\}$ form a feasible [with respect to (3.11)] control

$$\{u^*_{(n,i)} \ , \ \ i = 0,1,\ldots,n-1 \ ; \ \ x^*_{(n,i)} \ , \ \ i = 1,\ldots,n\} \in Q_n^M$$

Consequently,

$$\begin{aligned}&\sum_{i=1}^{n}(\lambda^M_{(n,i)})^T\gamma(x^M_{(n,i-1)}, u^M_{(n,i-1)}, t_{(n,i-1)})\Delta \geq \\ &\geq \sum_{i=1}^{n}(\lambda^*_{(n,i)})^T\gamma(x^*_{(n,i-1)}, u^*_{(n,i-1)}, t_{(n,i-1)})\Delta\end{aligned} \tag{3.31}$$

where $\{u^M_{(n,i)}\ ,\ i = 0,1,\ldots,n-1\ ,\ x^M_{(n,i)}\ ,\ i = 1,\ldots,n\}$, $\{\lambda^M_{(n,i)}\ ,\ i = 1,\ldots,n\}$ are the solution and its corresponding Lagrange vector to (3.11). By (3.22), we can write

$$\begin{aligned}&\sum_{i=1}^{n}(\lambda^*_{(n,i)})^T\gamma(x^*_{(n,i-1)}, u^*_{(n,i-1)}, t_{(n,i-1)})\Delta \geq \\ &\geq \int_0^T(\lambda^*(\tau))^T\gamma(x^*(\tau), u^*(\tau), \tau)d\tau - k_6\Delta\end{aligned} \tag{3.32}$$

and, from (3.18), it follows that

$$\begin{aligned}&\sum_{i=1}^{n}(\lambda^M_{(n,i)})^T\gamma(x^M_{(n,i-1)}, u^M_{(n,i-1)}, t_{(n,i-1)})\Delta \leq \\ &\leq \int_0^T(\lambda^M_n(\tau))^T\gamma(x^M_n(\tau), u^M_n(\tau), \tau)d\tau + k_3\Delta\end{aligned} \tag{3.33}$$

where $x^M_n(t)$, $\lambda^M_n(t)$ are the trajectories of the system (2.5) and its conjugate obtained by using the control $u^M_n(t)$.

Comparison of (3.31)–(3.33) leads to the relation

$$\begin{aligned}&\int_0^T(\lambda^M_n(\tau))^T\gamma(x^M_n(\tau), u^M_n(\tau), \tau)d\tau + (k_3 + k_6)\Delta \geq \\ &\geq \int_0^T(\lambda^*(\tau))^T\gamma(x^*(\tau), u^*(\tau), \tau)d\tau\end{aligned} \tag{3.34}$$

Similarly to (2.24), one can obtain

$$\tilde{G}^M_{n\epsilon} = G(x^M_n(T)) - \epsilon\int_0^T(\lambda^M_n(\tau))^T\gamma(x^M_n(\tau), u^M_n(\tau), \tau)d\tau + o(\epsilon) \tag{3.35}$$

with the term in $o(\epsilon)$ being uniform in n and being defined only by the values of the corresponding derivatives. According to Lemma 3.1, we also have

$$\begin{aligned}&|G(x^M_{(n,n)}) - G(x^M_n(T))| \leq \hat{L}_1\Delta \Rightarrow \\ &\Rightarrow G(x^M_n(T)) \leq G(x^M_{(n,n)}) + \hat{L}_1\Delta \overset{\Delta}{\leq} G^*_n + M\Delta + \hat{L}_1\Delta\end{aligned}$$

At the same time, from (3.28), it follows that

$$G^* + L_2\Delta \geq G_n^* \Rightarrow G(x_n^M(T)) \leq G^* + (\hat{L}_2 + \hat{L}_1 + M)\Delta \tag{3.36}$$

from which, taking into account (3.34), we can rewrite (3.35) in the form

$$\begin{aligned}\tilde{G}_{n\epsilon}^M \leq \ & G^* - \epsilon\int_0^T (\lambda^*(\tau))^T \gamma(x^*(\tau), u^*(\tau), \tau)\mathrm{d}\tau + o(\epsilon) + \\ & + \epsilon(k_3 + k_6)\Delta + (\hat{L}_2 + \hat{L}_1 + M)\Delta\end{aligned}$$

Assuming that

$$\epsilon(k_3 + k_6) \leq \hat{L}_3 = \text{constant}$$

and using the first result of Theorem 2.1, we obtain

$$\begin{aligned}& G_\epsilon^* \overset{\Delta}{\leq} \tilde{G}_{n\epsilon}^M \leq G^* - \epsilon\Phi^* + o(\epsilon) + (\hat{L}_2 + \hat{L}_3 + \hat{L}_1 + M)\Delta = \\ & = G_\epsilon^* + (L + M)\Delta + o(\epsilon)\end{aligned}$$

where $L \triangleq \hat{L}_2 + \hat{L}_3 + \hat{L}_1$.

The theorem allows us to exploit the special structure of the reduced problem since this is preserved under discretization.

As an example of the application of Theorem 3.1 we can construct the following algorithm to solve the problem (2.1)–(2.3):

(1) Estimate the constants M_0 and L in (3.14).
(2) Choose $M \geq M_0$ and n to satisfy the required solution precision.
(3) Find, by means of the feasible directions method, a feasible solution $\hat{x}^n$, $\hat{u}^n$ to the program (3.6)–(3.9) such that

$$G(\hat{x}_{(n,n)}) \leq G_n^* + M\Delta$$

(4) Starting from this solution as an initial approximation with the additional constraint

$$G(x_{(n,n)}) \leq G_n^* + M\Delta$$

find the solution to (3.11) by means of the feasible direction method.

Under the assumption that the solution to (2.9) is unique the results of Theorem 3.1 can be extended.

Theorem 3.2.

Let all the conditions of Theorem 3.1 be fulfilled, and let $u^*(t)$ be the unique (up to equivalence) solution to the auxiliary problem (2.9). Then

$$u_n^M(t) \underset{n\to\infty}{\overset{\text{measure}}{\longrightarrow}} u^*(t)\ ,\ x_n^M(t) \underset{n\to\infty}{\overset{c}{\longrightarrow}} x_{u^*}(t) \tag{3.37}$$

Proof. Let us show that

$$\lim_{n\to\infty} \int_0^T \|u_n^M(t) - u^*(\tau)\| d\tau = 0 \tag{3.38}$$

Assume this is not true. Then there exists a subsequence $\{n'\}$ and a number $\delta \geq 0$ such that

$$\int_0^T \|u_n^M(\tau) - u^*(\tau)\| d\tau \underset{n'\to\infty}{\geq} \delta \tag{3.39}$$

By the Helly theorem, one can state that a subsequence $\{n''\}$ of $\{n'\}$ exists such that

$$u_{n''}(t) \underset{n''\to\infty}{\longrightarrow} \bar{u}(t)\ ,\ \overset{T}{\underset{0}{V}}(\bar{u}) \leq \hat{k} \tag{3.40}$$

Similarly to Theorem 2.1, one can easily verify that

$$x_{n''}(t) \underset{n''\to\infty}{\overset{c}{\longrightarrow}} \bar{x}(t)\ ,\ \lambda_{n''}(t) \underset{n''\to\infty}{\overset{c}{\longrightarrow}} \bar{\lambda}(t)$$

where $\{x_{n''}(t), \lambda_{n''}(t)\}$, $\{\bar{x}(t), \bar{\lambda}(t)\}$ are the trajectories corresponding to $u_{n''}(t)$, $\bar{u}(t)$. By Lemma 3.1 and (3.36), we have

$$G(\bar{x}(T)) = \lim_{n''\to\infty} G(x_{n''}(T)) \leq \overline{\lim_{n\to\infty}}\ G(x^M_{(n,n)}) \triangleq$$
$$\triangleq \overline{\lim_{n\to\infty}}\ G_n^* \leq G^* \Rightarrow \bar{u}(t) \in \{u^\beta\}$$

i.e., $\bar{u}(t)$ is the solution to the reduced problem (2.4)–(2.6). Taking the limit in (3.34), we obtain

$$\int_0^T (\bar{\lambda}(\tau))^T \gamma(\bar{x}(\tau), \bar{u}(\tau), \tau) d\tau \geq \int_0^T (\lambda^*(\tau))^T \gamma(x^*(\tau), u^*(\tau), \tau) d\tau$$

It follows that $\bar{u}(\tau)$ is the solution to the auxiliary problem (2.9). Hence, in view of the uniqueness,

$$\bar{u}(t) = u^*(t) \tag{3.41}$$

On the other hand, taking the limit in (3.39) gives

$$\int_0^T \|\bar{u}(\tau) - u^*(\tau)\| d\tau \geq \delta > 0$$

which contradicts (3.41).

Thus, (3.38) is established. The statements of the theorem are direct consequences of (3.38). QED.

Theorem 3.2 indicates that the construction of approximate auxiliary problems may be considered a regularization procedure for the sequence of difference approximations to the reduced problem, which is incorrect in the sense of Tikhonov. This procedure, however, is essentially different from standard ones (see, e.g., [70, 80]). The difference is not a formal one and is closely connected with the formulation of the approach. In this approach the convergence (3.37) is due to the fact that the special structure of the problem (2.1)–(2.3) is taken into account.

6.4. Weak control (nonuniqueness of the reduced solution)

The problem of how to control dynamic systems if they are weakly controlled is of practical importance (as is noted, for example, in many works devoted to the control of spacecraft). We have studied such a situation in Section 6.1. However, we supposed there that the performance index depends on the control, and this allowed us to fix a zero-order approximation to the optimal control as the unique solution to the reduced problem. At the same time, the case in which the reduced problem does not depend on the control is also of great interest. Formally, the perturbed problem is described here as follows.

Find a measurable control

$$u(t) \in U \quad , \quad t \in [0, T] \tag{4.1}$$

which, on a fixed interval of time $[0, T]$, maximizes the performance index

$$\inf \{G(x_\epsilon(T))\} \tag{4.2}$$

subject to

$$\dot{x}_\epsilon(t) = f(x_\epsilon(t), t) + \epsilon\gamma(x_\epsilon(t), u(t), t) \quad , \quad x_\epsilon(0) = x_0 \tag{4.3}$$

The reduced system is, obviously, uncontrolled

$$\dot{x}(t) = f(x(t), t) \quad , \quad x(0) = x_0 \tag{4.4}$$

and an arbitrary control satisfying (4.1) may be taken as the optimal one to the reduced problem (i.e., the domain of feasible controls coincides with that of optimal controls).

Assume that the functions $f(\cdot)$, $f_x(\cdot)$, $\gamma(\cdot)$ are continuous and continuously differentiable over corresponding domains. Then the Cauchy problem (4.4) will be solvable and so will the conjugate Cauchy problem

$$\dot{\lambda}(t) = -f_x^T(x(t), t)\lambda(t) \quad , \quad \lambda(T) = -G_x(x(T)) \tag{4.5}$$

The auxiliary problem is reduced here to the maximization (for all $t \in [0, T]$) of the objective function

$$\Phi_t^* = \max\{\Phi_t(u) \,/\, u \in U\} \ , \ \Phi_t(t) \triangleq (\lambda(t))^T\gamma(x(t), u, t) \tag{4.6}$$

Suppose that there exists a measurable function

$$u^*(t) \triangleq \arg\max \Phi_t(t) \tag{4.7}$$

We shall also seek controls for the perturbed system (4.3) in the class of measurable functions. The results of Theorem 2.1 can now be stated more precisely.

Theorem 4.1.

Let the functions $f(\cdot)$, $f_x'(\cdot)$, $\gamma(\cdot)$, $G(\cdot)$, G_x' be continuous and have continuous derivatives with respect to x. Let the set U be compact, and let the function $u^*(t)$, defined by (4.7), be measurable. Then

(1) The value $\tilde{G}_\epsilon$ of the performance index, which is obtained by using $u^*(t)$ as a control in (4.3), is, to within $O(\epsilon^2)$, the optimal one

$$G_\epsilon^* = \tilde{G}_\epsilon + O(\epsilon^2)$$

(2) The expansion

$$G_\epsilon^* = G(x(T)) - \epsilon\int_0^T \Phi_t^* \mathrm{d}t + O(\epsilon^2)$$

is valid for the optimal value of the performance index.

(3) If the solution to problem (4.6) is almost everywhere unique, then for any control $\tilde{u}_\epsilon(t)$ such that

$$G(\tilde{x}_\epsilon(T)) = G_\epsilon^* + o(\epsilon) \tag{4.8}$$

the following relations are true:

$$\tilde{u}_\epsilon(t) \underset{\epsilon\to 0}{\overset{\text{measure}}{\longrightarrow}} u^*(t) \tag{4.9}$$

$$\tilde{x}_\epsilon(t) = x(t) + \epsilon x_1(t) + o(\epsilon) \tag{4.10}$$

where $\tilde{x}_\epsilon(t)$ is the solution to (4.3) corresponding to $\tilde{u}_\epsilon(t)$, and $x_1(t)$ is the solution to the Cauchy problem

$$\dot{x}_1(t) = f_x(x(t)\,,t)x_1(t) + \gamma(x(t)\,,u^*(t)\,,t) \quad , \quad x_1(0) = 0 \tag{4.11}$$

and where the error term in (4.10) is uniform with respect to t.

Proof. Let the function $\hat{u}_\epsilon(t)$ be a control, and let the corresponding trajectory of (4.3)

$$\hat{x}_\epsilon(t) = x_0 + \int_0^t [f(\hat{x}_\epsilon(\tau)\,,\tau) + \epsilon\gamma(\hat{x}_\epsilon(\tau)\,,\hat{u}_\epsilon(\tau)\,,\tau)]\mathrm{d}\tau \tag{4.12}$$

satisfy the condition

$$G(\hat{x}_\epsilon(T)) = G_\epsilon^* + O(\epsilon^2) \tag{4.13}$$

The first two results of the theorem will be proved if we show that

$$G(\hat{x}_\epsilon(T)) = G(\overline{x}_\epsilon(T)) + O(\epsilon^2) = G(x(T)) - \epsilon\int_0^T \Phi_t^* \mathrm{d}t + O(\epsilon^2) \tag{4.14}$$

where $\bar{x}_\epsilon(t)$ is the trajectory of (4.3) with $u^*(t)$ as a control

$$\bar{x}_\epsilon(t) = x_0 + \int_0^t [f(\bar{x}_\epsilon(\tau), \tau) + \epsilon\gamma(\bar{x}_\epsilon(\tau), u^*(\tau), \tau)]\mathrm{d}\tau \tag{4.14'}$$

Let us rewrite (4.4) in integral form and subtract it from (4.12). We obtain

$$\|\hat{x}_\epsilon(t) - x(t)\| \le L_1 \int_0^t \|\hat{x}_\epsilon(\tau) - x(\tau)\|\mathrm{d}\tau + L_2\epsilon$$

where L_1, L_2 are obviously defined constants. Applying the Gronwall–Bellman lemma, we have

$$\begin{aligned} &\|\hat{x}_\epsilon(t) - x(t)\| \le L_3\epsilon \;,\; L_3 = \text{constant} \Rightarrow \\ &\Rightarrow \hat{x}_\epsilon(t) = x(t) + \epsilon\hat{x}_1(t, \epsilon) \;,\; \|\hat{x}_1(t, \epsilon)\| \le L_3 \end{aligned} \tag{4.15}$$

Substitute (4.15) into (4.12) and take into account (4.4):

$$\begin{aligned} &\dot{\hat{x}}_1(t, \epsilon) = f_x(x(t), t)\hat{x}_1(t, \epsilon) + \gamma(x(t), \hat{u}_\epsilon(t), t) + O(\epsilon) \;,\; \hat{x}_1(0, \epsilon) = 0 \Rightarrow \\ &\Rightarrow G(\hat{x}_\epsilon(T)) = G(x(T)) + \epsilon G_x^T(x(T))\hat{x}_1(T, \epsilon) + O(\epsilon^2) \end{aligned}$$

where

$$G_x^T(x(T))\hat{x}_1(T, \epsilon) = -\lambda^T(T)\hat{x}_1(T, \epsilon) = -\int_0^T \lambda^T(\tau)\gamma(x(\tau), \hat{u}_\epsilon(\tau), \tau)\mathrm{d}\tau$$

This last arises from the equations

$$\begin{aligned} \frac{\mathrm{d}}{\mathrm{d}t}(\lambda^T(t)\hat{x}_1(t, \epsilon)) &= \dot{\lambda}^T(t)\hat{x}_1(t, \epsilon) + \lambda^T(t)\dot{\hat{x}}_1(t, \epsilon) \\ &= \lambda^T(t)\gamma(x(t), \hat{u}_\epsilon(t), t) + O(\epsilon) \end{aligned}$$

Thus,

$$G(\hat{x}_\epsilon(T)) = G(x(T)) - \epsilon\int_0^T \lambda^T(\tau)\gamma(x(\tau), \hat{u}_\epsilon(\tau), \tau)\mathrm{d}\tau + O(\epsilon^2) \tag{4.16}$$

Similarly, it is easy to show that

$$G(\bar{x}_\epsilon(T)) = G(x(T)) - \epsilon\int_0^T \lambda^T(\tau)\gamma(x(\tau), u^*(\tau), \tau)\mathrm{d}\tau + O(\epsilon^2) \tag{4.17}$$

However, by (4.13)

$$G(\hat{x}_\epsilon(T)) \leq G(\bar{x}_\epsilon(T)) + O(\epsilon^2) \tag{4.18}$$

At the same time, by definition,

$$\int_0^T \lambda^T(\tau)\gamma(x(\tau), \hat{u}_\epsilon(\tau), \tau)\mathrm{d}\tau \leq \int_0^T \lambda^T(\tau)\gamma(x(\tau), u^*(\tau), \tau)\mathrm{d}\tau \tag{4.19}$$

Comparison of (4.18), (4.19) and (4.16), (4.17) yields

$$\int_0^T \lambda^T(\tau)\gamma(x(\tau), \hat{u}_\epsilon(\tau), \tau)\mathrm{d}\tau = \int_0^T \lambda^T(\tau)\gamma(x(\tau), u^*(\tau), \tau)\mathrm{d}\tau + O(\epsilon) \tag{4.20}$$

Hence, (4.14) is established. To prove the third part, note that, similarly to (4.20), one can write

$$\int_0^T \lambda^T(\tau)\gamma(x(\tau), u^*(\tau), \tau)\mathrm{d}\tau - \int_0^T \lambda^T(\tau)\gamma(x(\tau), \tilde{u}_\epsilon(\tau), \tau)\mathrm{d}\tau \underset{\epsilon\to 0}{\longrightarrow} 0$$

or

$$\int_0^T [\Phi_\tau(u^*(\tau)) - \Phi_\tau(\tilde{u}_\epsilon(\tau))]\mathrm{d}\tau \underset{\epsilon\to 0}{\longrightarrow} 0$$

This means, in particular, that for arbitrary $\delta_1 > 0$

$$\text{mes}\,\{\Phi_\tau(u^*(\tau)) - \Phi_\tau(\tilde{u}_\epsilon(\tau)) \geq \delta_1\} \underset{\epsilon\to 0}{\longrightarrow} 0 \tag{4.21}$$

By the Luzin theorem [55], for any positive number μ there exists an open subset Θ of the interval $[0, T]$ such that:

$$\text{mes}\ \Theta \leq \mu \tag{4.22}$$

the vector-function $u^*(\tau)$ is continuous on the closed set

$$D \triangleq [0, T] / \Theta$$

and the solution to problem (4.6) is unique. Let δ be an arbitrary positive number. It is not difficult to show that there exists a positive number δ_1 (corresponding to δ) such that if

$$\|u - u^*(\tau)\| \geq \delta \ , \ u \in U \ , \ \tau \in D \tag{4.22'}$$

then

$$\Phi_\tau(u^*(\tau)) - \Phi_\tau(u) \geq \delta_1 > 0 \tag{4.22''}$$

Assume the contrary, i.e., assume that there exists a sequence $\tau_i \in D$ and an associated sequence $u_i \in U$ satisfying the conditions

$$\|u_i - u^*(\tau_i)\| \geq \delta \ , \ \Phi_{\tau_i}(u^*(\tau_i)) - \Phi_{\tau_i}(u_i) \underset{i \to \infty}{\longrightarrow} 0 \tag{4.23}$$

without loss of generality (taking into account the fact that the sets D and U are closed), we may assume that there exist the limits

$$\lim_{i \to \infty} \tau_i = \bar{\tau} \ , \ \lim_{i \to \infty} u_i = \bar{u} \in U \ , \ \lim_{i \to \infty} u^*(\tau_i) = u^*(\bar{\tau}) \tag{4.24}$$

Taking the limit in (4.23), we then have

$$\|\bar{u} - u^*(\bar{\tau})\| \geq \delta \ , \ \Phi_{\bar{\tau}}(u^*(\bar{\tau})) = \Phi_{\bar{\tau}}(\bar{u}) \ , \ \bar{u} \in U \ , \ \bar{\tau} \in D$$

This contradicts the uniqueness of the solution to (4.6), which, as was supposed, occurs in all points of D.

In view of (4.22′) and (4.22″), we can write

$$\begin{aligned} \text{mes}\ \{\|\tilde{u}_\epsilon(\tau) - u^*(\tau)\| \geq \delta\} &\leq \text{mes}\ \{\|\tilde{u}_\epsilon(\tau) - u^*(\tau)\| \geq \delta \, , \tau \in D\} + \mu \\ &\leq \text{mes}\ \{(\Phi_\tau(u^*(\tau)) - \Phi_\tau(\tilde{u}_\epsilon(\tau)) \geq \delta_1\} + \mu \end{aligned}$$

from which, by (4.21),

$$\overline{\lim_{\epsilon\to 0}} \text{ mes } \{\|\tilde{u}_\epsilon(\tau) - u^*(\tau)\| \geq \delta\} \leq \mu$$

Since μ is arbitrary, it follows that

$$\lim_{\epsilon\to 0} \text{ mes } \{\|\tilde{u}_\epsilon(\tau) - u^*(\tau)\| > \delta\} = 0$$

Similarly to (4.15),

$$\tilde{x}_\epsilon(t) = x(t) + \epsilon\tilde{x}_1(t,\epsilon) \quad , \quad \|\tilde{x}_1(t,\epsilon)\| \leq L_4 = \text{constant} \tag{4.25}$$

where $\tilde{x}_1(t,\epsilon)$ is the solution to the equation

$$\tilde{x}_1(t,\epsilon) = \int_0^T f_x(x(\tau),\tau)\tilde{x}_1(\tau,\epsilon)\mathrm{d}\tau + \int_0^T \gamma(x(\tau),\tilde{u}_\epsilon(\tau),\tau)\mathrm{d}\tau + O(\epsilon) \tag{4.26}$$

By virtue of the Lebesgue theorem [55] and the relation (4.9) just proved, we have

$$\int_0^t \gamma(x(\tau),\tilde{u}_\epsilon(\tau),\tau)\mathrm{d}\tau \underset{\epsilon\to 0}{\longrightarrow} \int_0^t \gamma(x(\tau),u^*(\tau),\tau)\mathrm{d}\tau$$

Rewriting (4.11) in integral form and subtracting it from (4.26), we obtain

$$\|\tilde{x}_1(t,\epsilon) - x_1(t)\| \leq k_4\int_0^t \|\tilde{x}_1(\tau,\epsilon) - x_1(\tau)\|\mathrm{d}\tau + \nu(\epsilon) \quad , \quad \nu(\epsilon) \underset{\epsilon\to 0}{\longrightarrow} 0$$

which leads, via the Gronwall–Bellman lemma, to

$$\|\tilde{x}_1(t,\epsilon) - x_1(t)\| = O(\nu(\epsilon))$$

QED.

In general, Theorem 4.1 does not assume that there exists a solution to the initial problem (4.1)–(4.3). It has the form of asymptotic sufficiency conditions for the approximate optimality of $u^*(t)$. However, if the existence of a solution to (4.1)–(4.3) is established or is postulated by the conditions, then with the help of Theorem 4.1 and the Pontriagin maximum principle, the statements of Theorem 4.1 can be extended by the following three theorems.

Theorem 4.2.

Let the conditions of Theorem 4.1 be fulfilled, and let the solution to problem (4.6) be almost everywhere unique on $[0, T]$. Let there also exist a solution $u_\epsilon^*(t)$, $x_\epsilon^*(t)$ to the problem (4.1)–(4.3). Then, for all t where the solution to the problem (4.6) is unique, one can use the estimate

$$\Phi_t(u^*(t)) - \Phi_t(u_\epsilon^*(t)) = O(\epsilon) \tag{4.27}$$

which is uniform with respect to $t \in [0, T]$, to evaluate the norm $\|u_\epsilon^*(t) - u^*(t)\|$.

Proof. By virtue of the maximum principle, there exists an absolutely continuous vector-function $\lambda_\epsilon^*(t)$ satisfying almost everywhere the equation

$$\begin{aligned} \dot{\lambda}_\epsilon^*(t) &= -[f_x^T(x_\epsilon^*(t), t) + \epsilon\gamma_x^T(x_\epsilon^*(t), u_\epsilon^*(t), t)]\lambda_\epsilon^*(t) \ , \\ \lambda_\epsilon^*(T) &= -G_x(x_\epsilon^*(T)) \end{aligned} \tag{4.28}$$

and such that, for almost all $t \in [0, T]$,

$$u_\epsilon^*(t) \in \arg\max_{u \in U} \{(\lambda_\epsilon^*(t))^T \gamma(x_\epsilon^*(t), u, t, \epsilon)\} \tag{4.29}$$

Using the representations (4.9)–(4.11) similarly to Theorem 4.1, it is easy to verify that, uniformly with respect to $t \in [0, T]$,

$$\lambda_\epsilon^*(t) = \lambda(t) + \epsilon\lambda_1(t) + o(\epsilon) \tag{4.30}$$

where $\lambda_1(t)$ is the solution to the Cauchy problem[†]

$$\dot{\lambda}_1(t) = -f_x^T(x(t), t)\lambda_1(t) + d(t) \ , \ \lambda_1(T) = -G_{xx}(x(T))x_1(T) \tag{4.31}$$

Thus, one can rewrite (4.29) in the form

$$\begin{aligned} &u_\epsilon^*(t) = \arg\max_{u \in U} \{\lambda^T(t)\gamma(x(t), u, t)\} + O(\epsilon) \Rightarrow \\ &\Rightarrow \Phi_t(u_\epsilon^*(t)) \geq \Phi_t(u^*(t)) + O(\epsilon) \end{aligned} \tag{4.32}$$

On the other hand, by definition,

[†] The measurable function $d(t)$ is obviously obtained via the substitution of (4.9), (4.10), and (4.30) into (4.28) and equating coefficients of like powers of ϵ.

$$\Phi_t(u^*(t)) \geq \Phi_t(u_\epsilon^*(t))$$

The last two inequalities complete the proof.

More detailed results can be obtained for some special classes of problems. If, for example, the system is linear in u

$$\gamma(x\,,u\,,t) = \gamma(x\,,t)u \tag{4.33}$$

then it is easy to establish the following result.

Theorem 4.3.

Assume that the conditions of Theorem 4.2 hold and that the system under consideration is linear in control, i.e., the expression (4.33) is valid, and the set U is a polyhedron. Then result (4.9) can be replaced by a stronger one:

$$\lim_{\epsilon \to 0} \text{mes}\,\{u_\epsilon^*(t) \neq u^*(t)\} = 0 \tag{4.34}$$

If, moreover, the solution to problem (4.6) is unique for all $t \in [0\,,T]$, then there exists $\epsilon_0 > 0$ such that, for $0 < \epsilon \leq \epsilon_0$,

$$u_\epsilon^*(t) = u^*(t) \tag{4.35}$$

Some interesting results may be adduced if the Hamiltonian $\Phi_t(u)$ is strongly concave in u. It is not difficult to show that in this case the following expansions

$$\begin{aligned} u_\epsilon^*(t) &= u^*(t) + \epsilon u_1(t) + \cdots + \epsilon^l u_l(t) + o(\epsilon^l) \\ x_\epsilon^*(t) &= x(t) + \epsilon x_1(t) + \cdots + \epsilon^l x_l(t) + o(\epsilon^{l+1}) \end{aligned} \tag{4.36}$$

are true if the functions entering into (4.1)–(4.3) are $(l+2)$-times continuously differentiable. However, we concentrate our efforts on another more convenient technique, which is close to that developed in Section 6.1.

Let

$$x_0(\epsilon\,,t) \triangleq x(t) \quad , \quad \lambda_0(\epsilon\,,t) \triangleq \lambda(t) \quad , \quad u_0(\epsilon\,,t) \triangleq u^*(t)$$

Let $x_1(\epsilon\,,t) \triangleq \overline{x}_\epsilon(t)$ be the solution to the Cauchy problem (4.14′), let $\lambda_1(\epsilon\,,t)$ be the solution to the Cauchy problem

$$\begin{aligned} \dot{\lambda}_1(\epsilon,t) &= -[f_x^T(x_1(\epsilon,t),t) + \epsilon\gamma_x^T(x_1(\epsilon,t),u_0(\epsilon,t),t)]\lambda_1(\epsilon,t) \\ \lambda_1(\epsilon,T) &= -G_x(x_1(\epsilon,T)) \end{aligned}$$

and let $u_1(\epsilon,t)$ be defined as

$$u_1(\epsilon,t) \triangleq \arg\max_{u \in U} \{\lambda_1^T(\epsilon,t),\gamma(x_1(\epsilon,t),u,t)\} \tag{4.37}$$

Let us now construct $x_i(\epsilon,t)$, $\lambda_i(\epsilon,t)$ as the respective solutions to the following Cauchy problems:

$$\begin{aligned} \dot{x}_i(\epsilon,t) &= f(x_i(\epsilon,t),t) + \epsilon\gamma(x_i(\epsilon,t),u_{i-1}(\epsilon,t),t) \quad , \quad x_i(\epsilon,0) = x_0 \\ \dot{\lambda}_i(\epsilon,t) &= -[f_x^T(x_i(\epsilon,t),t) + \epsilon\gamma_x^T(x_i(\epsilon,t),u_{i-1}(\epsilon,t),t)]\lambda_i(\epsilon,t) \\ \lambda_i(\epsilon,t) &= -G_x(x_i(\epsilon,T)) \end{aligned}$$

Finally, construct $u_i(\epsilon,t)$ as the solution to the mathematical program

$$u_i(\epsilon,t) \triangleq \arg\max_{u \in U} \{\lambda_i^T(\epsilon,t)\gamma(x_i(\epsilon,t),u,t)\} \tag{4.38}$$

Theorem 4.4.

Let the conditions of Theorem 4.1 be fulfilled and, for all sufficiently small ϵ, let there exist the solution $u_\epsilon^*(t)$, $x_\epsilon^*(t)$ to problem (4.1)–(4.3). Assume that the set U is convex and that the function $\Phi_t(u)$ is strongly (with respect to the parameter ρ) concave on U for all $t \in [0,T]$. Suppose, moreover, that the functions that define (4.1)–(4.3) are twice continuously differentiable with respect to x, u throughout the domain

$$\|x - x(t)\| \leq \hat{\delta} \quad , \quad u \in U \quad , \quad t \in [0,T]$$

where $\hat{\delta}$ is a positive number. Then there exist constants ϵ_0, $\tilde{k}$ ($\epsilon_0 < 1/\tilde{k}$) and constants $\bar{k}_0$, $\bar{k}_1$ such that

$$\begin{aligned} \|u_i(\epsilon,t) - u_\epsilon^*(t)\| &\leq \bar{k}_0(\tilde{k}\cdot\epsilon)^{i+1} \quad , \quad i = 0,1,\ldots \\ \|x_i(\epsilon,t) - x_\epsilon^*(t)\| &\leq \bar{k}_1(\tilde{k}\cdot\epsilon)^{i+1} \quad , \quad i = 0,1,\ldots \end{aligned}$$

Under the conditions of the theorem, the solution to problem (4.1)–(4.3) is in the class of continuous functions. This is a consequence of the fact that the function

$$\arg \max_{u \in U} \{(\lambda_\epsilon^*(t))^T \gamma(x_\epsilon^*(t), u, t)\}$$

is continuous. This continuous optimal control is unique and is the control denoted in the theorem as $u_\epsilon^*(t)$.

Proof. By the Gronwall–Bellman lemma, it is not difficult to obtain

$$\begin{aligned} &\max_{t \in [0,T]} \|x_i(\epsilon, t) - x(t)\| \le k\epsilon \ , \quad \max_{t \in [0,T]} \|\lambda_i(\epsilon, t) - \lambda(t)\| \le k\epsilon \\ &\max_{t \in [0,T]} \|x_\epsilon^*(t) - x(t)\| \le k\epsilon \quad , \quad \max_{t \in [0,T]} \|\lambda_\epsilon^*(t) - \lambda(t)\| \le k\epsilon \end{aligned} \tag{4.39}$$

from which, in particular,

$$\max_{t \in [0,T]} \|x_i(\epsilon, t) - x_\epsilon^*(t)\| \le 2k\epsilon \ , \quad \max_{t \in [0,T]} \|\lambda_i(\epsilon, t) - \lambda_\epsilon^*(t)\| \le 2k\epsilon \tag{4.40}$$

By virtue of the same lemma, it is also easy to obtain the inequalities

$$\begin{aligned} &\|x_i(\epsilon, t) - x_\epsilon^*(t)\| \le \epsilon k_1 \max_t \|u_{i-1}(t) - u_\epsilon^*(t)\| \\ &\|\lambda_i(\epsilon, t) - \lambda_\epsilon^*(t)\| \le \epsilon k_2 \max_t \|u_{i-1}(t) - u_\epsilon^*(t)\| \end{aligned} \tag{4.41}$$

The fact that the function $\Phi_t(u)$ is strongly concave in the parameter ρ means that there exists a positive number δ ($\hat{\delta} \ge \delta > 0$) such that the function $\lambda^T \gamma(x, u, t)$ will be strongly concave on U in the parameter $\rho/2$ if

$$|x - x(t)| \le \delta \ , \ |\lambda - \lambda(t)| \le \delta \ , \ t \in [0, T] \tag{4.42}$$

Taking this into account with (4.39), one can conclude that if

$$\epsilon \le \frac{\delta}{2k} \tag{4.43}$$

then the function $(\lambda_i^T(\epsilon, t)\gamma_i(x_i(\epsilon, t), u, t)$, $i = 1,2,\dots$ and also the function $(\lambda_\epsilon^*(t))^T \gamma(x_\epsilon^*(t), u, t)$ are strongly concave on U in the parameter $\rho/2$. This allows us to write [34]

$$\lambda_i^T(\epsilon, \tau)\gamma(x_i(\epsilon, \tau), u_i(\epsilon, \tau), \tau) - \lambda_i^T(\epsilon, \tau)\gamma(x_i(\epsilon, \tau), u_\epsilon^*(\tau), \tau) \ge \frac{\rho}{2} \|u_\epsilon^*(\tau) - u_i(\epsilon, \tau)\|^2$$

$$(\lambda_\epsilon^*(\tau))^T \gamma(x_\epsilon^*(\tau), u_\epsilon^*(\tau), \tau) - (\lambda_\epsilon^*(\tau))^T \gamma(x_\epsilon^*(\tau), u_i(\epsilon, \tau), \tau) \ge \frac{\rho}{2} \|u_\epsilon^*(\tau) - u_i(\epsilon, \tau)\|$$

The sum of these inequalities is

$$\begin{aligned}
&[\lambda_i^T(\epsilon,\tau)\gamma(x_i(\epsilon,\tau),u_i(\epsilon,\tau),\tau) - (\lambda_\epsilon^*(\tau))^T\gamma(x_\epsilon^*(\tau),u_i(\epsilon,\tau),\tau)] - \\
&- [\lambda_i^T(\epsilon,\tau)\gamma(x_i(\epsilon,\tau),u_\epsilon^*(\tau),\tau) - (\lambda_\epsilon^*(\tau))^T\gamma(x_\epsilon^*(\tau),u_\epsilon^*(\tau),\tau)] \geq \\
&\geq \rho\|u_\epsilon^*(\tau) - u_i(\epsilon,\tau)\|^2 \Rightarrow \\
&\lambda_i^T(\epsilon,\tau)[\gamma_x'(x_\epsilon^*(\tau),u_i(\epsilon,\tau),\tau) - \gamma_x'(x_\epsilon^*(\tau),u_\epsilon^*(\tau),\tau)](x_i(\epsilon,\tau) - x_\epsilon^*(\tau)) + \\
&+ (\lambda_i^T(\epsilon,\tau) - \lambda_\epsilon^*(\tau))^T[\gamma(x_\epsilon^*(\epsilon,\tau),u_i(\epsilon,\tau),\tau) - \gamma(x_\epsilon^*(\tau),u_\epsilon^*(\tau),\tau)] + \\
&+ M_1\|x_i(\epsilon,\tau) - x_\epsilon^*(\tau)\|^2 \geq \\
&\geq \rho\|u_\epsilon^*(\tau) - u_i(\epsilon,\tau)\|^2 \Rightarrow \\
&M_1\|x_i(\epsilon,\tau) - x_\epsilon^*(\tau)\| + M_2\|u_i(\epsilon,\tau) - u_\epsilon^*(\tau)\|\,\|x_i(\epsilon,\tau) - x_\epsilon^*(\tau)\| + \\
&+ M_3\|u_i(\epsilon,\tau) - u_\epsilon^*(\tau)\|\,\|\lambda_i(\epsilon,\tau) - \lambda_\epsilon^*(\tau)\| \geq \\
&\geq \rho\|u_i(\epsilon,\tau) - u_\epsilon^*(\tau)\|^2
\end{aligned} \tag{4.44}$$

where M_1, M_2, M_3 are constants defined by the values of the corresponding derivatives.
Using the bounds (4.40) and (4.41) in (4.44) we obtain, respectively,

$$M_1{\cdot}4k^2\epsilon^2 + (M_2{\cdot}2k\epsilon + M_3{\cdot}2k\epsilon)\Delta_i \geq \rho\Delta_i^2 \ , \ i = 0,1,\ldots \tag{4.45}$$

$$M_1k_1^2\epsilon^2\Delta_{i-1}^2 + (M_2k_1 + M_3k_2)\epsilon\Delta_{i-1}\Delta_i \geq \rho\Delta_i^2 \ , \ i = 1,2,\ldots \tag{4.46}$$

where

$$\Delta_i \triangleq \max_{t\in[0,T]} \|u_i(\epsilon,t) - u_\epsilon^*(t)\|$$

From (4.45) we now have

$$\Delta_i \leq \left[\left(\frac{M_2k}{\rho} + \frac{M_3k}{\rho}\right) + \sqrt{\left(\frac{M_2k}{\rho} + \frac{M_3k}{\rho}\right)^2 + \frac{4M_1k^2}{\rho}}\right]\epsilon$$

and, in particular,

$$\Delta_0 \leq k_3\epsilon \ , \ k_3 \triangleq \frac{(M_2 + M_3)k}{\rho} + \sqrt{\frac{(M_2 + M_3)^2k^2}{\rho^2} + \frac{4M_1k^2}{\rho}} \tag{4.47}$$

On the other hand, in view of (4.46),

$$\Delta_i \leq \epsilon\tilde{k}\Delta_{i-1} \ ,$$

$$\tilde{k} \triangleq \left(\frac{M_2}{2\rho}k_1 + \frac{M_3}{2\rho}k_2\right) + \sqrt{\left(\frac{M_2}{2\rho}k_1 + \frac{M_3}{2\rho}k_2\right)^2 + \frac{M_1}{\rho}k_1^2}$$

It follows, according to (4.47), that

$$\Delta_i \leq \tilde{k}_0(\tilde{k}\epsilon)^{i+1} \quad , \quad \tilde{k}_0 \triangleq \frac{k_3}{\tilde{k}}$$

The latter and (4.41) imply the statements of the theorem.

Let us consider, in conclusion, a differential game of perseption. Assume that the dynamics of the first participant (hunger) and the second (victim) are given by the respective equations

$$\dot{x}_\epsilon = f(x_\epsilon, t) + \epsilon\gamma(x_\epsilon, u, t) \quad , \quad u \in U \quad , \quad x_\epsilon(0) = x_0 \tag{4.48}$$

$$\dot{y}\epsilon = g(y_\epsilon, t) + \epsilon\beta(y_\epsilon, v, t) \quad , \quad v \in V \quad , \quad y_\epsilon(0) = y_0 \tag{4.49}$$

where x, y are the state variable vectors; $f(\cdot)$, $g(\cdot)$, $\gamma(\cdot)$, $\beta(\cdot)$ are vector-functions of corresponding dimension; and $u(m_1 \times 1)$, $v(m_2 \times 1)$ are control vectors.

In accordance with the standard problem statement [128], the goal of the first participant is to minimize the distance between himself and his victim at the moment T:

$$F_1^* = \inf_{u \in U} \{\sup_{v \in V} \|x_\epsilon(T) - u_\epsilon(T)\|^2\} \triangleq \inf_{u \in U} \hat{J}_\epsilon^*(u) \tag{4.50}$$

when $\|\cdot\|$ is the Euclidean norm. Conversely, the goal of the second participant is to make the distance as large as possible:

$$F_2^* = \sup_{v \in V} \{\inf_{u \in U} \|x_\epsilon(T) - y_\epsilon(T)\|^2\} = \sup_{v \in V} \check{J}_\epsilon^*(v) \tag{4.51}$$

Suppose that both participants seek their controls in the class of measurable functions. Suppose also that the functions $u^*(t)$, $v^*(t)$ belong to this class, where they are defined as solutions to the parametric problems

$$\max_{u \in U} \lambda^T(t)\gamma(x(t), u, t) \tag{4.52}$$

$$\max_{v \in V} \mu^T(t)\beta(y(t), v, t) \tag{4.53}$$

and where $x(t)$, $y(t)$, $\lambda(t)$, $\mu(t)$ are the solutions to the Cauchy problems

$$\dot{x} = f(x, t) \quad , \quad x(0) = x_0 \tag{4.54}$$

$$\dot{y} = g(y, t) \quad , \quad y(0) = y_0 \tag{4.55}$$

$$\dot{\lambda} = -f_x^T(x(t), t)\lambda(t) \quad , \quad \lambda(T) = -2(x(T) - y(T)) \tag{4.56}$$

$$\dot{\mu} = -g_y^T(y(t), t)\mu(T) \quad , \quad \mu(T) = -2(x(T) - y(T)) \tag{4.57}$$

Theorem 4.5.

Let the functions $f(\cdot)$, $g(\cdot)$, $\gamma(\cdot)$, $\beta(\cdot)$, $f_x(\cdot)$ be continuously differentiable, and let U, V be compact sets. Then

(1) If the first participant uses the control $u^*(t)$, the value of his performance index (4.50) will be within $O(\epsilon^2)$ of the optimal one.
(2) If the second participant uses the control $v^*(t)$, the value of his performance index will be within $O(\epsilon^2)$ of the optimal one.
(3) The following is valid:

$$\begin{aligned} F_1^* &= F_2^* + O(\epsilon^2) \\ &= (x(T) - y(T))^2 + \epsilon\int_0^T [\mu^T(t)\beta(y(t), v^*(t), t) - \\ &\quad - \lambda^T(t)\gamma(x(t), u^*(t), t)]dt + O(\epsilon^2) \end{aligned} \tag{4.58}$$

(4) If the solution to (4.52) is almost everywhere unique, then there exists the limit

$$u_\epsilon(t) \overset{\text{measure}}{\longrightarrow} u^*(t) \tag{4.58'}$$

where $u_\epsilon(t)$ is an arbitrary feasible control such that the corresponding value of the performance index (4.50) is within $o(\epsilon)$ of the optimal one. In particular, the limit (4.58′) will exist for the optimal control if it exists. Similarly, if the solution to (4.53) is almost everywhere unique, then there exists the limit

$$v_\epsilon(t) \underset{\epsilon \to 0}{\overset{\text{measure}}{\longrightarrow}} v^*(t)$$

where $v_\epsilon(t)$ delivers a $o(\epsilon)$-optimum to (4.51).

We shall give here only an outline of the proof since the details are completely analogous to those in Theorem 4.1. First of all, it is necessary to establish that the relations

$$x_\epsilon(t) = x(t) + \epsilon x_1(t) + O(\epsilon^2) \tag{4.59}$$

$$y_\epsilon(t) = y(t) + \epsilon y_1(t) + O(\epsilon^2) \tag{4.60}$$

are fulfilled uniformly with respect to $t \in [0, T]$ and all admissible controls, where

$$\dot{x}_1 = f_x(x(t),t)x_1(t) + \gamma(x(t),u,t) \quad , \quad x_1(0) = 0 \tag{4.61}$$

$$\dot{y}_1 = g_y(y(t),t)y_1(t) + \beta(y(t),v,t) \quad , \quad y_1(0) = 0 \tag{4.62}$$

It follows that the estimates of the performance indices (4.50) and (4.51) can be written in the form

$$\begin{aligned} F_1^* &= \inf_{u\in U} \sup_{v\in V} (x(T) + \epsilon x_1(T) - y(T) - \epsilon y_1(T))^2 + O(\epsilon^2) \\ &= (x(T) - y(T))^2 + \epsilon \inf_{u\in U} \{2(x(T) - y(T))^T x_1(T)\} + \\ &\quad + \epsilon \sup_{v\in V} \{-2(x(T) - y(T))^T y_1(T)\} + O(\epsilon^2) \ ; \end{aligned} \tag{4.62'}$$

$$\begin{aligned} F_2^* &= (x(T) - y(T))^2 + \epsilon \inf \{2(x(T) - y(T))^T x_1(T)\} + \\ &\quad + \epsilon \sup_{v\in V} \{-2(x(T) - y(T))^T y_1(T)\} + O(\epsilon^2) \Rightarrow \end{aligned} \tag{4.63}$$

$$F_1^* = F_2^* + O(\epsilon^2) \tag{4.64}$$

Thus, problems (4.50) and (4.51) are reduced to the separate optimization of system (4.61) with the performance index

$$\Phi_1^* = \inf_{u\in U} \{2(x(T) - y(T))^T x_1(T)\} \tag{4.65}$$

and of system (4.62) with the performance index

$$\begin{aligned} \Phi_2^* &= \sup_{v\in V} \{-2(x(T) - y(T))^T y_1(T)\} \\ &= -\inf_{v\in V} \{2(x(T) - y(T))^T y_1(T)\} \end{aligned} \tag{4.66}$$

It is easy to verify that the solutions to the problems (4.61), (4.65) and (4.62), (4.66) exist and are determined according to (4.52), (4.53) with

$$\Phi_1^* = -\int_0^T \lambda^T(t)\gamma(x(t),u^*(t),t)dt$$

$$\Phi_2^* = \int_0^T \mu^t(t)\beta(y(t),v^*(t),t)dt$$

This leads to the first three statements of the theorem. The fourth is established similarly to that in Theorem 4.1.

6.5. Aggregation in a singular perturbed problem

This section is closely connected with Section 5.2 and is devoted to the study of systems

$$\dot{x} = (A_0 + \epsilon A_1)x + (B_0 + \epsilon B_1)u \tag{5.1}$$

$$u \in U \tag{5.2}$$

which lose the property of complete controllability when ϵ tends toward zero. Unlike Section 5.2, we shall suppose that the feasible set of controls is bounded, that the time interval is finite, and that the performance index is of general form. (Singularly perturbed LQ problems on a finite time interval have been investigated by many authors (see e.g., [10, 104].) We shall consider only a critical case characterized by the following conditions:

(A) The matrix A_0 $(n \times n)$ is singular and its zero eigenvalue is semisimple and k-multiple.

(B) The matrix V_L $(n \times k)$ satisfies the relation

$$V_L^T B_0 = 0 \tag{5.3}$$

where V_L is defined as in Section 5.2 (i.e., its columns constitute a basis for the left eigensubspace of A_0 corresponding to the zero eigenvalue).

It is obvious that, if $\epsilon = 0$, k linear combinations (aggregates) of state variables remain constant

$$V_L^T x(t) = V_L^T x_0 \triangleq z_0$$

no matter what control is chosen. If $\epsilon > 0$, the aggregates change slowly, this variation being finite at a time of the order of $1/\epsilon$ (that is, it does not tend toward zero together with ϵ). Therefore, we shall optimize on the interval $[0, T/\epsilon]$, $T =$ constant. One of the reasons for calling such problems "singular" is that, in general, the optimal reduced control is not suboptimal for the perturbed problem.

For example, if the performance index is of the form

$$\inf G(V_L^T x(T/\epsilon))$$

then the reduced problem is invariant with respect to the choice of control, while the perturbed one is essentially dependent on it. As was established in Section 5.2, the singularly perturbed system (5.1) can be written in the form (2.21) of Chapter 5 by means of the transformation (2.6) and (2.7).

Let us again use the heuristic concerning the behavior of "fast" and "slow" subsystems. Suppose that the "fast" subsystem is stable (i.e., that the matrix Λ_1 is stable). It is natural to suggest that this subsystem reaches a quasistationary state in a period of time that is negligible from the viewpoint of its influence on the "slow" subsystem. This "quasistationary state" is defined by the control applied at the same time

$$\Lambda_1 \xi_1 \simeq -D_1 u \Rightarrow \xi_1 \simeq -\Lambda_1^{-1} D_1 u \tag{5.3'}$$

and, consequently, the "slow" subsystem can be transformed to

$$\dot{\xi}_2 = \epsilon\{\Lambda_{22}\xi_{21} + (D_{21} - \Lambda_{21}\Lambda_1^1 D_1)u\}$$

In other words, one may hope that the system motion on the whole is approximately described by an independent consideration of the simplified "fast" subsystem

$$\dot{\xi}_1 = \Lambda_1 \xi_1 + D_1 u$$

and the "slow" subsystem described above.

To abbreviate the notation, we shall henceforth use η, z, M instead of ξ_1, ξ_2, Λ_{22}. Then the simplified system is written as

$$\dot{z} = \epsilon\{Mz + (D_{21} - \Lambda_{21}\Lambda_1^{-1} D_1)u\} \tag{5.4}$$

$$\dot{\eta} = \Lambda_1 \eta + D_1 u \tag{5.5}$$

where, according to (2.9),

$$D_{21} = V_L^T B_1 \ , \ \Lambda_{21} = V_L^T A_1 W_R \ , \ D_1 = W_L^T B_0 = W_R^T B_0 \tag{5.6}$$

and where the initial conditions are of the form

$$z(0) = z_0 \triangleq V_L^T x_0 \ , \ \eta(0) = \eta_0 \triangleq W_L^T x_0 = W_R^T (E - V_R V_L^T) x_0$$

Our goal in this section is to investigate formally whether the optimization of (5.1) can be reduced to the separate optimizations of (5.4) and (5.5). Note that a number of works are devoted to this matter in connection with the optimization of singular perturbed systems (see for an overview [104]), presented in the form (2.22′) in Chapter 5, i.e.,

$$\epsilon \frac{\mathrm{d}x^{(1)}}{\mathrm{d}\tau} = a_{11}x^{(1)} + a_{12}x^{(2)} + b_1 u$$
$$\frac{\mathrm{d}x^{(2)}}{\mathrm{d}\tau} = a_{21}x^{(1)} + a_{22}x^{(2)} + b_2 u$$

However, one can obtain the equivalent representation in the form (5.1). Then the simplified system can be written as

$$\dot{z} = \epsilon\{(a_{22} - a_{21}a_{11}^{-1}a_{12})z + (b_2 - a_{22}a_{11}^{-1}b_1)u\} \ , \ z(0) = x_0^{(2)}$$
$$\dot{\eta} = a_{11}\eta + b_1 u \ , \ \eta(0) = x_0^{(1)} + a_{11}^{-1}a_{12}x_0^{(2)}$$

The majority of the works mentioned in [104] deal with the case in which the optimal control is a smooth function of the primal and conjugate variables. Under this condition the problem of the approximate optimization of (5.6) is solvable by constructing boundary layer corrections for the solutions of the corresponding boundary value problems. The technique used is close to that developed in [9, 79, 99] for the Cauchy problem. Here we shall follow another approach based on classical perturbation theory for linear finite-dimensional operators.

Let us first of all establish the connection between the solutions of (5.1), (5.4) and (5.5) given that the control is fixed.

Theorem 5.1.

Let the matrix Λ_1 be stable. Then for any feasible bounded control

$$\|u(t)\| \leq L = \text{constant} \tag{5.7}$$

the conditions

$$x_u(t) = V_R z_u(t) + W_R \eta_u(t) + O(\epsilon) \tag{5.8}$$
$$V_L^T x_u(t) = z_u(t) + O(\epsilon) \tag{5.9}$$

are fulfilled uniformly with respect to $t \in [0, T/\epsilon]$ where $x_u(t)$, $z_u(t)$, $\eta_u(t)$ are the solutions to (5.1), (5.4), and (5.5) corresponding to the control $u(t)$. The proof of the theorem uses the following lemma.

Lemma 5.1. The matrix exponent associated with the perturbed system (5.1) may be written in the form

$$e^{(A_0+\epsilon A_1)t} = V_R(\epsilon)e^{\epsilon M(\epsilon)t}V_L^T(\epsilon) + W_R(\epsilon)e^{\Lambda_1(\epsilon)t}W_L^T(\epsilon) \tag{5.10}$$

where

$$\begin{aligned} M(\epsilon)(k \times k) &= M + \epsilon Q_1 + \epsilon^2 Q_2 + \cdots , \\ \Lambda_1(\epsilon)(n-k \times n-k) &= \Lambda_1 + \epsilon \Lambda_1' + \epsilon^2 \Lambda_1'' + \cdots , \\ W_L(\epsilon)(n \times n-k) &= (E - V_L V_R^T) W_R + \epsilon W_L' + \cdots , \\ W_R(\epsilon)(n \times n-k) &= W_R + \epsilon W_R' + \cdots , \\ V_R(\epsilon)(n \times k) &= V_R + \epsilon V_R' + \cdots , \\ V_L(\epsilon)(n \times k) &= V_L + \epsilon V_{L1} + \cdots \end{aligned} \tag{5.11}$$

and V_{L1} satisfies the relation

$$V_{L1}^T B_0 = -V_L^T A_1 W_R \Lambda_1^{-1} W_R^T B_0 \tag{5.12}$$

Proof. As noted above, the columns of the matrix W_R constitute a basis for the right eigensubspace of the matrix A_0 corresponding to nonzero eigenvalues. Hence,

$$\det \{ V_R , W_R \} \neq 0$$

As follows from general perturbation theory [36], the eigenvalues of $A_0 + \epsilon A_1$ are decomposed into two groups. The first group, of total multiplicity k, consists of eigenvalues asymptotically (in $\epsilon \to 0$) close to zero, and the second one, of total multiplicity $n - k$, consists of eigenvalues that are isolated from zero. The associated projectors on the right eigensubspaces of $A_0 + \epsilon A_1$ can be developed into the power series

$$\begin{aligned} P^1(\epsilon) &= P_0^1 + \epsilon P_1^1 + \cdots \\ P^2(\epsilon) &= P_0^2 + \epsilon P_1^2 + \cdots \end{aligned} \tag{5.13}$$

where P_0^1, P_0^2 are the projectors on the right eigensubspaces of A_0 corresponding to zero and nonzero eigenvalues.

By virtue of (5.13), one can write

$$\begin{aligned} V_R(\epsilon) &\triangleq P^1(\epsilon) V_R = P_0^1 V_R + \epsilon P_1^1 V_r + \cdots = V_R + \epsilon V_{R1} + \cdots \\ W_R(\epsilon) &\triangleq P^2(\epsilon) W_R = P_0^2 W_R + \epsilon P_2^1 W_r + \cdots = W_R + \epsilon W_{R1} + \cdots \end{aligned} \tag{5.14}$$

$$\det \{ V_R(\epsilon) , W_R(\epsilon)\} = \det \{ V_R , W_R \} + O(\epsilon) \neq 0 \tag{5.15}$$

Let us now define the matrices $V_L(\epsilon)$, $W_L(\epsilon)$ such that

$$\begin{vmatrix} V_L^T(\epsilon) \\ W_L^T(\epsilon) \end{vmatrix} \triangleq \{ V_R(\epsilon) , W_R(\epsilon)\}^{-1} \Rightarrow \tag{5.16}$$

$$V_R(\epsilon) V_L^T(\epsilon) + W_R(\epsilon) W_L^T(\epsilon) = E_n \tag{5.17}$$

$$\begin{aligned} V_L^T(\epsilon) V_R(\epsilon) &= E_k \ , \quad V_L^T(\epsilon) W_R(\epsilon) = 0 \\ W_L^T(\epsilon) V_R(\epsilon) &= 0 \ , \quad W_L^T(\epsilon) W_R(\epsilon) = E_{n-k} \end{aligned} \tag{5.18}$$

Taking into account (5.14) and (5.15), it is easy to show that

$$V_L(\epsilon) = V_L + \epsilon V_{L1} + \cdots \;, \quad W_L(\epsilon) = W_L + \epsilon W_{L1} + \cdots \tag{5.19}$$

Since the columns of $V_R(\epsilon)$ constitute a basis of an invariant subspace of $A_0 + \epsilon A_1$, we have

$$\begin{aligned}
&V^R(\epsilon)\, V_L^T(\epsilon)(A_0 + \epsilon A_1)\, V_R(\epsilon) = (A_0 + \epsilon A_1)\, V_R(\epsilon) \Rightarrow \\
&V_R(\epsilon)[\, V_L^T(\epsilon)(A_0 + \epsilon A_1)\, V_R(\epsilon)]^r\, V_L^T(\epsilon) = (A_0 + \epsilon A_1)^r\, V_R(\epsilon)\, V_L^T(\epsilon) \;, \quad r = 0,1,2,\ldots \Rightarrow \\
&V_R(\epsilon) e^{V_L^T(\epsilon)(A_0 + \epsilon A_1) V_R(\epsilon) t}\, V_L^T(\epsilon) = e^{(A_0 + \epsilon A_1)t}\, V_R(\epsilon)\, V_L^T(\epsilon)
\end{aligned}$$

Similarly,

$$W_R(\epsilon) e^{W_L^T(\epsilon)(A_0+\epsilon A_1) W_R(\epsilon) t}\, W_L^T(\epsilon) = e^{(A_0+\epsilon A_1)t}\, W_R(\epsilon)\, W_L^T(\epsilon)$$

from which, by (5.17),

$$\begin{aligned}
e^{(A_0+\epsilon A_1)t} &= e^{(A_0+\epsilon A_1)t}\, V_R(\epsilon)\, V_L^T(\epsilon) + e^{(A_0+\epsilon A_1)t}\, W_R(\epsilon)\, W_R^T(\epsilon) \\
&= V_R(\epsilon) e^{\epsilon M(\epsilon) t}\, V_L^T(\epsilon) + W_R(\epsilon) e^{\Lambda_1(\epsilon) t}\, W_L^T(\epsilon)
\end{aligned}$$

where

$$\begin{aligned}
\epsilon M(\epsilon) &\triangleq V_L^T(\epsilon)(A_0 + \epsilon A_1)\, V_R(\epsilon) \\
\Lambda_1(\epsilon) &\triangleq W_L^T(\epsilon)(A_0 + \epsilon A_1)\, W_R(\epsilon)
\end{aligned}$$

The relations obtained above with (5.14) and (5.19) lead to (5.10) and (5.11).

To obtain (5.12), note first that, in view of (5.18), the columns of the matrix $V_L(\epsilon)$ form a basis for the left eigensubspace of $A_0 + \epsilon A_1$ corresponding to eigenvalues that are asymptotically close to zero. Consequently, we can write

$$\begin{aligned}
&V_L(\epsilon)\, V_R^T(\epsilon)(A_0^T + \epsilon A_1^T)\, V_L(\epsilon) = (A_0^T + \epsilon A_1^T)\, V_L(\epsilon) \Rightarrow \\
&(V_L + \epsilon V_{L1} + \cdots)(\epsilon M^T + \epsilon^2 Q_1^T + \cdots) = (A_0^T + \epsilon A_1^T)(V_L + \epsilon V_{L1} + \cdots) \Rightarrow \\
&V_L M^T = A_1^T V_L + A_0^T V_{L1}
\end{aligned} \tag{5.20}$$

Let us construct V_{L1} as

$$V_{L1} = V_L L_V + W_R L_W \tag{5.21}$$

where L_V, L_W are matrices of appropriate dimension. Substituting (5.21) into (5.20), we obtain

$$V_L M^T - A_1^T V_L = A_0^T W_R L_W \Rightarrow$$
$$-W_R^T A_1^T V_L = \Lambda_1^T L_W \Rightarrow$$
$$L_W = -\Lambda_1^{-T} W_R^T A_1^T V_L \Rightarrow$$
$$V_{L1}^T B_0 = L_W^T W_R^T B_0 = -V_L^T A_1 W_R \Lambda_1^{-1} W_R^T B_0$$

Proof of the theorem. According to (5.10), the solution of (5.1) may be written in the form

$$\begin{aligned} x_u(t) = {} & e^{(A_0+\epsilon A_1)t} x_0 + \int_0^t e^{(A_0+\epsilon A_1)(t-s)}(B_0 + \epsilon B_1)u(s)ds \\ = {} & V_R(\epsilon) e^{\epsilon M(\epsilon)t} V_L^T(\epsilon) x_0 + \\ & + \int_0^t V_R(\epsilon) e^{\epsilon M(\epsilon)(t-s)} V_L^T(\epsilon)(B_0 + \epsilon B_1)u(s)ds + \\ & + W_R(\epsilon) e^{\Lambda_1(\epsilon)t} W_L^T(\epsilon) x_0 + \\ & + \int_0^t W_R(\epsilon) e^{\Lambda_1(\epsilon)(t-s)} W_L^T(\epsilon)(B_0 + \epsilon B_1)u(s)ds \end{aligned} \tag{5.22}$$

By virtue of Lemma 5.1, the relations

$$V_R(\epsilon) e^{\epsilon M(\epsilon)t} V_L^T(\epsilon) x_0 = V_R e^{\epsilon Mt} V_L^T x_0 + O(\epsilon) \tag{5.23}$$

$$\begin{aligned} & \int_0^t V_R(\epsilon) e^{\epsilon M(\epsilon)(t-s)} V_L^T(\epsilon)(B_0 + \epsilon B_1)u(s)ds = \\ & = \epsilon \int_0^t V_R e^{\epsilon M(t-s)}(V_L^T B_1 + V_{L1}^T B_0)u(s)ds + O(\epsilon) = \\ & = \epsilon \int_0^t V_R e^{\epsilon M(t-s)}(V_L^T B_1 + V_L^T A_1 W_R \Lambda_1^{-1} W_R^T B_0)u(s)ds + O(\epsilon) \end{aligned} \tag{5.24}$$

hold where the estimate is uniform in $t \in [0, T/\epsilon]$. At the same time, we have

$$\begin{aligned} \|e^{\Lambda_1(\epsilon)t} - e^{\Lambda_1 t}\| & = \|e^{\Lambda_1 t}\| \cdot \|e^{[\epsilon\Lambda_1' + \epsilon^2 \Lambda_1'' + \cdots]t} - E_n\| \\ & \leq \|e^{\Lambda_1 t}\| \cdot [\epsilon t \|\Lambda_1'(\epsilon)\| + \frac{\epsilon^2 t^2 \|\Lambda_1'(\epsilon)\|^2}{2!} + \cdots] \\ & \leq \|e^{\Lambda_1 t}\| \cdot e^{\epsilon\|\Lambda_1'(\epsilon)\|t} \cdot \epsilon Lt \\ & \leq \|e^{\Lambda_1 t}\| e^{\epsilon Lt} \epsilon Lt \end{aligned} \tag{5.25}$$

where

$$\Lambda_1'(\epsilon) = \Lambda_1' + \epsilon\Lambda_1'' + \cdots \ , \quad \|\Lambda_1'(\epsilon)\| \leq L = \text{constant}$$

By the conditions of the theorem, the matrix Λ_1 is stable. Hence, there exist constants c, γ, p such that

$$\|e^{\Lambda_1 t}\| \leq ce^{-\gamma t}t^p$$

It follows that the inequality (5.25) can be strengthened

$$\|e^{\Lambda_1(\epsilon)t} - e^{\Lambda_1 t}\| \leq \epsilon(cL)t^{p+1}e^{-\gamma t}e^{\epsilon Lt}$$

Now let $0 < \tilde{\gamma} < \gamma$. Then, obviously, one can choose $t_1 > 0$ such that

$$t^{p+1}e^{-\gamma t} \leq t_1^{p+1}e^{-\gamma t_1} \triangleq L_1 \ , \quad 0 \leq t \leq t_1$$
$$t^{p+1}e^{-\gamma t} \leq e^{-\tilde{\gamma} t} \ , \quad t > t_1$$

from which, for $t \in [0\,, T/\epsilon]$,

$$\|e^{\Lambda_1(\epsilon)t} - e^{\Lambda_1 t}\| \leq \begin{cases} \epsilon(cLL_1)e^{LT} \ , & 0 < t \leq t_1 \\ \epsilon(cL)e^{-\tilde{\gamma} t}e^{LT} \ , & t > t_1 \end{cases} \tag{5.26}$$

As a consequence of (5.26) we have the following relations (uniformly for $t \in [0\,, T/\epsilon]$)

$$W_R(\epsilon)e^{\Lambda_1(\epsilon)t}W_L^T(\epsilon)x_0 = W_R e^{\Lambda_1 t}W_R^T(E_n - V_R V_L^T)x_0 + O(\epsilon) \tag{5.27}$$

$$\begin{aligned} &\int_0^t W_R(\epsilon)e^{\Lambda_1(\epsilon)(t-s)}W_L^T(\epsilon)(B_0 + \epsilon B_1)u(s)ds = \\ &= \int_0^t W_R(\epsilon)e^{\Lambda_1(\epsilon)s}W_L^T(\epsilon)(B_0 + \epsilon B_1)u(t-s)ds = \\ &= \int_0^t W_R e^{\Lambda_1 s}W_R^T B_0 u(t-s)ds + O(\epsilon) = \\ &= \int_0^t W_R e^{\Lambda_1(t-s)}W_R^T B_0 u(s)ds + O(\epsilon) \end{aligned} \tag{5.28}$$

Substituting (5.27), (5.28), (5.23) and (5.24) into (5.29), we obtain that, uniformly for $t \in [0, T/\epsilon]$,

$$x_u(t) = V_R e^{\epsilon M t} V_L^T x_0 + \epsilon \int_0^t V_R e^{\epsilon M(t-s)} (V_L^T B_1 - V_L^T A_1 W_R \Lambda_1^{-1} W_R^T B_0) u(s) ds + \\ + W_R e^{\Lambda_1^T} W_R^T (E_n - V_R V_L^T) x_0 + \int_0^t W_R e^{\Lambda_1 (t-s)} W_R^T B_0 u(s) ds + O(\epsilon)$$

which is equivalent to (5.8). After premultiplying (5.8) by V_L^T we come to (5.9).

As follows from Theorem 5.1, if a control is fixed, the trajectory of the perturbed system is expressed via the trajectories of the systems (5.4) and (5.5). This leads to the connection between the singularly perturbed optimal control problem

$$\begin{aligned} &\dot{x} = (A_0 + \epsilon A_1) x + (B_0 + \epsilon B_1) u \ , \ x(0) = x_0 \ , \ u \in U \\ &\inf \{ G(x(\frac{T}{\epsilon})) + \epsilon \int_0^{T/\epsilon} \varphi(x(\tau), u(\tau)) d\tau \} \triangleq G_\epsilon^* \end{aligned} \tag{5.29}$$

and the problem

$$\begin{aligned} &\dot{z} = \epsilon M z + \epsilon (D_{21} - \Lambda_{21} \Lambda_1^{-1} D_1) u \ , \ u \in U \\ &z(0) = z_0 \triangleq V_L^T x_0 \\ &\dot{\eta} = \Lambda_1 \eta + D_1 u \ , \ \eta(0) = \eta_0 \triangleq W_R^T (E_n - V_R V_L^T) x_0 \\ &\inf \{ G(V_R z(\frac{T}{\epsilon}) + W_R \eta(\frac{T}{\epsilon})) + \epsilon \int_0^{T/\epsilon} \varphi(V_R z(\tau) + W_R \eta(\tau), u(\tau)) d\tau \} \triangleq \bar{G}_\epsilon^* \end{aligned} \tag{5.30}$$

where the coefficient ϵ of the integral term makes it comparable in value with the terminal one. Note that in (5.30) the "slow" and "fast" subsystems are connected by the control influence only.

Theorem 5.2.

Let the matrix Λ_1 be stable, let U be a compact set, and let the functions $G(\cdot)$, $\varphi(\cdot)$ have bounded derivatives with respect to x. Then

$$G_\epsilon^* = \bar{G}_\epsilon^* + O(\epsilon) \tag{5.31}$$

Moreover, if $\bar{u}_\epsilon(t)$ is a control that, when used in (5.30), leads to a value of the performance index coinciding with the optimal one to within $O(\mu(\epsilon))$, then use of this control in (5.29) ensures a value differing from the optimal one by $O(\nu(\epsilon))$, where $\nu(\epsilon) \triangleq \max\{\epsilon, \mu(\epsilon)\}$ and $\mu(\epsilon)$ are arbitrary positive functions of ϵ tending together with it toward zero.

Proof. Let $\bar{u}_\epsilon(t)$ be a control satisfying the conditions of the theorem. By virtue of Theorem 5.1, the value $\tilde{G}_\epsilon$ of the performance index (5.29) obtained by using $\bar{u}_\epsilon(t)$ can be written in the form

$$\begin{aligned}
\tilde{G}_\epsilon &\triangleq G(x_{\bar{\mathbf{u}}_\epsilon}(\frac{T}{\epsilon})) + \epsilon \int_0^{T/\epsilon} \varphi(x_{\bar{\mathbf{u}}_\epsilon}(\tau), \bar{u}_\epsilon(\tau))d\tau \\
&= G(V_R z_{\bar{\mathbf{u}}_\epsilon}(\frac{T}{\epsilon}) + W_R \eta_{\bar{\mathbf{u}}_\epsilon}(\frac{T}{\epsilon})) + \\
&\quad + \epsilon \int_0^{T/\epsilon} \varphi(V_R z_{\bar{\mathbf{u}}_\epsilon}(\tau) + W_R \eta_{\bar{\mathbf{u}}_\epsilon}(\tau), \bar{u}_\epsilon(\tau))d\tau + O(\epsilon) \\
&\triangleq \bar{G}^*_\epsilon + O(\epsilon) + O(\mu(\epsilon)) \triangleq \bar{G}^*_\epsilon + O(\nu(\epsilon)) \Rightarrow \\
G^*_\epsilon &\leq \tilde{G}_\epsilon = \bar{G}^*_\epsilon + O(\nu(\epsilon))
\end{aligned} \tag{5.32}$$

In particular, for $\mu(\epsilon) = \epsilon$,

$$G^*_\epsilon \leq \tilde{G}_\epsilon = \bar{G}^*_\epsilon + O(\epsilon) \tag{5.33}$$

On the other hand, let $\hat{u}_\epsilon(t)$ be a control leading to the $O(\epsilon)$ optimal value of the performance index in (5.29):

$$G(x_{\hat{\mathbf{u}}_\epsilon}(\frac{T}{\epsilon})) + \epsilon \int_0^{T/\epsilon} \varphi(x_{\hat{\mathbf{u}}_\epsilon}(\tau), \hat{u}_\epsilon(\tau))d\tau \triangleq G^*_\epsilon + O(\epsilon) \tag{5.34}$$

Applying Theorem 5.1, we have

$$\begin{aligned}
&G(x_{\hat{\mathbf{u}}_\epsilon}(\frac{T}{\epsilon})) + \epsilon \int_0^{T/\epsilon} \varphi(x_{\hat{\mathbf{u}}_\epsilon}(\tau), \hat{u}_\epsilon(\tau))d\tau = \\
&= G(V_R z_{\hat{\mathbf{u}}_\epsilon}(\frac{T}{\epsilon}) + W_R \eta_{\hat{\mathbf{u}}_\epsilon}(\frac{T}{\epsilon})) + \epsilon \int_0^{T/\epsilon} \varphi(V_R z_{\hat{\mathbf{u}}_\epsilon}(\tau) + W_R \eta_{\hat{\mathbf{u}}_\epsilon}(\tau))d\tau + O(\epsilon) \geq \\
&\geq \bar{G}^* + O(\epsilon)
\end{aligned}$$

from which, by (5.34),

$$G_\epsilon^* \geq \bar{G}_\epsilon^* + O(\epsilon)$$

This last together with (5.33) and (5.32) yields the desired result.

Theorem 5.2 enables us to identify situations when one can use only a "slow" control determined from the aggregated "slow" subsystem to suboptimize the perturbed system. So the following statements are true.

Corollary 1. Let the conditions of Theorem 5.2 be fulfilled and, moreover, let the functions $G(\cdot)$, $\varphi(\cdot)$ that form the performance index in (5.29) be written as follows:

$$G(x) \triangleq \bar{G}(V_L^T x) \quad , \quad \varphi(x, u) \triangleq \bar{\varphi}(V_L^T x, u)$$

Let $u^*(t)$ be an optimal control for the problem

$$\begin{aligned} &\dot{z} = \epsilon M z + \epsilon(D_{21} - \Lambda_{21}\Lambda_1^{-1}D_1)u \quad , \quad z(0) = z_0 \\ &u \in U \\ &\inf \{\bar{G}(z(\frac{T}{\epsilon})) + \epsilon \int_0^{T/\epsilon} \bar{\varphi}(z(\tau), u)d\tau\} \end{aligned} \tag{5.35}$$

Then, by using this control in (5.29), the optimal performance index is achieved to within $O(\epsilon)$.

Corollary 2. Let the conditions of Theorem 5.2 be fulfilled, and let $B_0 = 0$. Let $\bar{u}^*(t)$ be the optimal control to the problem

$$\begin{aligned} &\dot{z} = \epsilon M z + \epsilon D_{21} u \quad , \quad z(0) = z_0 \\ &u \in U \\ &\inf \{G(V_R z(\frac{T}{\epsilon})) + \epsilon \int_0^{T/\epsilon} \varphi(V_R z(\tau), u(\tau))d\tau\} \end{aligned} \tag{5.36}$$

Then use of this control in (5.29) provides the value of the performance index, which is equal to the optimal one to within $O(\epsilon \ln 1/\epsilon)$.

For singularly perturbed sysetems presented in the form (2.22′) of Chapter 5, the condition of Corollary 1 is equivalent to the supposition that the performance index does not depend on the "fast" variables. The condition of Corollary 2 means that the "fast" variables are not directly controlled.

Let us consider further the case where the singularly perturbed problem can be written in the form

$$\begin{aligned} &\dot{x} = (A_0 + \epsilon A_1)x + (B_0 + \epsilon B_1)u \quad , \quad x(0) = x_0 \\ &u \in U \\ &\inf \{G(x(\frac{T}{\epsilon}))\} \end{aligned} \tag{5.37}$$

Consider also the problem

$$\begin{aligned}
&\dot{z} = \epsilon M z + \epsilon(D_{21} - \Lambda_{21}\Lambda_1^{-1}D_1)u \ , \ z(0) = z_0 \\
&u \in U \\
&\dot{\eta} = \Lambda_1 \eta + D_1 v \ , \ \eta(0) = \eta_0 \ , \ v \in U \\
&\inf G(V_R z(\frac{T}{\epsilon}) + W_R \eta(\frac{T}{\epsilon})) \triangleq \check{G}_\epsilon^*
\end{aligned} \tag{5.38}$$

Unlike (5.30), the subsystems here are controlled independently, and the connection between them exists via the performance index only.

Theorem 5.3.

Let the conditions of Theorem 5.2 be satisfied. Let $u^*(t)$, $v^*(t)$ be the optimal control to (5.38), and let α be a sufficiently large positive constant. Then the use of the control defined as

$$\check{u}(t) \triangleq \begin{cases} u^*(t) \ , & 0 < t \le \frac{T}{\epsilon} - \alpha \ln \frac{1}{\epsilon} \\ v^*(t) \ , & \frac{T}{\epsilon} - \alpha \ln \frac{1}{\epsilon} \le t \le \frac{T}{\epsilon} \end{cases}$$

yields a value of the performance index differing from the optimal one by $O(\epsilon \ln 1/\epsilon)$.

Proof. The values $\check{G}_\epsilon^*$, $\bar{G}_\epsilon^*$ of the performance indices of the problems (5.38) and (5.30) are connected by the inequality

$$\bar{G}_\epsilon^* \ge \check{G}_\epsilon^*$$

Hence, in view of Theorem 5.2, to prove the theorem it is sufficient to show that

$$G(V_R \check{z}(\frac{T}{\epsilon}) + W_R \check{\eta}(\frac{T}{\epsilon})) = \check{G}_\epsilon^* + O(\epsilon \ln \frac{1}{\epsilon}) \tag{5.39}$$

where

$$\check{z}(\frac{T}{\epsilon}) \triangleq e^{MT} Z_0 + \int_0^{T/\epsilon} e^{\epsilon M(T/\epsilon - s)}(D_{21} - \Lambda_{21}\Lambda_1^{-1}D_1)\check{u}(s)ds$$

$$\check{\eta}(\frac{T}{\epsilon}) \triangleq e^{\Lambda_1 T/\epsilon}\eta_0 + \int_0^{T/\epsilon} e^{\Lambda_1(T/\epsilon - s)} D_1 \check{u}(s)ds$$

For any constant α

$$
\begin{aligned}
\check{z}(\frac{T}{\epsilon}) &= \check{z}(\frac{T}{\epsilon} - \alpha \ln \frac{1}{\epsilon}) + O(\epsilon \ln \frac{1}{\epsilon}) \\
&= e^{M(T-\epsilon\alpha \ln 1/\epsilon)} z_0 + \\
&+ \int_0^{T/\epsilon-\alpha\ln 1/\epsilon} e^{\epsilon M(T/\epsilon-\alpha\ln 1/\epsilon - s)}(D_{21} - \Lambda_{21}\Lambda_1^{-1}D_1)\check{u}(s)\mathrm{d}s + \\
&+ O(\epsilon \ln \frac{1}{\epsilon}) \\
&= z_{u^*}(\frac{T}{\epsilon} - \alpha \ln \frac{1}{\epsilon}) + O(\epsilon \ln \frac{1}{\epsilon}) \\
&= z_{u^*}(\frac{T}{\epsilon}) + O(\epsilon \ln \frac{1}{\epsilon})
\end{aligned}
$$

i.e.,

$$
\check{z}(\frac{T}{\epsilon}) = z_{u^*}(\frac{T}{\epsilon}) + O(\epsilon \ln \frac{1}{\epsilon}) \tag{5.40}
$$

where $z_{u^*}(\tau)$ is the trajectory of the system (5.4) with $u^*(\tau)$ as a control. At the same time we can write

$$
\begin{aligned}
\check{\eta}(\frac{T}{\epsilon}) = \; & e^{\Lambda_1 T/\epsilon}\eta_0 + e^{\Lambda_1\alpha\ln 1/\epsilon} \int_0^{T/\epsilon-\alpha\ln 1/\epsilon} e^{\Lambda_1(T/\epsilon-\alpha\ln 1/\epsilon - s)} D_1\check{u}(s)\mathrm{d}s + \\
&+ \int_{T/\epsilon-\alpha\ln 1/\epsilon}^{T/\epsilon} e^{\Lambda_1(T/\epsilon - s)} D_1\check{u}(s)\mathrm{d}s
\end{aligned} \tag{5.41}
$$

Since the matrix Λ_1 is stable and the set U is bounded, the following inequalities hold for all t

$$
\begin{aligned}
&\|\int_0^t e^{\Lambda_1(t-s)} D_1 u(s)\mathrm{d}s\| \leq L = \text{constant} \Rightarrow \\
&\|e^{\Lambda_1\alpha\ln 1/\epsilon} \int_0^{T/\epsilon-\alpha\ln 1/\epsilon} e^{\Lambda_1(T/\epsilon-\alpha\ln 1/\epsilon-s)} D_1 u(s)\mathrm{d}s\| \leq \\
&\leq L\|e^{\Lambda_1\alpha\ln 1/\epsilon}\| \leq k\epsilon \;\;, \;\; \alpha \geq \alpha_0
\end{aligned}
$$

These inequalities allow us to rewrite (5.41) in the form

$$\check{\eta}(\frac{T}{\epsilon}) = \int\limits_{T/\epsilon-\alpha\ln 1/\epsilon}^{T/\epsilon} e^{\Lambda_1(T/\epsilon-s)} D_1\check{u}(s)ds + O(\epsilon) \triangleq$$

$$\triangleq \int\limits_{T/\epsilon-\alpha\ln 1/\epsilon}^{T/\epsilon} e^{\Lambda_1(T/\epsilon-s)} D_1 v^*(s)ds + O(\epsilon)$$

Similarly, it is easy to obtain

$$\eta_{v^*}(\frac{T}{\epsilon}) = \int\limits_{T/\epsilon-\alpha\ln 1/\epsilon}^{T/\epsilon} e^{\Lambda_1(T/\epsilon-s)} D_1 v^*(s)ds + O(\epsilon)$$

where $\eta_{v^*}(\tau)$ is the trajectory of system (5.5) corresponding to the control $v^*(\tau)$. The conjunction of the last two relations gives

$$\check{\eta}(\frac{T}{\epsilon}) = \eta_{v^*}(\frac{T}{\epsilon}) + O(\epsilon \ln \frac{1}{\epsilon})$$

which together with (5.40) prove the theorem.

Corollary 1. Let the conditions of Theorem 5.2 be fulfilled, and let the function $G(\cdot)$ be such that

$$G(V_R z + W_R \eta) = G_1(z) + G_2(\eta) \tag{5.42}$$

Representation (5.42) takes place, for instance, if $G(x)$ is a linear function of x or if

$$G(x) \triangleq x^T Q x$$

and

$$V_R^T Q W_R = 0$$

Then using in (5.37) the control

$$\check{u}(t) = \begin{cases} u^*(t) \ , & 0 \le t \le \dfrac{T}{\epsilon} - \alpha \ln \dfrac{1}{\epsilon} \\ v^*(t) \ , & \dfrac{T}{\epsilon} - \alpha \ln \dfrac{1}{\epsilon} \le t \le \dfrac{T}{\epsilon} \end{cases} \tag{5.43}$$

one can obtain a value of the performance index within $O(\epsilon \ln 1/\epsilon)$ of the optimal one, where $u^*(t)$ is the optimal control for the problem

$$\begin{aligned} &\dot{z} = \epsilon M z + \epsilon(D_{21} - \Lambda_{21}\Lambda_1^{-1}D_1)u \ , \ z(0) = z_0 \\ &u \in U \\ &\inf G_1(z(\frac{T}{\epsilon})) \end{aligned} \tag{5.44}$$

and $v^*(t)$ is the optimal control for the problem

$$\begin{aligned} &\dot{\eta} = \Lambda_1 \eta + D_1 v \ , \ \eta(0) = \eta_0 \\ &v \in V \\ &\inf G_2(\eta(\frac{T}{\epsilon})) \end{aligned} \tag{5.45}$$

If (5.42) does not hold, one can construct the solution via the successive approximations scheme, each cycle consisting of solving a problem of the (5.44) or (5.45) type.

In more detail, given the i-th iteration $\{u^i(t), z^i(t)\}$, $\{v^i(t), \eta^i(t)\}$, the $(i+1)$-th iteration is determined as follows: $\{u^{i+1}(t), z^{i+1}(t)\}$ is the solution to

$$\begin{aligned} &\dot{z} = \epsilon M z + \epsilon(D_{21} - \Lambda_{21}\Lambda_1^{-1}D_1)u \ , \ z(0) = z_0 \\ &u \in U \\ &\inf \{G(V_R z(\frac{T}{\epsilon}) + W_R \eta^i(\frac{T}{\epsilon}))\} \end{aligned}$$

and $\{v^{i+1}(t), \eta^{i+1}(t)\}$ is the solution to

$$\begin{aligned} &\dot{\eta} = \Lambda_1 \eta + D_1 v \ , \ \eta(0) = \eta_0 \ , \\ &v \in U \\ &\text{int} \{G(V_R z^{i+1}(\frac{T}{\epsilon}) + W_R \eta(\frac{T}{\epsilon}))\} \end{aligned}$$

It is easy to show that, if the set U and the function $G(t)$ are convex, the procedure converges to the solution of the problem (5.38) no matter what initial approximation $\{u^0(t), z^0(t)\}$, $\{v^0(t), \eta^0(t)\}$ is chosen.

Related Literature

Chapter 1.

In Section 1.1, we mentioned some works that discuss the general problems of decomposition and aggregation. All of them contain the classical schemes of Dantzig, Wolfe, Arrow, Hurwicz, and Uzawa as a basis. It is worth mentioning an approach presented by Geoffrion and Zipkin in papers [115–118], which are devoted to the *a priori* estimates of heuristic aggregation schemes. See also a review in [126]. Note also relatively new works by Ben-Israel *et al.* [157], Gauvin and Dubeau [158], and Lempio and Maurer [159] containing surveys.

The investigation of the dependence of mathematical program solutions on a small parameter has been a field of activity for many authors. From a formal point of view, the closest approach to that of Chapter 1 is contained in books and papers that solve the problem of differentiability of the objective with respect to the parameter (the so-called marginal values problem).

Beginning with the classical paper by Mills (reprinted in [144]), devoted to matrix games and linear programming, expansions of the (2.23)-type were constructed and proved under various suppositions. See, primarily Golshtein and Judin [21], where an error in the proof given by Mills was corrected, and also the most general result of Levitin [46].

However, in spite of the natural connection, results concerning the convergence between perturbed and unperturbed solutions to the auxiliary program had not been formulated and were published for the first time in the authors' papers [17, 18, 65, 149]. For the simplest case, where only the objective function contains a perturbation, similar results were independently obtained by Mangasarian and Meyer [119]. Some interesting results concerning perturbed linear programs are obtained by Gabassov and Kirillova in [127].

Chapter 2.

The material is original in approach and based on the investigation of the authors and their colleagues. Partial results were published in the papers [17, 18, 65, 66]. Some were presented in the book [64] at the heuristic level. At the same time, there are certain connections with the papers by Volkonskii [12] and Dudkin and Ershov [27, 31]. Moreover, there are no real barriers separating approximation methods and well-known schemes oriented toward an exact solution. Some approaches presented in Chapter 2 coincide with the initial iterations of exact procedures based on duality theory (references were given in the text).

Chapter 3.

The results are completely original. Partial results were published in the authors' papers [13, 14, 15, 16]. Some results of Section 3.4 are closely connected with those of papers devoted to the aggregation of Markov processes. First of all, this is referred to in the works of Koroljuk and his colleagues (see, e.g., [43, 44]). The investigation of the control of Markov chains was developed by Delebecque and Quadrat in France (essentially later but, as far as we know, independently).

Chapter 4.

The material is, as a whole, original. The results were obtained by a development of corresponding results for mathematical programs. The introductory section is similar to the paper by Arbuzova [3]. The formulation of problems adopted in Section 4.2 is partially analogous to those investigated by Kolmanovskii and Cheronous'ko [41, 42, 131], Fleming [94], and Holand [98] in the theory of stochastic differential equations but the techniques are essentially different. The results of Sections 4.4 and 4.5, published in the authors' papers [19, 66], were partially predicted in [64].

Chapter 5.

The LQ problem is the favorite subject of the perturbation method. The main results were obtained by Bayley and Ramapriyan [87] for the regular case and by Kokotovic, O'Malley, and their colleagues for singularly perturbed systems [100–108].

A new element introduced by the authors is the establishment of connections between singularity and degeneracy, i.e., the loss of complete controllability and (or) complete observability. This allows the unification of results relating to simplification of the LQ problem by decomposition and aggregation. In particular, it is now clear that the heuristic schemes of aggregation suggested by Aoki in [86] and other papers can only be efficient if the problems are close to completely observable ones (see also the authors' paper [135]). The examples of applications presented in Sections 5.3 are original. The first one is of practical interest and closely connected with the theory of linear production and inventory systems developed in the book [64].

The material of Section 5.4 is also original, pieces of which were published in [136, 143]. However, the control problem for the pure oscillatory system was investigated earlier by Larin in [132], who used the factorization scheme and obtained an explicit result for the one-degree-of-freedom system. References relating to singularly perturbed LQG problems are presented in Section 5.5 directly. There is a close connection between these problems and LQG problems for systems with an infinite number of eigenvalues; see Pervozvanskii and Solonina [161].

It is worth noting that almost all the proofs presented in Chapter 5 are original. Only the stationary case is considered here, and this allowed the use of the simplest algebraic approach.

Chapter 6.

Because the perturbation method plays the main role in the analytical theory of nonlinear optimal control, the number of publications is immense and our reference are very incomplete here. Section 6.1 is, principally a brief review of the main results obtained by many authors. In this connection works by Novoselov [58], Kirillova [37], Moisseev [52], Chernous'ko [85], Kolmanovskii [39, 40], Plotnikov [129, 130], Garrard [96], Lukes [106], and Nishikawa *et al.* [109] should be mentioned foremost.

Sections 6.2, 6.3 and 6.4 contain original results. Partial results were published in the authors' papers [138, 140].

The last section is devoted to singularly perturbed linear systems with a non-quadratic objective. The material is closely connected with other works in this field. One can obtain information about this field by means of the survey [160] or see, e.g., the papers by Kokotovic and his colleagues mentioned above and papers by Soviet authors, especially by Vasiljeva and Dmitriev [10, 24, 25]. All these authors apply the boundary layer techniques developed earlier by Tichonov [79], Vasiljeva [9, 142] and Hoppenstedt [99] to the analysis of ordinary differential equations. However, in Section 6.5, another technique is used that is simpler and more efficient for linear systems. A new technique and results for nonlinear singularly perturbed systems are described in [162].

Bibliography

Russian-language periodicals

Avtomatika i telemekhanika (Automatics and remote control)
Diff. uravnenija (Differential equations)
Dinamika sistem (Dynamics of systems)
Dokl. AN SSSR (Reports of the USSR Academy of Science)
Ekon. i mat. metody (Economics and mathematical methods)
Izv. AN SSSR, Tekhn. kibernetika (Proceedings of the USSR Academy of Science, Technical cybernetics)
J. vychisl. mat. i mat. fiz. (Journal of computing mathematics and mathematical physics)
Kibernetika (Cybernetics)
Mat. sbornik (Mathematical transactions)
Mat. metody reshenija ekon. zadach (Mathematical methods for economic problems)
Metody optimiz. (Optimization methods)
Planovoje khoz. (Planned economy)
Prikl. mat. i mekh. (Applied mathematics and mechanics)
Problemy rasshir. vozmogn. avtomatov (Problems of automata theory and applications)
Trudy sckoly po mat. program. [Proceedings of the school in mathematical programming (Drogobych)]
Uspekhi mat. nauk (Progress in mathematics)
Vestnik MGU (Proceedings of Moscow University)
Voprosy sovershi. planirov. (Planning problems)

---

[1] Andreev, Ju.N., *Control of Linear Finite-Dimensional Systems*, Moscow, Nauka, 1976 (in Russian).

[2] Anissimov, V.V., Asymptotic aggregation of random processes, *Kibernetika*, 3, 1973 (in Russian).

[3] Arbuzova, N.I., On a stochastic stability of dual linear programs, *Ekon. i mat. metody*, 2, 4, 1966 (in Russian).

[4] Babakov, I.M., *Theory of Vibrations*, Moscow, Nauka, 1968 (in Russian).

[5] Bellman, R., *Introduction to Matrix Analysis*, New York-Toronto-London, McGraw-Hill, 1960.

[6] Bratus, L.S., Iterative method of the solution of Bellman equations for the optimal control of stochastic systems, *Prikl. mat. i mekh.* 39, 2, 1975 (in Russian).

[7] Budak, B.N., Berkovich, E.N., and Solovjeva, E.I., Difference approximations in optimal control problems, *Vestnik MGU*, Series 1, 2, 1968 (in Russian).

[8] Vainberg, M.M. and Trenogin, V.A. *Bifurcation Theory of Nonlinear Equations*, Moscow, Nauka, 1969 (in Russian).

[10] Vasiljeva, A.B. and Dmitriev, M.G., Singular perturbations and some optimal control problems, *Proceedings of the Seventh World Congress of IFAC*, Helsinki, 1978.

[11] Vilchevskii, N.O., Mutual relations of a center and executives in optimal planning, in *Theory of Hierarchical Systems*, Irkutsk, SOAN SSSR, 1978 (in Russian).

[12] Volkonskii, V.A., Utilization of special structure of matrice in planning problems, *Ekon. i. mat. metody*, 2, 2, 1966 (in Russian).

[13] Gaitsgori, V.G. and Pervozvanskii, A.A., Aggregation of Markov chains with weak interaction, *Kibernetika*, 3, 1975 (in Russian).

[14] Gaitsgori, V.G. and Pervozvanskii, A.A., Separation of motions in Markov systems, *Dinamika sistem*, 3, 1975 (in Russian).

[15] Gaitsgori, V.G. and Pervozvanskii, A.A., Hierarchical procedure of optimal control of finite Markov chains, *Trudy sckoly po mat. program.*, 7, 1974 (in Russian).

[16] Gaitsgori, V.G., Iterative aggregation as a tool for the solution of incorrect linear programs, *Voprosy soversh. planirov.*, 1976 (in Russian).

[17] Gaitsgori, V.G. and Pervozvanskii, A.A., Small nonlinear perturbations in optimal problems, *Ekon. i mat. metody*, 13, 1, 1977 (in Russian).

[18] Gaitsgori, V.G. and Pervozvanskii, A.A., On a simplified solution of optimal planning problems, *Mat. metody reshenijy ekon. zadach*, 7, Moscow, Nauka, 1977 (in Russian).

[19] Gaitsgori, V.G. and Pervozvanskii, A.A., On approximate optimization of the sliding planning, *Avtomatika i telemekhanika*, 10, 1977 (in Russian).

[20] Gantmacher, F.R., *Theory of Matrices*, Moscow, Nauka, 1967 (in Russian).

[21] Golshtein, E.G. and Judin, D.B., *New Directions in Linear Programming*, Moscow, Sov. Radio, 1966 (in Russian).

[22] Grosdovskii, G.L., Ivanov, Ju.N., and Tokarev, V.V., *Mechanics of Cosmic Flights with the Small Thrust*, Moscow, Nauka, 1966 (in Russian).

[23] Dantzig, G.B., *Linear Programming and Extensions*, Princeton, NJ, Princeton University Press, 1963.

[24] Dmitriev, M.G., On continuity of solutions to Mayer problems under singular perturbations, *J. vychisl. mat. i mat. fiz.*, 3, 1972 (in Russian).

[25] Dmitriev, M.G., On a class of singular perturbated problems of optimal control, *Prikl. mat. i mekh.*, 42, 2, 1978 (in Russian).

[26] Dorofejuk, A.A., Algorithms of automatic classification, *Problemy rasshir. vozmogn. avtomatov*, 1, 1971 (in Russian).

[27] Dudkin, L.M. and Ershov, E.B., Interindustrial balance and material balances of products, *Planovoje khoz.*, 5, 1965 (in Russian).

[28] Zangwill, W.I., *Nonlinear Programming: A Unified Approach*, Englewood Cliffs, NJ, Prentice-Hall, 1969.

[29] Egorov, A.J., *Optimal Control of Thermal and Diffusional Processes*, Moscow, Nauka, 1978 (in Russian).

[30] Esipova, V.A., Asymptotic solution of minimal time linear problem, *Diff. uravnenija*, 3, 1973 (in Russian).

[31] Ershov, E.B., On detection and using of structural properties of matrices in planning problems, *Ekon. i mat. metody, 2, 2, 1966 (in Russian).*

[32] Ioslovich, I.B. and Makarenkov, Ju. M., On methods of reduction of dimension in linear programming, *Ekon. i mat. metody*, 11, 3, 1975 (in Russian).

[33] Ioffe, A.D. and Tichomirov, V.M., *Theory of Extremal Problems*, Moscow, Nauka, 1974 (in Russian).

[34] Karmanov, V.G., *Mathematical Programming*, Moscow, Nauka, 1975 (in Russian).

[35] Kartvelishvili, N.A. and Galaktionov, Ju.I., *Idealization of Complex Dynamic Systems*, Moscow, Nauka, 1976 (in Russian).

[36] Kato, T., *Perturbation Theory for Linear Operations*, Berlin, Springer-Verlag, 1966.

[37] Kirillova, F.M., On continuous dependence of solutions of some optimal control problems on initial values and parameters, *Uspekhi mat. nauk*, 17, 4, 1962 (in Russian).

[38] Kolmogorov, A.N. and Fomin, S.V. *Elements of the Theory of Functions and Functional Analysis*, Moscow, Nauka, 1972 (in Russian).

[39] Kolmanovskii, V.B., Application of perturbation method to some problems of optimal control, *Prikl. mat. i mekh.*, 39, 5, 1975 (in Russian).

[40] Kolmanovskii, V.B., Optimal control of some weakly controllable systems, *Kibernetika*, 2, 1976 (in Russian).

[41] Kolmanovskii, V.B. and Chernous'ko, F.L., Optimal control problems with non-complete information, *Trudy sckoly po mat. program.*, 4, 1971 (in Russian).

[42] Kolmanovskii, V.B., On approximate synthesis of some stochastic quasilinear systems, *Avtomatika i telemekhanika*, 1, 1975 (in Russian).

[43] Koroljuk, V.S., Polishchuk, L.I., and Tomusjak, A.A., On a limit theorem for semi-Markov processes, *Kibernetika*, 4, 1969 (in Russian).

[44] Koroljuk, V.S. and Turbin, A.F., *Semi-Markov Processes and Their Applications*, Kiev, Naukova dumka, 1976 (in Russian).

[45] Lee, E.B. Markus, L., *Foundations of Optimal Control Theory*, New York, John Wiley, 1967.

[46] Levitin, E.S., On differentiability with respect to parameter of optimal value of parametric mathematical programs, *Kibernetika*, 1, 1976 (in Russian).

[47] Levitin, E.S., On estimates of solutions' stability in unconstrained optimal problems, *J. Vychisl. mat. i mat. fiz.*, 3, 1976 (in Russian).

[48] Lasdon, L., *Optimization Theory for Large-Scale Systems*, New York, Macmillan, 1970.

[49] Mine, H. and Osaki, S., *Markovian Decision Processes*, New York, American Elsevier, 1970.

[50] Meerov, M.V., *Synthesis of Structure of High-Precision Control Systems*, Moscow, Fizmatgiz, 1958 (in Russian).

[51] Meerov, M.V., and Litvak, B.L., *Optimization of Multi-Connected Control Systems*, Moscow, Nauka, 1972 (in Russian).

[52] Moisseev, N.N., *Asymptotic Methods of Nonlinear Mechanics*, Moscow, Nauka, 1969 (in Russian).

[53] Moisseev, N.N., *Numerical methods in the Theory of Optimal Systems*, Moscow, Nauka, 1971 (in Russian).

[54] Mishchenko, E.F. and Rozov, N.H., *Differential Equations with a Small Parameter and Relaxation Oscillations*, Moscow, Nauka, 1975 (in Russian).

[55] Natanson, I.P., *Theory for Functions of Real Variable*, Moscow, Nauka, 1975 (in Russian).

[56] Nebesnov, V.I. and Plotnikov, V.A., *Mathematical Methods for Research of Naval Complexes*, Moscow, MMF, 1977 (in Russian).

[57] Nikaido, H., *Convex Structures and Economic Theory*, New York and London, Academic Press, 1968.

[58] Novoselov, V.S., Method for approximate analytic solution of optimization problems, *Proceedings of Symposium in Applied Mathematics and Cybernetics*, Gorki, 1967 (in Russian).

[59] Novosselov, V.S., *Analytic Theory for Optimization in Gravitational Fields*, Leningrad, LGU Press, 1972 (in Russian).

[60] Pervozvanskaja, T.N., Recurrent relations of dynamic programming for solution of block linear programs, *Trudy sckoly po mat. program.*, 4, 1971 (in Russian).

[61] Pervozvanskaja, T.N. and Pervozvanskii, A.A., Decentralization of optimal planning for complex system, *Avtomatika i telemekhanika*, 7, 1968 (in Russian).

[62] Pervozvanskii, A.A., On a connection of basic theorems for mathematical programming and maximum principle, *Izv. AN SSR, Tekhn. kibernetika*, 3, 1966 (in Russian).

[63] Pervozvanskii, A.A., *Methods for Search*, Moscow, Nauka, 1970 (in Russian).

[64] Pervozvanskii, A.A., *Mathematical Models for Industrial Management*, Moscow, Nauka, 1975 (in Russian).

[65] Pervozvanskii, A.A., Optimization of weakly connected systems, *System Science*, 2, 2, 1976 (in Russian).

[66] Pervozvanskii, A.A. and Gaitsgori, V.G., Operative control of industrial activity as a problem of stochastic programming, *System Science*, 3, 1, 1977.

[67] Pervozvanskii, A.A. and Gaitsgori, V.G., Approximate optimization, decomposition, aggregation, *Proceedings of the Seventh USSR Congress for Control Problems*, Minsk, 1977 (in Russian).

[68] Pizarevskii, V.A., Two-level optimal planning with the minimal expenditure criterium, *Mat. metody reshenija ekon. zadach*, 7, 1977 (in Russian).

[69] Plotnikiv, V.A., *Asymptotic Methods for Optimal Control Problems*, Odessa, Odessa University Press, 1976 (in Russian).

[70] Poluektov, R.A. (ed.), *Dynamic Theory of Biologic Populations*, Moscow, Nauka, 1974 (in Russian).

[71] Pontrjagin, L.S., *Ordinary Differential Equations*, Moscow, Nauka, 1974 (in Russian).

[72] Prochorov, Ju.V. and Rozanov, Ju.A., *Probability Theory*, Moscow, Nauka, 1973 (in Russian).

[73] Pliskin, L.G., Optimization for industrial processes, *Energija*, 1975 (in Russian).

[74] Pontrjagin, L.S., Boltjanskii, V.G., Gamkrelidze, P.V., and Mishchenko, E.F., *Mathematical Theory of Optimal Processes*, Moscow, Fizmatgiz, 1961 (in Russian).

[75] Rakitskii, Ju.V., Ustinov, S.M., and Chernorutskii, I.G., *Numerical Methods for Solution of Stiff Ordinary Differential Equation*, Leningrad, LPI Press, 1977 (in Russian).

[76] Romanovskii, Ju.N., Stepanova, N.V., and Chernavskii, D.S., *Mathematical Modeling in Biophysics*, Moscow, Nauka, 1975 (in Russian).

[77] Soljanik, A.I. and Chernous'ko, F.L., Approximate synthesis of optimal control for stochastic systems, *Prikl. mat. i mekh.*, 36, 5, 1972 (in Russian).

[78] Ter-Krikorov, A.M., *Optimal Control and Mathematical Economics*, Moscow, Nauka, 1977 (in Russian).

[79] Tichonov, A.N., Systems of differential equations with small parameters before derivatives, *Mat. sbornik*, 71(73), 3, 1952 (in Russian).

[80] Tichonov, A.N. and Arsenin, V.Ja., *Methods for Solution of Incorrect Problems*, Moscow, Nauka, 1974 (in Russian).

[81] Fiacco, A.V. and McCormick, G.R., *Nonlinear Programming: Sequential Unconstrained Minimization Techniques*, New York, John Wiley, 1968.

[82] Khomjakov, V.A., Generalization of a proof for convergence of iterative aggregation process of linear systems solution, *Avtomatika i telemekhanika*, 3, 1973 (in Russian).

[83] Hadley, G., *Nonlinear and Dynamic Programming*, Reading, MA, Addison-Wesley, 1964.

[84] Howard, R.A., *Dynamic Programming and Markov Processes*, New York, John Wiley, 1960.

[85] Chernous'ko, F.L., Some optimal control problems with a small parameter, *Prikl. mat. i mekh.*, 32, 1, 1968 (in Russian).

[86] Aoki, M., Control of large-scale dynamic systems by aggregation, *IEEE Trans. on A.C.*, 13, 3, 1968.

[87] Bayley, F.N. and Ramapriyan, H.K., Bound on suboptimality in the control of linear dynamic systems, *IEEE Trans. on A.C.*, 18, 5, 1973.

[88] Binding, P., Singularly perturbed optimal control problems 1: Convergence, *SIAM J. on Control and Optimization*, 14, 4, 1976.

[89] Benveniste, A., Bernard, P., and Cohen, G., On the decomposition of stochastic control problems, *IFAC Symposium on Large-Scale Systems Theory and Applications*, Udine, 1976.

[90] Chenery, H.B. and Clark, P.G., *Interindustry Economics*, New York, John Wiley, 1959.

[91] Clamroch, N.Y., Duality and bounds for the matrix Riccati equation, *J. of Math. Appl.*, 25, 1969, pp. 622–627.

[92] Evers, J., *Linear Programming over an Infinite Horizon*, Tilburg, Tilburg University Press, 1973.

[93] Fisher, W.D., *Clustering and Aggregation in Economics*, Baltimore, MD, Johns Hopkins Press, 1969.

[94] Fleming, W.H., Stochastic control for small noise intensities, *SIAM J. on Control*, 9, 3, 1971.

[95] Freedman, H.I. and Kaplan, J.L., Singular perturbations of two point boundary value problems, *SIAM J. on Control and Optimization,* 14, 2, 1976.

[96] Garrard, W.L., Suboptimal feedback control for nonlinear systems, *Automatica*, 8, 2, 1972.

[97] Hoffman, A.J., On approximate solutions of systems of linear inequalities, *Journal of Research of the National Bureau of Standards*, 49, 4, 1952.

[98] Holand, C.J., Small noise open loop control, *SIAM J. on Control*, 12, 3, 1974.

[99] Hoppenstedt, F., Properties of solutions of ordinary differential equations with small parameters, *Com. Pure Appl. Math.*, 24, 1971, pp. 807–840.

[100] Kokotovic, P.V., Feedback design of large linear systems, in J.B. Gruz (ed.), *Feedback Systems*, New York, McGraw-Hill, 1972.

[101] Kokotovic, P.V. and Yackel, R.A., Singular perturbations of linear regulators: basic theorems, *IEEE Trans. on A.C.*, 17, 1, 1972.

[102] Kokotovic, P.V. and Wilde, R.R., Optimal open and closed loop control of singularly perturbed linear systems, *IEEE Trans. on A.C.,* 18, 6, 1973.

[103] Kokotovic, P.V. and Haddad, A.M., Controllability and time-optimal control of systems with slow and fast modes, *IEEE Trans. on A.C.*, 20, 1, 1975.

[104] Kokotovic, P.V., O'Malley, R.E., and Sanuti, P., Singular perturbations and order reductions in control theory — an overview, *Automatica*, 12, 1976, pp. 123–132.

[105] Kwong, R.H., Chong, C.Y., and Athans, M., On stochastic control system design methods for weakly coupled large systems, *Proc. JACC*, 1972, pp. 220–229.

[106] Lukes, D.L., Optimal regulation of nonlinear dynamical systems, *SIAM J. on Control,* 7, 1, 1969.

[107] O'Malley, R.E., *Introduction to Singular Perturbations*, New York, Academic Press, 1974.

[108] O'Malley, R.E. and Kung, C.E., The singularly perturbed linear state regulators problem, *SIAM J. on Control*, 13, 2, 1975.

[109] Nishikawa, Y., Sannomiya, N., and Itakura, N., A method for suboptimal design of nonlinear feedback systems, *Automatica*, 7, 1971, pp. 703–712.
[110] Özgüner, Ü., Analysis and control of interconnected systems which have a multi-time scale property, *IFAC Workshop on Control and Management of Integrated Industrial Complexes*, Toulouse, 1977.
[111] *Stiff Differential Systems*, Proceedings of an International Symposium, New York, John Wiley, 1974.
[112] Wilde, D.J., *Optimum Seeking Methods*, Englewood Cliffs, MJ, Prentice-Hall, 1964.
[113] Wismer, D.A. (ed.), *Optimization Method for Large-Scale Systems*, New York, McGraw-Hill, 1971.
[114] Rockafellar, R.T., *Convex Analysis*, Princeton, NJ, Princeton University Press, 1970.
[115] Geoffrion, A.M., *Aggregation Theory and Its Application to Models in Mathematical Programming*, Western Management Science Institute, University of California, Working Paper 278, December 1979.
[116] Geoffrion, A.M., Objective function approximations in mathematical programming, *Mathematical Programming*, 13, 1.
[117] Geoffrion, A.M., A priori error bounds for procurement commodity aggregation in logistics planning models, *Naval Res. Log. Quart., 24, 2.*
[118] Zipkin, P., Aggregation in linear programming, Ph.D. thesis, School of Organization and Management, Yale University, December, 1977.
[119] Mangasarian, O.L. and Meyer, R.R. Nonlinear perturbation of linear programs, *SIAM J. Control and Optimization*, 17, 6, 1979.
[120] Delebecque, F. and Quadrat, J.P., Asymptotic problems for control of Markov chains with strong and weak interactions, *Automatica*, 1980.
[121] Kwakernaak, H. and Sivan, R., *Linear Optimal Control Systems*, New York, John Wiley, 1972.
[122] Sandell, N.R., Jr. and Teneketzis, D., Linear regulator design for stochastic systems by a multiple time scalea method, *IEEE Trans. on A.C.*, 22, 4, pp. 615–621, 1977.
[123] Haddad, A.H. and Kokotovic, P.V., Stochastic control of linear singularly perturbed systems, *IEEE Trans. on A.C.*, 22, 5, pp. 815–821, 1977.
[124] Kushner, H.J., *Stochastic Stability and Control*, New York, Academic Press, 1967.
[125] Golstein, E.G., *Duality Theory in Mathematical Programming and Its Application*, Moscow, Nauka, 1971 (in Russian).
[126] Dudkin, L.M. (ed.), *Iterative Aggregation and Its Application for Planning*, Moscow, Ekonomika, 1979 (in Russian).
[127] Gabassov, R. and Kirillova, F.M., *Linear Programming*, part III, Special problems, Minsk, BGU Press, 1980 (in Russian).
[128] Krasovskii, N.N., *Game Problems of Trajectory Intersections*, Moscow, Nauka, 1970 (in Russian).
[129] Plotnikov, V.A., Method of partial averaging for optimal control problems, *Diff. uravnenija*, 2, 1978 (in Russian).

[130] Plotnikiv, V.A., Averaging method for differential inclusions and its applications in optimal control theory, *Diff. uravnenija*, 3, 1978 (in Russian).

[131] Chernous'ko, F.L. and Kolmanovskii, V.B., *Optimal Control of Stochastic Systems*, Moscow, Nauka, 1979 (in Russian).

[132] Larin, V.B., On a weak control of weakly damped systems, *Prikl. mat. i mekh.*, 42, 6, 1978 (in Russian).

[133] Rakitskii, Ju.V., Ustinov, S.M., and Chernorutskii, I.G., *Numerical Methods for Solution of Stiff Systems*, Moscow, Nauka, 1979 (in Russian).

[134] Pervozvanskii, A.A. and Gaitsgori, V.G., *Decomposition, Aggregation and Approximate Optimization*, Moscow, Nauka, 1979 (in Russian).

[135] Gaitsgori, V.G. and Pervozvanskii, A.A., On aggregation of linear control systems, *Avtomatika i telemekhanika,* 3, 1980 (in Russian).

[136] Gaitsgori, V.G. and Pervozvanskii, A.A., Control of linear systems which are close to conservative ones, *Avtomatika i telemekhanika,* 10, 1980 (in Russian).

[137] Lur'e, A.I., *Some Nonlinear Problems of Control Theory*, Moscow and Leningrad, Gostechizdat, 1951 (in Russian).

[138] Pervozvanskii, A.A. and Gaitsgori, V.G., Suboptimization, decomposition and aggregation, *Proceedings of the Seventh Congress of IFAC*, pp. 1057–1061, 1978.

[139] Pervozvanskii, A.A. and Gaitsgori, V.G., Perturbation method for LQ problems: duality in degenerate cases, *Proceedings of the Second Warsaw Workshop on Multi-Level Control*, 1978, pp. 141–164.

[140] Gaitsgori, V.G. and Pervozvanskii, A.A., Perturbation method in the optimal control problems, *System Science*, 5, 1, 1979.

[141] Sandell, R.S., Varaja, P., Athans, M., and Safonov, M.G., Survey of decentralized control methods for large-scale systems, *IEEE Trans. on A.C.*, 23, 2, 1978.

[142] Vasiljeva, A.B. and Butuzov, V.F., *Singularly Perturbed Equations: Critical Cases*, MGV Press, 1978 (in Russian).

[143] Gaitsgori, V.G., Small parameter method for singular LQ problems. A case of pure imaginary eigenvalues, VINITI, 2488, 1979 (in Russian).

[144] Kuhn, H.W. and Tucker, A.W. (eds.), *Linear Inequalities and Related Systems*, Princeton, NJ, Princeton University Press, 1956.

[145] Golstein, E.G. and Tretjakov, N.V., Modified Lagrange functions, *Ekon. i mat. metody*, 10, 3, 1074 (in Russian).

[146] Gaitsgori, V.G. and Pervozvanskii, A.A., On the optimization of weakly controlled stochastic systems, *Soviet Math. Dokl.*, 21, 2, 1980.

[147] Pervozvanskii, A.A., On the "freezed coefficient" method for the optimal synthesis of linear control, *Dokl. AN SSSR*, 268, 5, 1983 (in Russian).

[148] Kalman, R.E., Mathematical description of linear dynamic systems, *SIAM J. on Control and Optimization*, Series, A, 1, 1963.

[149] Gaitsgori, V.G. and Pervozvanskii, A.A., Perturbation method for optimization problems, *Dinamika sistem*, 14, 1978 (in Russian).

[150] Khalil, H.K., Control of linear singularly perturbed systems with colored noise disturbance, *Automatica*, 14, March 1978.

[151] Chernous'ko, F.L., Akulenko, L.D., and Sokolov, B.I., *Control of Oscillations*, Moscow, Nauka, 1980 (in Russian).

[152] Mangasarian, O.L. and Fromovitz, S., The Fritz John necessary optimality: conditions in the presence of equality and inequality constraints, *Journal of Mathematical Analysis and Applications*, 17, 1, 1967.

[153] Gauvin, J. and Tolle, J.W., Differential stability in nonlinear programming, *SIAM J. on Control and Optimization*, 15, 2, 1977.

[154] Gaitsgori, V.G. and Pervozvanskii, A.A., Decomposition and aggregation in problems with a small parameter, *Izv. AN SSSR, Tekhn. kibernetika*, 1, 1983 (in Russian).

[155] Gaitsgori, V.G. and Pervozvanskii, A.A., Perturbation method for mathematical programming problems, *Ekon. i mat. metody*, 4, 1983 (in Russian; English version in *Mathecon.*, 20, 3, 1984).

[156] Gaitsgori, V.G. and Pervozvanskii, A.A., Perturbation theory for mathematical programming, *Journal of Optimization Theory and Applications*, 3, 1985.

[157] Ben-Israel, A., Ben-Tal, A., and Zlobec, S., Optimality in convex programming: a feasible directions approach, *Math. Programming Study*, 19, 1982, pp. 16–38.

[158] Gauvin, J. and Dubeau, F., Differential properties of the marginal function in mathematical programming, *Math. Programming Study*, 19, 1982, pp. 101–119.

[159] Lempio, F. and Maurer, H. Differentiability of optimal values in infinite-dimensional nonlinear programming, *Applied Mathematics and Optimization*, 6, 1980, pp. 139–152.

[160] Saksena, V.R., O'Reilly, J., and Kokotovic, P.V., Singular perturbations and time-scale methods in control theory: survey 1976–1983, *Automatica*, 20, 3, pp. 273–293, 1984.

[161] Pervozvanskii, A.A. and Solonina, N.V., Suboptimal finite-dimensional control law for distributed plants, *Avtomatika i telemekhanika*, 4 (part I), 7 (part II), 1984 (in Russian).

[162] Gaitsgori, V.G. Theorems on the existence and structure of limiting set for outputs of a control system and their applications to suboptimization of singular perturbed control systems (an overview), *Metody optimiz.* (Proceedings of VNISI), Moscow, 1984 (in Russian).

Index